INTRODUCTION TO
TECHNOLOGY

Third Edition

Alan J. Pierce
Dennis Karwatka

Mc Graw Hill **Glencoe**

New York, New York   Columbus, Ohio   Chicago, Illinois   Peoria, Illinois   Woodland Hills, California

Safety Notice

The reader is expressly advised to consider and use all safety precautions described in this book or that might also be indicated by undertaking the activities described herein. In addition, common sense should be exercised to help avoid all potential hazards.

Publisher and Authors assume no responsibility for the activities of the reader or for the subject matter experts who prepared this book. Publisher and Authors make no representation or warranties of any kind, including but not limited to, the warranties of fitness for particular purpose or merchantability, nor for any implied warranties related thereto, or otherwise. Publisher and Authors will not be liable for damages of any type, including any consequential, special or exemplary damages resulting, in whole or in part, from reader's use or reliance upon the information, instructions, warnings or other matter contained in this book.

Brand Disclaimer

Publisher does not necessarily recommend or endorse any particular company or brand name product that may be discussed or pictured in this text. Brand name products are used because they are readily available, likely to be known to the reader, and their use may aid in the understanding of the text. Publisher recognizes that other brand name or generic products may be substituted and work as well or better than those featured in the text.

The McGraw·Hill Companies

Send all inquiries to:
Glencoe/McGraw-Hill
3008 W. Willow Knolls Drive
Peoria, IL 61614

ISBN 0-07-861219-5 (Student Edition)
Printed in the United States of America
3 4 5 6 7 8 9 10   058   08 07 06 05

Contents in Brief

Lane Beard
Agriscience/Industrial Maintenance
Instructor
St. James Parish Career & Technology Center
Lutcher, Louisiana

Chuck Bridge, DTE
Technology Education Teacher
Chisholm Trail Middle School
Round Rock, Texas

LaMarr Brooks
Industrial Technology & Engineering Teacher
Pendleton High School
Pendleton, South Carolina

Oscar L. Cockerham
Technology Application Instructor
Amite County Vocational Center
Liberty, Mississippi

Oscar L. Dean
Technology Teacher
Dunbar Magnet Middle School
Little Rock, Arkansas

Mike Fischer
Teacher
Lawrence County High School
Lawrenceburg, Tennessee

Lynne M. Grimes, MLS
Freelance Editor & Indexer
Science Reviewer
Bedford, Texas

Pamela G. Hall
Technology Instructor
Gilmer Middle School
Ellijay, Georgia

Ed Hamilton
Wood Tech./Construction Tech.
Berryhill High School
Tulsa, Oklahoma

Richard H. Koblin
Technology Teacher
Manhasset Middle School
Manhasset, New York

Jeffrey Parkman
Teacher
Mt. Greylock Regional School District
Williamstown, Massachusetts

Mark K. Smedley
Technology Instructor
Billingsley School
Billingsley, Alabama

James W. Starling
Industrial Education Department Head
Dunedin High School
Dunedin, Florida

Daniel B. Stout
Technology Teacher
Michael Grimmer Middle School
Schereville, Indiana

David R. Vigil
CAD/Carpentry Instructor
Taos High School
Taos, New Mexico

Brooks R. Wadsworth
Engineering Technology Teacher
North Edgecombe High School
Leggett, North Carolina

Michael W. Warwick
Career & Technical Education Instructor
West High School
Knoxville, Tennessee

Ronald D. Yuill, DTE
Department Chair & Teacher
Tecumseh Middle School
Lafayette, Indiana

Our lives are becoming more and more dependent on technology. For that reason technology literacy is very important for everyone. Several years ago the International Technology Education Association (ITEA) initiated the Technology for All Americans Project. This project ultimately developed standards for what students should know and be able to do in order to be technologically literate. These standards are titled *Standards for Technological Literacy: Content for the Study of Technology* and are commonly referred to as *Technology Content Standards*.

Introduction to Technology is written to specifically address the *Technology Content Standards*.

Each chapter of this text is divided into sections. At the beginning of each section is a list of the standards covered in that section, including the section review. With these standards as a basis of study, students will have a better understanding of technology and its impact on our world after using this text.

ABOUT THE AUTHORS

Dr. Alan J. Pierce is a retired college professor. He has served as an educator at the elementary, middle, and college levels for 39 years. Dr. Pierce has authored numerous articles in the field of education, co-authored curriculum projects, and served as a technical consultant on a children's technology book. He is an editor and writer of the middle school AgBio curriculum for the NSF-funded TECHknow project at North Carolina State University.

Dr. Pierce created the "Technology Today" column for *Tech Directions* magazine and has written this column since 1995.

Dennis Karwatka is a Professor in the Department of Industrial Education and Technology at Morehead State University. He taught for many years in the high school Upward Bound Program. He has written two middle/high school textbooks, five technical history books, and many articles.

Mr. Karwatka created the monthly "Technology's Past" column in *Tech Directions* and has written that column since 1980. He is a Registered Professional Engineer with experience in the Apollo lunar landing program and with jet engine development.

Contents

Unit 2—Engineering Design

Unit 3—Information & Communication Technologies

Unit 4—Biotechnologies

Unit 5—Manufacturing Technologies

Unit 6—Construction Technologies

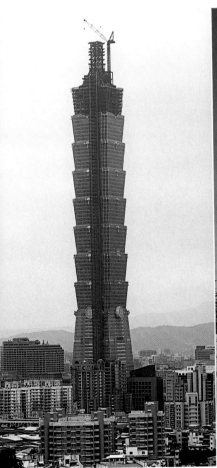

Unit 7—Transportation Technologies

Highlighting the Career Skills Needed for Success

TECHNOLOGY CHALLENGE ACTIVITY

ETHICS in ACTION

LINK TO THE PAST

LOOK TO THE FUTURE

How to Use This Book

Technology is all around you. Without it, you would not have food, clothing, or a home to live in. You need technology, but it can be complicated or confusing. To be a smart consumer, a good citizen, a successful worker, and a healthy person, you need to understand technology and use it wisely.

You know that *literacy* means being able to read and write. Literate people are comfortable with the use of words. Do you know what *technological literacy* means? It means knowing about technology and being comfortable using it. **Introduction to Technology** will help you become technologically literate. You will learn how technology is designed, made, and used. You will also learn about the impact of technology on society and where technology is going in the future.

This book is grouped into seven units. Each unit will help you understand and allow you to work with the different aspects of technology.

- **Nature of Technology** describes the basics of technology, including why we study technology and important concepts of technology.

- **Engineering Design** explains how technology works. It covers design, problem solving, drafting, and modeling.

- The remaining five units each cover one area of technology: **information and communication**, **biotechnology**, **manufacturing**, **construction**, and **transportation**.

As shown on the next few pages, this book helps you study technology in a variety of ways. Not only will you learn *about* technology, you will also learn to *do* technology.

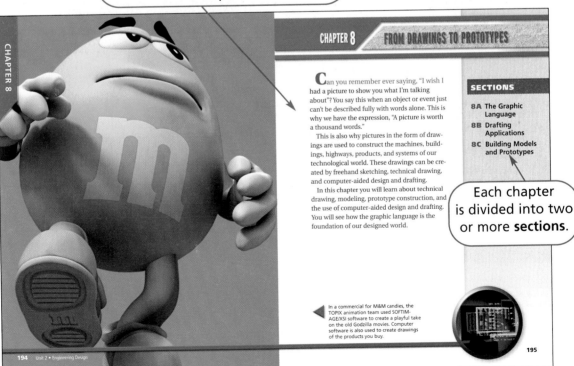

Chapter Introduction focuses your attention on the main topics that will be explored in the chapter.

Each chapter is divided into two or more **sections**.

Section Openers tell what you will learn in each section of the chapter.

Questions under headings help you focus on the most important concepts.

Objectives describe what you'll be able to do.

Terms to Learn are important words used in technology. These key terms are defined in the margins of the pages where they are first used.

Standards for Technological Literacy are identified for each section.

SECTION 8C

Objectives

▸▸ **Tell** the purpose of models and prototypes.
▸▸ **Describe** rapid prototyping.
▸▸ **Explain** the purpose of scientific and engineering visualization.

Terms to Learn

- model
- prototype
- rapid prototyping
- visualization software

Standards

■ Characteristics & Scope of Technology
■ Relationships & Connections
■ Engineering Design
■ Design Process
■ Use & Maintenance

model
replica of a proposed product that looks real but doesn't work

prototype
working model of a proposed product

Building Models and Prototypes

▸ What are models and prototypes and how are they used?

Many projects require construction of models and prototypes. A **model** doesn't actually work. It is created to determine how a product, building, or system will look. A **prototype** is usually a model that works. Models and prototypes can be crafted by hand, by machine, or with a computer system.

If a prototype is an accurate representation, it is used to test such things as product reliability and safety. These tests all take place before the product is mass-produced. The drawings that will be used to build the product might be changed many times to reflect improvements discovered during testing and inspection.

At one time, airplane manufacturers had no choice but to build wood models of their airplane designs. They would fly these models in wind tunnels. If the model had good flying characteristics, the next step was to build a prototype that actually worked.

Some designers still like to build models out of clay. See Figure 8-14. However, computers have made virtual three-dimensional models and prototypes possible. For example, NASA uses the oldest and newest technologies to mold the shape of future flight. See Figure 8-15. On a tour of NASA, you would see wood models, high-tech computer work stations, and machines that can make prototypes.

Figure 8-14 Designers make clay models of cars that you may be driving one day.

212   Unit 2 • Engineering Design

SECTION REVIEW 8C

Answers to Section Reviews are located in the *Teacher Resource Guide.*

Recall ▸▸

1. What is rapid prototyping?
2. Why are models and prototypes created?
3. What is the purpose of scientific and engineering visualization?

Think ▸▸

4. Name several manufactured products that you think might have required a model or prototype.
5. Why do you think the appearance of a product is important?

Apply ▸▸

6. **Construction** Build a model airplane of your own design and test it in a wind tunnel of your own design.
7. **Data Analysis.** Use computer software to graph the data from the model plane activity. What conclusions can you draw about your model?

Section Reviews ask you to recall, think about, and apply what you've learned.

CHAPTER 2

Summary Activity

For each numbered blank, pick the answer from the list on the right that makes the most sense in the entire passage. Write your answer on a separate sheet of paper. No answer will be used more than once.

Most of our __1__ developed because individuals had ideas that they were able to turn into useful devices. The __2__ resources of technology include people, information, tools and machines, capital, time, materials, and energy.

Today, large companies buy all the __3__ needed to create new technology. They hire __4__ with the knowledge and skill needed to use tools and machines to convert materials into __5__ products. The company supplies the energy to run the machines and gives its people the time they need to create new ideas.

Technology has developed many systems and subsystems. A car engine is a __6__ of the automobile. An automobile is a complete system, but it is also a subsystem of our transportation __7__.

In an open-loop system __8__ is the information, ideas, and activities needed to plan for production. Process is the construction stage, where machines, labor, and materials __9__ the product. Output is what the system has produced. An open-loop system changes into a closed-loop system when __10__ is introduced.

Products are designed to meet the needs and __11__ of consumers. Marketing firms use advertising to create desire for new products. Product designers determine what features will make the products sell, taking into consideration a product's __12__ and constraints. They try to make the most of its positive features while reducing its negative features. Tradeoffs may be needed to turn a design into a new product that can be built at a __13__ cost.

To protect people and the environment we need to monitor new technology to determine possible __14__ outcomes. As a technologically __15__ person, you must weigh the positive outcomes against the negative outcomes.

Answer List
• build
• criteria
• feedback
• input
• literate
• negative
• new
• people
• reasonable
• resources
• seven
• subsystem
• system
• technology
• wants

> **Chapter Reviews** provide summary activities, comprehension checks, and critical thinking questions to help you review the main points of each chapter.

Comprehension Check

1. What is technology?
2. What is a primary tool? Name two examples.
3. What is the difference between a tool and a machine?
4. Is the fabric in a 100 percent cotton shirt a processed material or a manufactured material? Explain your answer.
5. Is a table lamp a closed-loop system or an open-loop system? Explain your answer.

Critical Thinking

1. **Infer.** Why is skill important when creating technology?
2. **Analyze.** How might tools affect the amount of time needed to make a product?
3. **Compare.** The ZAP Company makes batteries that cost only one-third as much as other brands, but they last only two-thirds as long. Is this a good tradeoff? Why or why not?
4. **Evaluate.** Describe three positive and three negative impacts of computers.
5. **Design.** Suppose your company is designing a toy truck for children twelve to eighteen months old. Write safety specifications for the toy.

> **Visual Engineering** gives you a unique way to explore a topic covered in the chapter.

Visual Engineering

Communication devices for vehicles have changed dramatically over the years. First it was radio, then tape and CD players, phones, DVD players, and Global Positioning Systems (GPS). What might be next? Develop your own idea for a new communication device for vehicles. Describe to the class what the device would do and how it would work.

TECHNOLOGY CHALLENGE ACTIVITY

Programming a Computer to Control a Machine

Equipment and Materials

- LEGO® Mindstorms™ Invention System
- computer system

FIGURE A

FIGURE B

Background ▶ The United States landed the first people on the moon using less computer power than is found in today's automobiles. Today, computers control the fuel system, engine, and other parts of your car as well as many other technology systems. Have you ever programmed a computer to control a motor-powered machine?

Goal ▶ For this activity you will build a motorized robot machine using a LEGO® Mindstorms™ Invention System. The motors and sensors that are part of your machine will be controlled by a computer program that you will write. See Figures A, B, and C.

Criteria and Constraints

» You will use the problem-solving process to create your robot machine.

» Your computer program must be thoroughly tested before downloading.

» You will download your program into your robot machine using the RCX transmitter.

266   Unit 3 • Information & Communication Technologies

Technology Challenge Activities let you apply what you have learned about technology.

FIGURE C

FIGURE D

Design Procedure

❶ Read the user's guide that comes with the LEGO Mindstorms Invention System. You might want to complete the control experiment before trying to build and program a machine of your own.

❷ Each user's guide provides a step-by-step procedure for a different computer-controlled project. Keep in mind that you should use the problem-solving process as you work. Define your problem, generate ideas, select a solution, test the solution, make the item, evaluate it, and present the results.

❸ Pick one machine and follow the directions for assembly.

❹ Test motors and sensors following the user's guide.

❺ Use the LEGO manuals to learn the programming language for the RCX controller. You must use different commands for controlling the motors and sensors on your machine. The programming language lets you "talk" to the computer using phrases that the machine (through the software) can understand.

❻ Plan out the sequence of commands that will tell the computer how to control your machine.

❼ Test each command one at a time before downloading to your machine's microprocessor. Troubleshoot solutions to any malfunctions.

❽ Use the infrared transmitter to transfer your program to your machine.

❾ Demonstrate the operation of your machine to the class.

Evaluating the Results

1. In computer control systems you often find that the machine under control is equipped with optic sensors, touch sensors, and motors. How do these subsystems play a part in the control of the machine?

2. If you were to design and program another machine, what would you do differently?

Chapter 10 • Computer Technologies   267

⚠ SAFETY

Reminder

Throughout this book you will be using tools and machines to work on various activities. Be sure to always follow appropriate safety procedures and rules. Remember, safety is an attitude that you must develop and maintain at all times.

Safety notes remind you to follow all safety rules and procedures.

CAREERS

Careers features describe a technology career. Features emphasize the skills needed to succeed in the workplace.

Computer Technician

Computer technician wanted for large doctor's office. Our computer system currently handles medical, scheduling, and billing information. The successful candidate must have a strong background in computer technologies and a willingness to learn and grow as computer systems change.

Willingness to Learn

Having a willingness to learn will help you throughout your life, whether you're in school, at work, or at home.

Employers need people who are willing to improve and update their skills. They often promote workers who show a willingness to learn. Fields in which you must obtain a license or certificate often require more education to renew them. Continued learning is especially important in computer technology. New programs and hardware appear almost daily. Are you interested in a career in which you must stay up to date? Here's how to find out:

✔ **Look up the career on the Web site for the Occupational Outlook Handbook. Check under Education.**

✔ **Talk to someone who is currently working in the field. Ask the person how important updating your skills is to success.**

SuperStars OF TECHNOLOGY

INVESTIGATION

Research the life of another important technologist and report your findings to the class. Here are some names to start with: Granville Woods, Lee DeForest, Margaret Bourke-White, Erastus Bigelow, Elijah McCoy, Lillian Gilbreth, Augustus Fruehauf, Robert Goddard, Igor Sikorsky.

Lewis Latimer
(1843-1928)

The only African-American of Edison's research team, Latimer studied Thomas Edison's work with the light bulb and made improvements on the incandescent lamp, resulting in the Latimer lamp. Latimer's greatest technological achievement was when he patented the process for manufacturing carbon filaments in light bulbs. His process created a better filament that lasted much longer than other filaments of the time that were made of paper. He also assisted in drawings for the patent of Alexander Graham Bell's telephone.

Steven Jobs (1955-)
Stephen Wozniak (1950-)

In 1977, the two Steves designed and assembled the world's first commercially successful personal computer, the Apple II. To get enough money for the project, Jobs sold his Volkswagen bus and Wozniak sold his calculator. Then they built the computer in Jobs' garage. They chose the name "Apple" because Jobs had once spent a summer picking apples in Oregon. The Macintosh, which came out in 1984, was named after Wozniak's favorite apple, the McIntosh. By the time both left in 1985, Apple was a large, successful company. Jobs rejoined the company in 1996.

48   Unit 1 • The Nature of Technology

Elisha Graves Otis
(1811-1861)

The safety of early elevators relied on the strength of the rope used to haul them up and down. If the rope broke, the elevator crashed. Otis invented the safety elevator in 1852 for a New York company. He used wooden guide rails with teeth cut into them. The elevator was designed in such a way that, if the rope broke, the teeth caught the elevator and locked it into place.

As a scientist in Germany, von Braun helped develop a liquid-fueled rocket engine used during World War II. After the war, he became a U.S. citizen and the first director of the Marshall Space Flight Center in Huntsville, Alabama. He helped build the Saturn V booster rocket used to put Americans on the moon.

Wernher von Braun
(1912-1977)

Catharine Beecher
(1800-1878)

The most influential American house designer of the 19th century, Beecher was an architect, drafter, and efficiency expert. Her house plans included storage areas, use of heating fuels, and ways to perform household tasks more efficiently. She also developed revolutionary floor plans for apartments, churches, and schools.

Chester Carlson
(1906-1968)

Until Chester Carlson came along, only photographic techniques could make exact duplicates of images. Carlson developed the xerographic process in New York in 1938. However, no one realized its importance at first, and it took Carlson a while to find a company interested in using it. The first Xerox copy machine was made available to the public in 1959.

49

> **SuperStars** features in each unit tell about people who helped shape the world of technology.

ETHICS in ACTION — Honoring Copyright

Have you ever seen a funny story or joke on an Internet Web site and wanted to copy it? The Internet makes it easy to copy and transmit such material. However, when material is copyrighted, it means it should not be reproduced without permission. You must assume that everything you see on the Internet is copyrighted. Using copyrighted material without permission is stealing. Lawmakers are now working on laws to regulate copying material found on the Internet.

ACTIVITY Research the laws being considered to protect material found on the Internet.

> **Ethics in Action** features emphasize the importance of using technology appropriately.

LOOK TO THE FUTURE

More Information, Less Space

In 2002, researchers at IBM demonstrated a new technology for storing computer data. Millipede stores data in the form of tiny indentations in a thin plastic film—1 trillion bits of information in one square inch. That's equal to the information on 25 million printed pages or 25 DVDs. The first commercial applications will probably be in cell phones, watches, and PDAs (personal digital assistants). As the technology improves, Millipede's storage capacity will grow even larger.

> **Look to the Future** features focus on technologies that are in development now or that may one day exist.

TECH CONNECT
LANGUAGE ARTS

Powerful Prefix

Gas turbine engines have the prefix *turbo* in their names. The prefix means "driven by a turbine." Turbochargers use a turbine to increase the power of an engine.

ACTIVITY Find out about turbosails used to power sailing ships. Give an oral presentation about them to the class.

> **Tech Connect** features emphasize how technology relates to other subjects: mathematics, language arts, science, and social studies. **Activities** at the end extend your learning.

TECH CONNECT
SCIENCE

Which Came First?

Throughout history, the "how" of technology often came before the "why" of science. People used special "seeds" to make bread rise for thousands of years before science discovered the seeds were actually yeast, a type of fungus.

ACTIVITY Obtain a packet of yeast used to make bread and sprinkle it into a bowl of lukewarm water along with a pinch of sugar. Observe what happens over the next twenty minutes.

TECH CONNECT
MATHEMATICS

Computer Algebra

George Boole, an English mathematician, developed a system for processing logical statements in a mathematical (symbolic) way. His system is called Boolean algebra and is the basis for the way computers work.

ACTIVITY Find out how words can be as important as numbers in mathematics. What three key words in Boolean algebra allow computers to work?

TECH CONNECT
SOCIAL STUDIES

An Inventive President

Our presidents are known for many things. One of our presidents was an inventor who had his invention patented. (A patent is a government document granting a person the right to produce or sell his invention. No one else may copy it.)

ACTIVITY Research past presidents and see if you can find which one was granted a patent for his invention.

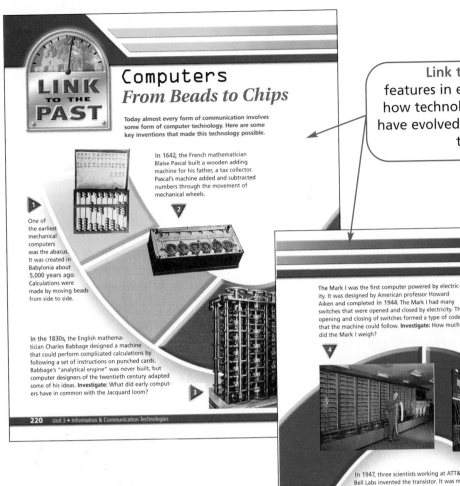

Computers
From Beads to Chips

Today almost every form of communication involves some form of computer technology. Here are some key inventions that made this technology possible.

In 1642, the French mathematician Blaise Pascal built a wooden adding machine for his father, a tax collector. Pascal's machine added and subtracted numbers through the movement of mechanical wheels.

1 One of the earliest mechanical computers was the abacus. It was created in Babylonia about 5,000 years ago. Calculations were made by moving beads from side to side.

In the 1830s, the English mathematician Charles Babbage designed a machine that could perform complicated calculations by following a set of instructions on punched cards. Babbage's "analytical engine" was never built, but computer designers of the twentieth century adapted some of his ideas. **Investigate:** What did early computers have in common with the Jacquard loom?

The Mark I was the first computer powered by electricity. It was designed by American professor Howard Aiken and completed in 1944. The Mark I had many switches that were opened and closed by electricity. The opening and closing of switches formed a type of code that the machine could follow. **Investigate:** How much did the Mark I weigh?

In 1946, two Americans, J. Presper Eckert, Jr., and John Mauchly, completed the ENIAC. It used vacuum tubes instead of the mechanical switches of the Mark I. The ENIAC could perform 5,000 mathematical operations per second. However, it had to be rewired for each new job, and that took days. **Investigate:** What did the letters ENIAC stand for?

In 1947, three scientists working at ATT&T's Bell Labs invented the transistor. It was much smaller, used less power, worked faster, and lasted longer. The first computer designed to use transistors was built in 1956. By 1960, all computers were using transistors instead of vacuum tubes.

In the 1960s, engineers developed a way to put dozens of transistors onto a single chip called an integrated circuit. The first chips were used in calculators.

In 1969, the Intel Development Corporation produced the first programmable computer chip. This technology made possible the design of personal computers. During the 1970s and 1980s, personal computers became more powerful, less expensive, and easier to operate. Today, almost half of all U.S. households have at least one computer. **Investigate:** Research and report on early PCs, such as Honeywell's "Kitchen Computer."

> **Link to the Past** features in each unit explore how technologies of the past have evolved into what we use today.

This book has been carefully designed for you, the technology education student. Regardless of what career you choose, what you learn from this book will be of value to you in your future. Technology will always be part of your life. ***Introduction to Technology*** will give you a better understanding of the technological world.

The first unit, "The Nature of Technology," begins on the next page with a Link to the Past. This feature shows how technology has evolved over thousands of years. Chapter 1, which begins on page 28, will tell you how technology changes to meet people's needs.

Chances are excellent that technology will change faster and faster in your lifetime. To prepare yourself for those changes, use this book to learn more about the exciting world of technology.

Technology
How It All Began

Human use of technology began about 2.5 million years ago when the first human changed a natural material to make it more usable. People use technology to extend their abilities and satisfy their wants and needs.

1

The first real civilizations were based on agriculture. When the plow was developed around **4000** B.C.E., civilization took a giant leap forward.

Although no one knows for sure, some experts believe the wheel was also invented about the same time as the plow. Until the invention of the first true automobile almost **6,000 years later**, the wagon remained the basic form of transportation.

Investigate: *Who invented the first true automobile and how was it powered?*

2

Construction methods, too, were slow to change, and early buildings were very simple in design. Then around **1700** B.C.E., construction technology began to change rapidly. Simple rectangular structures were transformed with columns and beams, as seen in Greek architecture.

3

The Romans, too, were engineers. They developed bridges, roads, tunnels, and aqueducts, some of which are still being used today. Similar achievements were also taking place in China and Central America.

Investigate: *Which people developed a well-maintained system of roads in South America, and how many miles did it cover?*

5 Energy is essential to most technology, and, since prehistoric times, humans had used energy in the form of muscle power and fire. **The Middle Ages** saw a development in wind power, not only for ships, but also for manufacturing.

6 Then, during the **18th century**, the steam engine was developed. The industrial age began.

Investigate: *What famous revolution did the steam engine start and which technologies were affected?*

7 Equally important to the development and growth of technology was communication. What moved communication forward at lightning speed was the development of movable type **in 1440**. Information spread rapidly, especially information about all kinds of technology.

What is technology? Let's answer that question by asking another: What *isn't* technology? Look out a window. Perhaps you see trees, plants, rocks, or other natural items. That's what technology isn't: anything created by nature. Sand, as it exists in nature, doesn't involve technology, and neither does water.

People take those natural materials and turn them into useful products. Trees are turned into lumber to make houses and furniture. Plants such as asparagus and tomatoes are put into cans and sold as food. Special rocks like iron ore are turned into metal, and sand can be changed into glass. Water can be used to make soft drinks.

Technology is usually involved in all the products, inventions, and discoveries made by people. How much do you know about technology? Do you know how inventors invent? Do you know what makes your computer or bike work? Do you understand how technology can hurt the environment? In this book you are going to learn the answers to these questions and many more *about* technology. You will also learn to *do* technology. Best of all, you will probably learn to *enjoy* technology.

Mirrors and optical lenses, when combined with a laser beam, can do amazing things in the world of technology.

Objectives

▶▶ **Define** technology.

▶▶ **Give** reasons for studying technology.

▶▶ **Explain** the advantages of being technologically literate.

Terms to Learn

- **technologically literate**
- **technology**

Standards

- Cultural, Social, Economic & Political Effects
- Role of Society
- Information & Communication Technologies

technology
the practical use of human knowledge to extend human abilities and to satisfy human needs and wants

Technology and You

▶ Do you see any products of technology around you?

Technology is the practical use of human knowledge to extend human abilities and to satisfy human needs and wants. Technology provides us with most of the things we use in our society.

Lift your eyes from this page and look around. What are some of the things that you can see?

- You might see a computer, a cell phone, or a DVD player. Those are all products of communication technology. See Figure 1-1.

- You might see a car, a commuter train, or an airplane. Those are products of transportation technology. See Figure 1-2.

- Perhaps you see cereal on a kitchen table or vitamins on a shelf. Those are products of biotechnology.

- If the items you see were made in factories, they are also considered products of manufacturing technology.

- When you look out the window, do you see a house, a bridge, or a skyscraper? Those are products of construction technology.

Figure 1-1 Did you realize that a home entertainment center is a communication system? What do you think is being communicated?

Figure 1-2 This is a concept car—a designer's idea of a car for the future. How are the seats different from those in today's cars?

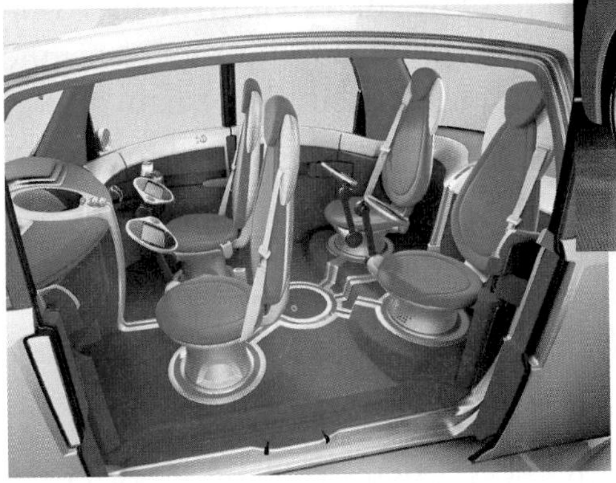

All this technology has improved over the years. When your grandparents were your age, each home probably had one rotary-dial telephone like the one shown in Figure 1-3. When your parents were growing up, they might have had one or two keypad telephones. In your home today, you might have access to several keypad telephones and maybe a cell phone with an Internet connection or one that sends and receives pictures. Advances in telephone design have been almost continuous since Alexander Graham Bell developed his model in 1876. You could say similar things about almost any other item.

Think about how cars are different from what they were like several years ago. Modern houses are also different, as are the foods you buy and your forms of entertainment. Medical equipment and procedures are other examples. What other changes in technology can you name?

Figure 1-3 Have you ever used a rotary-dial phone? It seems very slow compared to keypad phones.

Why We Study Technology

▶ Why do you think the average person needs to understand technology?

Why study technology? That's an easy question to answer. Technology is fun, rewarding, and exciting.

Technology is fun because you get to work with your hands and mind using tools and materials. Instead of only reading about bridges, you might build a model of one. Instead of only reading about electricity, you might make a small electric motor. Instead of only reading about wood, you might make something useful out of it. See Figure 1-4. Technology is always an up-to-date subject. You'll learn how to make and repair many things that can help you, not only in the future, but also right now.

Technology is rewarding because you can see the results of your work. People who put space shuttles together have a direct connection to the success of the mission. They are the ones who tighten the bolts, assemble the electronic equipment, and test the controls. A successful flight frequently brings tears of joy to many of the workers. The shuttle means more to them than just a piece of equipment.

Technology is exciting because each day brings new ideas and new challenges. Some days you and your class

Figure 1-4 Learning how to build a wood project yourself can be very rewarding. Do you know what machine these people are using?

Selecting a Rewarding Career

Finding a career suited to your needs and wants is very important.

Web designers often feel satisfied at seeing something they designed being displayed on the Internet for millions to see and use.

Being a web designer may not interest you, but you probably have some ideas about the kind of work you would like to do some day. That's good, because it's not too early to begin thinking about a future career. For example, consider your activities during a typical week:

✔ **Which of those activities did you enjoy the most?**
✔ **What makes those activities especially enjoyable?**
✔ **How do those activities reflect your interests and values?**

Answering questions like these will point you toward a career you will find both satisfying and rewarding.

might work on computer projects and other days on engine projects. See Figure 1-5 on page 34. It will be that way if you become a full-time technologist. No two days will be the same. Think how dull your life would be if you always did the same thing. That simply doesn't happen in the field of technology.

Studying technology will also help you develop your problem-solving skills. You will learn to identify a problem and come up with a solution. Technology is important in solving problems. For example, technology is working to provide pure water to an ever-increasing population. Technology also lets us maintain communication links with fire departments and hospitals. More efficient automobiles extend our limited fuel supplies. New construction techniques can quickly replace older buildings. The list is almost endless.

You will also find that technology is related to other subjects you study in school. Mathematics and science are

Figure 1-5 Does working on a small engine sound difficult to you? With a basic understanding of technology, you may not find it as hard as you think.

TECH CONNECT
MATHEMATICS

Use Your Seatbelt

Do you wear your safety belt when traveling in a car? What states require the use of seatbelts? Has seatbelt technology paid off? Statistics can tell you.

ACTIVITY Research and compare the percentage of people who receive serious injuries when not wearing seatbelts to those who receive serious injuries while wearing them. Do you like those odds?

technologically literate
term used to describe someone who is informed about technology and feels comfortable with it

two examples. You may find you enjoy all your classes more after you begin to see this relationship.

Being Technologically Literate

▷ Do you know what *technologically literate* means?

Technology is often part of the news. A journalist might report on a particular electrical power plant, a food additive, or a safety device on an automobile. It is important that you understand the importance of the subject. For example, automobile air bags have saved many lives in collisions. However, they inflate so rapidly that they have caused injury and death in some cases. As a result, the federal government lets car owners install an on/off switch for the air bags. Do you think the government should let people do this?

Understanding technology can help you answer that question. However, there is usually no one correct answer. You will have to evaluate each situation and make an informed decision. When you can do this, you will be **technologically literate**. You will be comfortable with the use of technology. You will not be afraid of it, and you will not think it has all the answers either.

Why Are Ethics Important?

Ethics are moral principles and values. Although something that is unethical may not be against the law, most people will probably agree that it is wrong.

Some professions, such as medicine, follow a code of ethics that regulates behavior. For example, it would not be ethical for a doctor to promote a treatment that had been proven not to work.

Ethics have an impact on us and on our society. Unethical people are seldom trustworthy and may have trouble keeping friends or a job. They may have trouble feeling good about themselves.

The development and use of technology sometimes poses ethical issues. For example, people can eavesdrop on cell phone users. Is this ethical? Some of these issues will be discussed in this course. What effect do you think ethics has or should have on technology?

ACTIVITY Interview three friends or family members. Ask them to give an example of what ethics means to them.

SECTION REVIEW 1A

Recall ▸▸

1. Name two communication products.
2. Name two transportation products.
3. Name two construction products.
4. Name one biotechnology product.
5. What does it mean to be technologically literate?

Think ▸▸

6. How are cars today different from those made fifty years ago? How did society's demands and values shape these differences?

7. How are grocery store foods today different from those made fifty years ago? How did society's demands and values shape these differences?

Apply ▸▸

8. **Research** Investigate automobile air bags and list their advantages and disadvantages. Develop an opinion about the use of air bags and write a paragraph defending your opinion.

Objectives

▸▸ **Identify** the workers who do technology.

▸▸ **Tell** how science, engineering, and technology are linked.

▸▸ **Explain** how teens have contributed to technology.

Terms to Learn

• engineering
• science

Standards

▪ Relationships & Connections
▪ Influence on History
▪ Manufacturing Technologies

Making Technology Happen

⯈ What are some ways in which technology involves people?

More than anything else, technology involves people. People make things happen. People apply the technology.

The People Who Do Technology

⯈ Who are the people who do technology?

The word *technology* comes from the ancient Greek word *techne*, which means "art." You might think that art means only paintings or sculpture. To the Greeks, it meant much more. They felt it took a real artist to make useful products from natural materials. They never thought that technology abused nature. They believed instead that technology linked nature to the human spirit. Many centuries later, Walter Chrysler, the man who started the car company, expressed the same idea. He said, "Someday I'd like to show a poet how it feels to design and build a railroad locomotive."

Several terms have been used to refer to people who work with technology. Early workers in technology called themselves *artisans*. It's still a name we hear from time to time. It referred to a highly skilled worker or craftsperson. During the 1800s, the unusual name of *mechanician* was used. From this word came our word *mechanic.* The word *mechanician* also came from a Greek word—*mechanikos*, meaning "machine."

Two modern names for people who work in technology are *technician* and *technologist.* It is customary to use either word to describe a person who works in technology. See Figure 1-6.

Figure 1-6 This technologist is working on complex machinery. He knows how every component (part) works.

Science, Engineering, and Technology

▶ How are the areas of science, engineering, and technology related?

You may notice that the words *science* and *technology* are often used together. A classmate might say, "I want to study science and technology in school." Your dictionary no doubt uses the word *science* in its definition of the word *technology*. Although science and technology are related, they're not the same. **Science** explains how things happen. Technology *makes* things happen.

Scientific effort created electronic microchips. Technology used those microchips to make automatic cameras, DVD players, and many other electronic devices. Scientists discovered nuclear reactions before technologists built nuclear power plants for generating electricity.

People didn't always need formal science to do technology, however. People made objects out of bronze 5,000 years ago, long before there was a branch of science known as metallurgy. Before the sciences of chemistry and biology were identified, people used certain plants to help cure injuries and illnesses.

Another important profession is **engineering**. It fits between science and technology. Using their knowledge of science and mathematics, engineers determine *how to make* things. See Figure 1-7. For example, chemical engineers working with mechanical engineers design machines that produce plastics and other materials. Civil engineers design bridges and roads. Electronic engineers design communication systems. Chemical, civil, electronic, and mechanical technologists build the products engineers design.

science
knowledge covering general truths or laws that explain how something happens

engineering
profession that involves designing products or structures so that they are sound and deciding how they should be made and what materials should be used

Figure 1-7 Today's engineers do most of their design work on computers. Some computer programs can also analyze a design and find any flaws in it.

Teens Contribute to Technology

▷ How have teenagers contributed to the development of technology?

Believe it or not, teens have sometimes made contributions to technology. Have you ever heard of George Westinghouse? He started a company that still uses his name. He is best known for inventing improved brakes for trains. However, when he was only nineteen, he patented a new type of steam engine. It wasn't successful, but he was on his way.

George Washington, our first president, was a self-taught surveyor long before he became a military leader. He assisted in surveying five million acres for the largest landowner in Virginia and helped lay out the city of Alexandria. He was appointed official surveyor of Culpeper County, Virginia, when he was only seventeen. See Figure 1-8.

At fourteen, Elmer Sperry invented a swiveling headlamp for locomotives so that the engineer could see around curves. Although the headlamp wasn't successful, the gyrocompass he later invented was remarkable. Sperry's gyrocompass is still used in all ship and airplane guidance systems.

Good inventions are still being created by young people today. Students at Hampshire College in Massachusetts created the Grease Car, which turns used cooking oil into fuel. See Figure 1-9. Other students there invented a hand braking system for wheelchairs and a scooter-bicycle combination.

Would you like to be an inventor? Ask your instructor for more information about organizations like the National Collegiate Inventors and Innovators Alliance and about competitions like the Odyssey of the Mind and West Point's Engineering Design Contest.

Figure 1-8 As a young man, George Washington surveyed millions of acres of land. The instrument he used was a circumferentor set.

Figure 1-9 This inventor who worked on the Grease Car probably got his start in a technology class like yours.

LOOK TO THE FUTURE

What's in Your Future?

In a feature like this one within every chapter, you will read about exciting advances in technology. These advances may make important changes in our lives. Do you think you can predict (forecast) what some of those changes might be? For example, how could the Grease Car change our lives?

SECTION REVIEW 1B

Recall »

1. What is the meaning of the Greek word *techne*?
2. What is the difference between science and technology?
3. Who decides how big a gear should be: a scientist, an engineer, or a technologist?
4. Who makes the gear: a scientist, an engineer, or a technologist?

Think »

5. The Greeks thought that technology linked nature to the human spirit. What do you think that means?

6. Give two examples of technological advances that did not first require science. Don't use any example from this section of the book.

Apply »

7. **Manufacturing** Manufacture useful products out of raw materials. For example:
   - Tree branches into small wooden items
   - Locally obtained rocks into jewelry or other decorative items
   - Dyed sand into paintings, as done by Native Americans
   - Grass or straw into woven hot pads

How Technology Changes

▷ **How does technology change to meet people's needs?**

Throughout history, technology has helped change societies and cultures. It has influenced politics and economies. In turn, technology itself has been influenced by changes in civilization. See Table 1-A.

Building on the Past

▷ What are some examples of today's technology that are based on pre-existing technology?

In everything we do, we build on the efforts of people who came before us. Your teachers teach you a bit more or a bit better than they were taught. We continue to build on the English language. We often add brand-new words such as *spam* (unwanted e-mail) and *blog* (Web log).

Isaac Newton was a famous British scientist during the 1700s who investigated the motion of the planets. He once said, "If I have seen further [than others], it is by standing upon the shoulders of giants." Newton didn't really stand on anyone's shoulders. He meant that his accomplishments were based on the earlier work of other people. That's something you share with Newton. Everyone builds on what came before. Sometimes knowledge or something new developed for one purpose can be used for something else. In technology, we build on the accomplishments of early artisans, mechanicians, technicians, engineers, and scientists.

Technology is continually evolving, which means it is growing and changing. Thomas Edison invented phonograph records, but he knew nothing about tape recordings. Tapes have been replaced by compact disc recordings. This evolutionary process is normal and never stops. See Figure 1-10 on page 42.

Boeing 707 airplanes were designed by people who used slide rules for calculations. Today's airplanes are designed by people who use pocket calculators and desktop com-

TABLE 1-A History of Technology

Iron Age—Beginning about 1200 B.C.E.	Iron replaced metals such as copper and bronze in tools. Iron was plentiful and its use spread over much of the known world. Iron agricultural tools made permanent farming settlements more desirable.
Middle Ages—Beginning about 500	The waterwheel, the windmill, the horseshoe, papermaking, mechanical clocks, and faster ships were all developed during this time. All had important effects on the societies of the period.
The Renaissance—Beginning about 1300	This period saw a rebirth in the arts and humanities. Canal construction and architecture also flourished. The microscope, the telescope, and many other devices and processes were invented.
The Industrial Revolution—Beginning about 1750	Manufacturing, transportation, communication, and construction all advanced rapidly during this time. Education was improved, and people had more leisure time.
The Information Age—Beginning about 1900	This period began with the invention of machines that could process, store, and exchange data electronically. Computers and other electronic devices have brought many advances.

puters. Technology continually moves forward by adapting, so that each new product is an improvement over existing products. New technologies often create new processes as well.

For example, a new technology you might have heard about is **nanotechnology**. It is the science of working with the atoms or molecules of materials to develop very small machines. (The term comes from the word *nano*, meaning "one billionth.") Years ago noted physicist Richard Feynman predicted that we would one day be able to build

nanotechnology
the science of working with the atoms or molecules of materials to develop very small machines

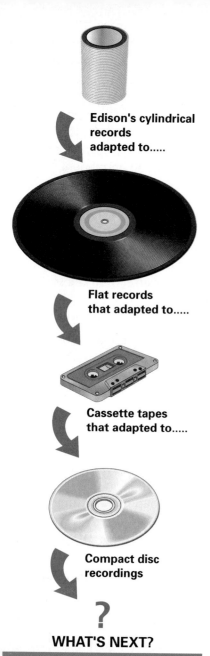

Edison's cylindrical records adapted to.....

Flat records that adapted to.....

Cassette tapes that adapted to.....

Compact disc recordings

?

WHAT'S NEXT?

Figure 1-10 Voice and music recording media have evolved from Edison's records to today's CDs. What recording medium do you think will be popular ten years from now?

a machine so small that it would be the size of just a few thousand atoms. His prediction came true. Tiny machines can be built by assembling molecules and atoms as building blocks. They have gears so small that you need a microscope to see them. See Figure 1-11. Someday these machines may be injected into the human body or used in other places people can't normally reach.

Technology is also changing more rapidly today than ever before. One reason for this is better communication. Centuries ago, news traveled slowly. Today, information travels around the world in seconds.

Technology in the United States

▷ What happened when technology met democracy?

Since the 1700s, this country has developed a reputation as a place where an intelligent and energetic person could be successful. During the early years, social conditions here were unlike any found in other countries. Americans had new ideas about work, community, and success. Under democracy, they felt free to try different ways of doing things. These beliefs encouraged technical advances and the establishment of businesses.

After the United States first entered the technology arena in the field of **machine tools**, many other products were developed here. Some of them include electronic computers, industrial robots, liquid-fueled rockets, reliable suspension bridges, photocopy machines, diesel engines for locomotives, electronic television, electronic flash for photography, the metal-framed skyscraper, and the practical helicopter. The heritage of American technology extends far back, and its roots are long and deep. You have benefited from and inherited a powerful technological tradition. It's up to you to continue that tradition. See Figure 1-12.

machine tool
a machine used for shaping or finishing metals and other materials

Figure 1-11 This tiny machine is so small it can be seen only with the aid of a microscope.

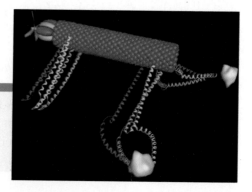

Figure 1-12 Through design, experimentation, and construction of models like this, you too can add to the technological tradition you have inherited.

Recall »

1. Who was Isaac Newton?
2. With what type of product did the United States first enter the arena of technology?
3. Name five products that were originally developed in the United States.

Think »

4. Give an example from your own life of how you have built on the efforts of someone else.

5. Give an example of the evolution of a technological device.

Apply »

6. **Research** Investigate an interesting career in technology. Make a display.

Summary Activity

For each numbered blank, pick the answer from the list on the right that makes the most sense in the entire passage. Write your answer on a separate sheet of paper. No answer will be used more than once.

Technology is the practical use of human knowledge. It involves turning __1__ items into useful products. We study technology because it is fun, __2__, and exciting. If you are technologically __3__, you are comfortable with technology. You are able to evaluate each situation and make informed decisions.

The word technology comes from the ancient Greek word *techne*, which means __4__. In the 1800s, workers in technology called themselves mechanicians, which is very similar to our modern-day word __5__.

The areas of science, technology, and engineering are related but different. __6__ try to explain how things happen; they conduct experiments in laboratories. Engineers figure out how to make things; they __7__ parts for machines and structures. Technologists make things by operating machines and assembling parts. All of these people work together to create and produce the __8__ we need.

Teenagers also can contribute to the development of technology. For example, as a teen __9__ invented a new kind of steam engine. Good inventions are being created by young people today. Students at Hampshire College in Massachusetts created the Grease Car, which turns used oil into fuel.

In everything we do, we build on the efforts of people who came before us. The famous scientist __10__ once said, "If I have seen further [than others], it is by standing upon the shoulders of giants." Technology moves forward by __11__ so that each new product is an improvement over existing products. New inventions are almost always based on earlier accomplishments.

One reason technology has thrived in the United States is because people felt free under __12__ to try different ways of doing things. The United States first entered the technology arena in the field of __13__. This laid the foundation for all modern manufacturing methods, including the use of industrial __14__.

Answer List

- adapting
- art
- democracy
- design
- George Westinghouse
- Isaac Newton
- literate
- machine tools
- mechanic
- natural
- products
- rewarding
- robots
- scientists

Comprehension Check

1. Define technology.
2. Give at least three reasons for studying technology.
3. Name at least one famous teen mentioned in this chapter who contributed to technology.
4. How does technology change to meet people's needs?
5. How did democracy influence the development of technology in the U.S.?

Critical Thinking

1. **Analyze.** How was the design for the automobile based on the design of a horse-drawn carriage?
2. **Evaluate.** Do you know someone who is technologically literate? What is your evidence for thinking so?
3. **Infer.** Think of a technological device. How do you think scientific knowledge helped make that device possible?
4. **Compare.** Select an earlier period in history from Table 1-A. How do you think your use of technology differs from the use of technology by a person of that period?
5. **Evaluate.** Access the Internet Public Library to learn what's available. Do you think it might be helpful during this course? How?

Visual Engineering

Future Engineering. In the past, engineers almost always drew their designs on paper. Today, they may still sketch ideas on paper, but, for final drawings, a computer is used. The same computer can be programmed to automatically check the many mathematical equations used in engineering. It can also evaluate the design for safety and other factors.

With two or three other students, try to imagine what tools engineers might be using in the next ten, twenty, or fifty years. What kinds of projects do you think they'll be working on? Present your ideas to the class.

TECHNOLOGY CHALLENGE ACTIVITY

Building a High-Tech Paper Airplane

Equipment and Materials

- paper
- pencil
- ruler
- colored markers
- yardstick, meterstick, or tape measure
- stopwatch

F4F Wildcat — **FIGURE A**

1/2"

Fold in half with 1/2" flap inside

1. 2.

3.

Draw wing design and cut along line

Fold wing tips up

Fold tail fins down

Bend inside nose up

Background You probably already know how to fold a simple paper airplane. The ordinary pointed-nose style has been around for a long time. However, you may not have had the opportunity to fold a high-technology paper airplane (HTPA). An HTPA takes careful planning and folding.

Several years ago, *Scientific American* magazine held its first International Paper Airplane Competition. There were almost 12,000 entries from twenty-eight countries. The plans in this activity are based on those used in the competition.

Goal Your task is to make two different HTPAs and see which one flies farther and stays up longer. The ordinary pointed-nose style usually flies about 15 feet and can stay up for about 4 seconds. One winning time in the *Scientific American* competition was 10.2 seconds. One winning distance was 91 feet. Of course, many other distances and times are possible. A breeze from an open window or a heating vent, for example, could help or hurt a flight.

Criteria and Constraints

During this activity, you will use a simple *systems technique*. A systems technique breaks down a complex project into basic elements. Here are the basic elements for each HTPA construction:

▶▶ Select an HTPA design.
▶▶ Draw the plans.
▶▶ Construct the HTPA.
▶▶ Operate and fly the HTPA.
▶▶ Collect flight information.
▶▶ Evaluate the flight information.

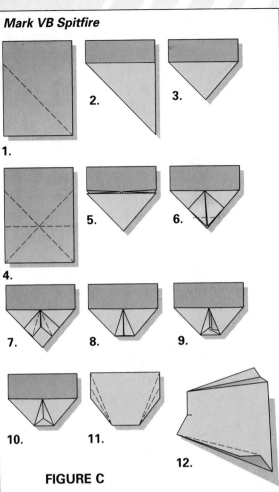

Mark VB Spitfire

1. 2. 3.

4. 5. 6.

7. 8. 9.

10. 11. 12.

FIGURE C

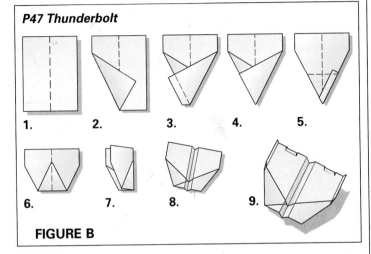

P47 Thunderbolt

1. 2. 3. 4. 5.

6. 7. 8. 9.

FIGURE B

Design Procedure

1. Select a paper airplane design from the plans shown in Figures A, B, and C.
2. Carefully draw the plans for your design on a sheet of ordinary notebook paper.
3. Fold the airplane according to the plans. Be sure your folds make sharp edges. A key to a good paper airplane is to have straight sharp folds.
4. Decorate your airplane with colored markers so that you can easily identify it.
5. Fly your HTPA several times.
6. Use the stopwatch to measure the length of each flight.
7. Use a measuring device to determine the straight-line distance from the point where the HTPA was launched to the point where it stopped.
8. If your plane doesn't fly well at first, try placing a very small weight at the nose. A piece of tape or small paper clip sometimes helps.
9. Repeat the preceding steps with another paper airplane design.

Evaluating the Results

1. Which HTPA flew farther and stayed up longer? Why do you think one was better than the other?
2. Which HTPA was more fun to fly? Why do you think one was more fun than the other?

SuperStars of TECHNOLOGY

INVESTIGATION ////

Research the life of another important technologist and report your findings to the class. Here are some names to start with: Granville Woods, Lee DeForest, Margaret Bourke-White, Erastus Bigelow, Elijah McCoy, Lillian Gilbreth, Augustus Fruehauf, Robert Goddard, Igor Sikorsky.

Lewis Latimer
(1843-1928)

The only African-American of Edison's research team, Latimer studied Thomas Edison's work with the light bulb and made improvements on the incandescent lamp, resulting in the Latimer lamp. Latimer's greatest technological achievement was when he patented the process for manufacturing carbon filaments in light bulbs. His process created a better filament that lasted much longer than other filaments of the time that were made of paper. He also assisted in drawings for the patent of Alexander Graham Bell's telephone.

Steven Jobs (b. 1955)
Stephen Wozniak (b. 1950)

In 1977, the two Steves designed and assembled the world's first commercially successful personal computer, the Apple II. To get enough money for the project, Jobs sold his Volkswagen bus and Wozniak sold his calculator. Then they built the computer in Jobs' garage. They chose the name "Apple" because Jobs had once spent a summer picking apples in Oregon. The Macintosh, which came out in 1984, was named after Wozniak's favorite apple, the McIntosh. By the time both left in 1985, Apple was a large, successful company. Jobs rejoined the company in 1996.

Elisha Graves Otis

(1811-1861)

The safety of early elevators relied on the strength of the rope used to haul them up and down. If the rope broke, the elevator crashed. Otis invented the safety elevator in 1852 for a New York company. He used wooden guide rails with teeth cut into them. The elevator was designed in such a way that, if the rope broke, the teeth caught the elevator and locked it into place.

Catharine Beecher

(1800-1878)

The most influential American house designer of the 19th century, Beecher was an architect, drafter, and efficiency expert. Her house plans included storage areas, use of heating fuels, and ways to perform household tasks more efficiently. She also developed revolutionary floor plans for apartments, churches, and schools.

Chester Carlson

(1906-1968)

Until Chester Carlson came along, only photographic techniques could make exact duplicates of images. Carlson developed the xerographic process in New York in 1938. However, no one realized its importance at first, and it took Carlson a while to find a company interested in using it. The first Xerox copy machine was made available to the public in 1959.

As a scientist in Germany, von Braun helped develop a liquid-fueled rocket engine used during World War II. After the war, he became a U.S. citizen and the first director of the Marshall Space Flight Center in Huntsville, Alabama. He helped build the *Saturn V* booster rocket used to put Americans on the moon.

Wernher von Braun

(1912-1977)

If a time machine could take you back to ancient times, would you be smarter, stronger, or faster than the people of that time period? Suppose you cannot take with you any of the things that have been created by technology.

Would you know how to survive in the wilderness? You wouldn't be able to run as fast, see as far, or hear as well as the predators who would view you as a possible dinner. To avoid being eaten, you would need to do a lot of inventing or hiding. Would you be able to adapt to find food and shelter in that environment? You would have to improvise and use your knowledge to do all these things.

What resources and methods did our earliest ancestors use to create the tools and weapons they needed to protect themselves and keep themselves alive? What resources and methods do we use today to create the tools and devices that make our lives easier and more enjoyable? In this chapter you will learn the answers to these questions.

Using drawings was one way people communicated in ancient times. Would you be able to survive in their time using only the resources that they used?

Objectives

▶▶ **Identify** and analyze the seven technology resources.

▶▶ **Explain** how people's skills and creativity lead to new ideas and inventions.

▶▶ **Describe** the six simple machines.

Terms to Learn

- **capital**
- **machine**
- **primary tool**
- **resource**
- **skill**
- **tool**

Standards

- Characteristics & Scope of Technology
- Relationships & Connections
- Cultural, Social, Economic & Political Effects
- Role of Society
- Information & Communication Technologies
- Influence on History
- Design Process
- Manufacturing Technologies

resource
a source of information, capital, materials, equipment, or support

Technology Resources

▶ **What resources do you need to create technology?**

Technology requires knowledge, skill, raw materials, tools, and energy to create the products and services that we want and need. The same resources used in ancient days are still being used today to develop new technology.

Our early ancestors knew very little compared to what people know today. However, they were still able to use their limited knowledge and their hands to form raw materials, such as stones, into useful tools. Their tools were simple and crude by our standards. See Figure 2-1. Our tools will seem simple and crude to people in the twenty-third century.

It was our ability to create technology that led to our powerful position on our planet. Our early weapons turned us into hunters rather than the hunted. Our telescopes and microscopes gave us the eyes to see what had once been invisible. Our telephones and satellites gave us the ability to hear even whispers over fantastic distances. The computer gave us the ability to recall the smallest details and solve problems in seconds.

To create new technology today, people use the same **resources** that were used by our earliest ancestors. They include all of the following:

- People
- Information
- Tools and machines
- Capital
- Time
- Materials
- Energy

Figure 2-1 Stone-headed axes like this one were used by people in prehistoric times.

People

▷ Why can't we create technology without people?

Any list of resources needed for technology must start with people. People are creators of technology. If we subtract humans from our list of resources, what do we have left? Who would create technology to meet human needs and wants?

Do you like to make something new or do things in a new way? That means you're creative. New technology has always been developed by people who used their imaginations to find new solutions to existing problems. Do you use a DVD player to watch movies? Its inventor came up with a way to reproduce movies that improved on the image quality of videotapes.

Can you define the term **skill**? Skill is what you develop when you combine knowledge and practice in order to perform an activity well. Technology also requires people for their skills. They use these skills to convert their ideas into real products, systems, or processes. Are there activities, such as sports, in which you have already developed a high level of skill?

People have not only learned to create new tools, but they have also passed their inventions on to future generations. Each generation has the opportunity to benefit from the accomplishments of the past.

skill
the combination of knowledge and practice that enables you to do something well

LOOK TO THE FUTURE

More Information, Less Space

In 2002, researchers at IBM demonstrated a new technology for storing computer data. Millipede stores data in the form of tiny indentations in a thin plastic film—1 trillion bits of information in one square inch. That's equal to the information on 25 million printed pages or 25 DVDs. The first commercial applications will probably be in cell phones, watches, and PDAs (personal digital assistants). As the technology improves, Millipede's storage capacity will grow even larger.

Figure 2-2 Skilled people are needed in the technological world.

TECH CONNECT
LANGUAGE ARTS

Pass It On

You weren't born knowing how to draw or write—someone had to teach you. Without some form of language to pass skills on, would we still be able to accomplish things we do today? Communicating information is the first step in teaching someone how to do something.

ACTIVITY Write basic instructions about how to perform a task, keeping in mind the knowledge and experience of your audience.

People are also the users of the products that their technology has built. Between the designer and the user are many jobs that must be done by people. People build the tools and machines, set up the factories, run the machines, and finally package and ship the products. Other people are in the service area of technology. They sell, install, and repair these products. See Figure 2-2.

Information

▷ Why do we need information to create technology?

Information can lead to knowledge, learning, scholarship, understanding, and wisdom. All are needed to create technology.

We use information, skill, and natural resources to meet our needs and wants. If a chimpanzee takes a branch (natural resource) and moves an object into its reach (skill), it is using technology to get food (need). A bird's nest is another example. It is a complex construction project for which a bird has used technology.

When our early ancestors used a stick to gather food, they used elementary technology similar to that used by a chimp. This basic tool was refined by each generation and passed down to us. People learned (gained knowledge) that a stone attached to the stick improved its performance. Others learned that the reaching stick could be thrown as a weapon.

Tools and Machines

▷ Why are tools and machines needed to develop technology?

Did your teacher ever complain that you came to school "without the tools of learning"? Your teacher considered pencils, pens, and books your tools. People consider all devices that help them perform their job as "tools of their trade." **Tools** increase our ability to do work. If learning is your work, then pens, pencils, and books are your tools.

Today all people use tools to help them perform their jobs better. Can you think of an occupation that is performed without the use of any tools?

The first tools invented were all hand held and muscle powered. These early tools were used to construct things that met early human wants and needs. These tools were also used to make other tools. Without the development of these original **primary tools**, more complex technology would never have developed. Primary tools increased a person's ability to hold, cut, drill, bend, and hammer materials. See Figure 2-3.

Machines are often referred to as tools. A tool becomes a mechanical machine when a power system takes advantage of certain scientific laws and makes the tool work better. You will learn more about power and power systems later in this book.

tool
an instrument or apparatus that increases your ability to do work

primary tool
a basic tool that is hand held and muscle powered

machine
a tool with a power system that takes advantage of certain scientific laws that enable the tool to work better

Figure 2-3 Several primary tools are built into a circular saw. Can you identify how they are used?

Small Machines, Big Achievements

Some machines of the future will have gears so small that you'll need a microscope to see them. These tiny machines could build materials completely from scratch by rearranging atoms. Raw carbon, for example, could be transformed, atom by atom, into a perfect diamond!

All mechanical power systems use one or more of the following six simple machines to change direction, speed, or force. See Figure 2-4. Very often complex machines will use a number of these simple machines in combination:

- **Wheel and axle.** The best-known simple machine is the wheel. It is round and connected to an axle, which is the center shaft. Gears and cams are related to the wheel and axle. The gear is a wheel with teeth around its circumference. The teeth allow gears to mesh (fit together) without any chance of slipping. Your bicycle has a gear that you turn by rotating your legs with your feet on the pedals. This gear meshes with the chain that meshes with the gear that drives your rear wheel. The cam uses the principle of the wheel in combination with the principle of the inclined plane. Most cams look like wheels that aren't perfectly round.

- **Pulley.** The pulley uses the principle of the wheel in combination with a rope or chain to lift heavy objects. In a one-pulley system, the full weight of the object must be lifted by pulling the rope. In a two-pulley system, the person pulling the rope feels as if the object weighs one-half its actual weight.

- **Lever.** The lever is a bar that turns on a fixed point and allows you to lift something heavy. When you were younger, you probably played on a seesaw. This playground toy consists of a long board that is raised off the ground and fastened securely at its middle so that each end can swing up and down. If you place heavier people closer to the middle of the board, lighter people can easily lift them.

- **Inclined plane.** The inclined plane is an angled ramp that makes it easy to raise things by rolling them uphill. Cars driven into parking garages often travel along an inclined plane.

- **Wedge.** The wedge is actually a small inclined plane that is used to spread things apart. Its shape converts downward movement into a force that separates things. The axe is a wedge on a stick. Scissors are two wedges joined together. The plow is one of the most important wedge-shaped tools ever invented.

- **Screw.** The screw is actually an inclined plane that runs around a metal rod. Notice how the ramp in a parking garage looks like a giant screw.

FIGURE **2-4**

Simple Machines

WHEEL

RESISTANCE
AXLE
EFFORT

PULLEY

EFFORT
RESISTANCE

CLASS 3 LEVER

RESISTANCE
FULCRUM
EFFORT

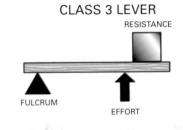

INCLINED PLANE

RESISTANCE
EFFORT

WEDGE

EFFORT
RESISTANCE

SCREW

RESISTANCE
EFFORT

Not all machines have mechanical power systems. Some machines use electronic power systems. The computer is an example of an electronic machine. Its power transfer system has no moving parts. It works by pushing electrons through a conducting material (see Chapter 5, "Electricity to Electronics").

Still other machines are biological in nature. You are a perfect example of a biological machine. In Chapter 15, "Agricultural Biotechnologies," you will learn how scientists are turning cells into machines that can manufacture needed chemicals.

capital
accumulated wealth, which may be money, credit, or property

Capital

▷ Why do we need capital to develop technology?

Capital is money, credit, or property—in other words, accumulated wealth. At the dawn of technology, primitive inventors probably created their tools without the help of others. Our earliest ancestors didn't need capital.

Later inventors couldn't get all the necessary tools and materials without the assistance of others. If they had to purchase or trade for tools, materials, and labor, they were using some form of capital. Capital's importance grew.

Using investment capital, people can now buy the other resources needed to create technology. Today, a team approach is used to develop most new ideas. Under the direction of large corporations, experts are hired. The company obtains materials, tools, information, and skilled and creative people. These companies construct the needed laboratories, mix all the resources together, and then wait for team effort to deliver useful products.

However, spending great sums of money won't guarantee success. You can still find independent, underfinanced inventors at work in the world. Can a basement or garage inventor still succeed? Such a person might create the next invention that will spur the growth of a multibillion-dollar business. See Figure 2-5.

Figure 2-5 Young people are inventors, too. Toys may provide inspiration for new inventions.

Time

▷ How does time affect the development of all technology?

Can you name something that takes place instantly without the passing of time? Everything takes time, even if the time is measured in millionths of a second.

People are paid for the time they work. Products for which people supply most of the labor are usually more expensive than products made by machines that are controlled by other machines.

All food recipes call for measured amounts of different ingredients. After the ingredients are put together, you must stir, mix, heat, or cool the contents for a specific amount of time. Too much or too little of any ingredient, including time, could ruin your results. Whether you are making a cake, building a car, or designing a new product, your results will take shape over time.

CAREERS

Technology Editor
Publisher of technology textbooks is looking for a person to develop and edit books about technology. The applicant need not be an expert in any one area but should show a general understanding of several technology areas. Must have good writing and speaking skills. Applicant must be willing to learn and be able to follow directions.

Following Directions

Making sure to follow directions will help you save time, be more efficient, and work well with others.

Finishing your formal education doesn't mean you are finished with learning. Even if you have been trained in school to do a certain job, you will still have lots to learn. You will be asked to complete tasks and follow directions. To be able to follow directions is a very important job skill. You might want to practice it now. Here are a few suggestions to help in following directions at school and at home:

✔ **Listen very carefully, even if you think you already know how something is done.**
✔ **Take notes. They will help you recall the directions later.**
✔ **Try to visualize the steps in the directions.**
✔ **Ask questions. If you do not understand the directions, ask questions until you understand the procedure.**

TABLE **2-A**

Material Resources

Raw materials are materials in their natural state. They are found on or in our land and sea and in the air. They include rocks, metal ores, crude oil, coal, sand, soil, clay, animals, and plants.

Processed materials are natural resources that have been changed into a more useful form. They include lumber from trees, leather from animals, and stone from rock quarries. When you look at a processed material, you can usually tell where it came from.

Manufactured materials are created when natural resources are altered by processes that do more than merely change their size or shape. Examples include gasoline, paper, concrete, and metals. Manufactured materials are so changed in form that you can't recognize where they came from.

Synthetic materials are created artificially. Industrial diamonds, human-made rubber, and plastics are all synthetic materials.

Materials

▷ **Why do we need materials to develop technology?**

Materials are needed to create the products and processes of technology. People have also learned to create new materials by combining or refining natural resources in ways not done by nature. Material resources can be classified by how they were formed and include raw, processed, manufactured, and synthetic materials. See Table 2-A. Sources for materials are first located. Then they are gathered by means of such processes as harvesting, drilling, and mining.

Energy

▶ Why must we use energy to create technology?

After a hard workout in a gym, did you ever feel like you had run out of energy? Your muscles use a great deal of energy to perform the tasks that you do daily. Even when you are at rest, you use energy to breathe, think, and pump blood throughout your body.

Energy is the source of power that runs all of our technological systems, too. There are many different sources of energy. These sources may be natural or synthetic and are in limited or unlimited supply. You will learn more about this important resource of technology in Chapter 4, "Energy and Power for Technology."

SECTION REVIEW 2A

Recall ▶▶

1. What seven resources do we need to create new technology?
2. How do imagination and skill relate to technology?
3. Name the six simple machines that are used to change direction, speed, or force.

Think ▶▶

4. If you traveled back in time, in what ways would you be different from our earliest ancestors?
5. Rate the level of importance of each resource of technology. Indicate your reason for placing the resources in this order.
6. Name and describe an occupation in which people don't use any tools at all.

Apply ▶▶

7. **Teamwork** Work in teams of three or four. Using a maximum of five sheets of paper, build a paper bridge that spans nine inches. The paper can be cut, bent, folded, taped, or glued. However, you can't use staples or any other material not listed above. Record the weight of your bridge. Determine the strongest bridge by adding weights to the center of each to see which holds the most.
8. **Construction** Build an object using only raw materials, only processed materials, only manufactured materials, or only synthetic materials. Your teacher will assign the type of materials that you or your team can use. You may use glue to hold the parts together. Your object may be useful or just an artistic design.

Objectives

▶▶ **Describe** systems and subsystems.

▶▶ **Identify** the difference between open- and closed-loop systems.

▶▶ **Explain** how systems play a part in technology.

Terms to Learn

- closed-loop system
- feedback
- input
- open-loop system
- output
- process
- subsystem
- system

Standards

■ Core Concepts of Technology

■ Design Process

system
a group of parts that work together in an organized way to complete a task

subsystem
a system that is part of another, larger system

Technology Systems
▶ What are systems and subsystems?

When we talk about a **system**, we are talking about an organized way of doing something. A system is made of parts that work together to complete a task. Systems are one of the building blocks of technology.

Your body digests food through the combined efforts of the organs that make up your digestive system. Many of your parents get to work using a highway system or public transportation system.

In mathematics you are taught how to use different systems to solve addition, subtraction, division, and word problems. In science you are taught the scientific method, which is a system used to solve science problems. In social studies you study our system of government. Our country has formed many governmental, military, legal, and educational subsystems to carry out the basic ideas of our system of freedom and democracy.

Subsystems are smaller systems that exist within larger systems. A subsystem can't usually function properly without its surroundings. For example, the digestive system is a subsystem of any living organism. The jet engine is one of many subsystems of the airplane. However, some systems can be both. The airplane is a complete system, but it is also a subsystem of the transportation system.

Technology systems turn ideas, facts, and principles into the things that we want and need. This is done through the skilled use of people, information, capital, tools, machines, materials, energy, and time.

Diagramming Systems
▷ Why do people diagram their plans?

Football coaches often diagram plays to help team members understand what they are going to do. Technology uses a method of diagramming, originally developed by engineers, that helps people understand how any system operates. It shows how one part of a system relates to the other parts. This same diagram can also help people organize their plans when developing new ideas.

Open-Loop Systems

▶ What is an open-loop system?

When a system includes no way of measuring or controlling its product, it is called an **open-loop system**. Old-fashioned bathtubs, stoves, and traffic lights are all examples of open-loop systems. What makes them open-loop systems? What is not being controlled?

These devices cannot shut themselves down at the appropriate time. A human being must do it. The bathtub overflows. The stove burns the food. A red traffic light stops the flow of traffic in lanes with cars, while at the same time giving a green light to empty lanes. Can you think of other open-loop systems?

The open-loop system includes three parts: input, process, and output. See Figure 2-6. In our system diagram, **input** includes the resources, ideas, and activities that determine what we need to accomplish. For example, suppose you want to run for school president. You and your friends decide to make campaign posters and buttons that will display your picture and a slogan. All the steps that lead up to the idea of creating these posters and buttons are part of input.

The **process** is the conversion of ideas or activities into products through the use of machines and labor. Designing your buttons and posters and determining the steps involved in making them are part of the process

open-loop system
a system with no way of controlling or measuring its product

input
whatever is put into a system

process
the conversion of a system's input into a useful product

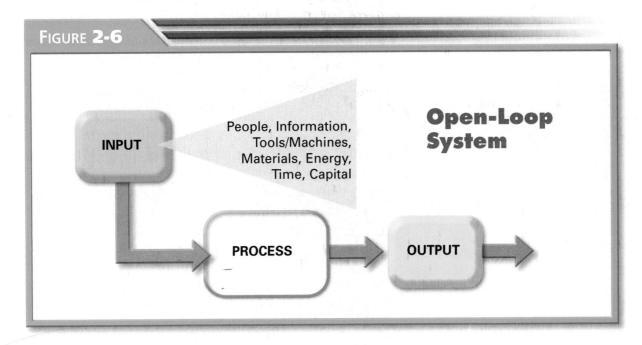

FIGURE **2-6**

INPUT

People, Information, Tools/Machines, Materials, Energy, Time, Capital

Open-Loop System

PROCESS

OUTPUT

FIGURE **2-7**

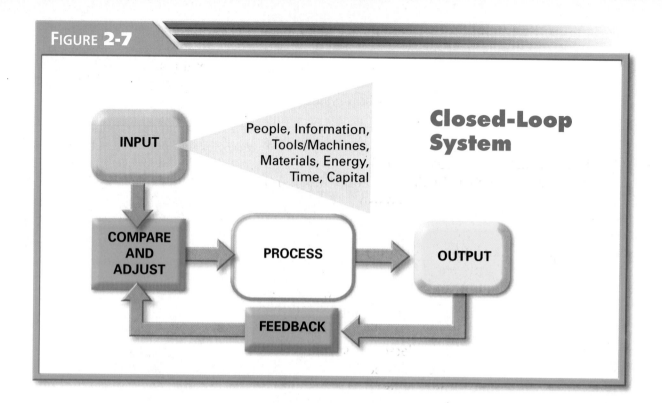

Closed-Loop System

shown on our system diagram. The production of your posters and buttons would also be part of the process. Different products and different technologies involve different kinds of processes.

Output is simply what the system produces. Your posters and buttons would be the output of your election planning.

Simply stated, the three parts of an open-loop system diagram contain an idea (input), which leads to an action (process), which leads to an outcome (output).

Can an open-loop system measure effectiveness? Will you be able to tell if your buttons and posters accurately convey your message to the other students at school? How would you add a controlling device to regulate an old-fashioned bathtub, stove, or traffic light? How would you measure the effectiveness of your buttons and posters?

Closed-Loop Systems

▷ What is a closed-loop system?

When an effort is made to control quality in an open-loop system, you are able to get information about your end product (output). If you knew that your posters were turning students off, what would you do? You would change your posters to correct the problem.

output
what a system produces

Feedback is the name given to the part of a system that measures and controls the outcomes of the system. When feedback is used to change an open-loop system in some way, the feedback becomes a kind of input. This "closes" the loop and makes the system a **closed-loop system**. See Figure 2-7. It is more stable than an open-loop system.

Can you name examples of closed-loop systems? The heater in a fish tank warms the water in the tank and shuts off when the water reaches the desired temperature. If it didn't shut off, you would have cooked fish.

Many other technological devices use feedback loops. A traffic light that has metal detectors in the intersection automatically stays green for lanes that have traffic. It remains red for empty lanes.

This same kind of feedback is used to open doors in some public buildings. In one system a motion detector placed above the door senses your movement in the same way that a police radar system detects a speeding car. Ovens that have cooking probes are another example of closed-loop systems. When the food is cooked, the probe tells the oven to shut down. Complex systems have many layers of feedback and control. See Figure 2-8.

Technological systems can be connected to one another. Some houses, for example, have computer systems that control the oven and other household systems. One system can also be part of another, larger system. The timer on an oven is a small system that is part of the larger oven system.

feedback
the part of a closed-loop system that provides control or measurement of the product

closed-loop system
a system that has a way of controlling or measuring its product

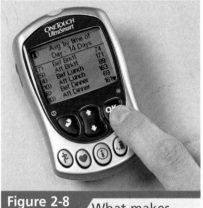

Figure 2-8 What makes this glucose-monitoring machine for diabetics part of a closed-loop system?

SECTION REVIEW 2B

Recall ▸▸

1. What are systems and subsystems?
2. Why do people diagram plans?
3. What is an open-loop system?
4. What is a closed-loop system?

Think ▸▸

5. What is the simplest operating system that you can think of?

6. When would an open-loop system be preferred to a closed-loop system?

Apply ▸▸

7. **Assemble** Construct two LEGO® objects. Make one of them an example of an open-loop system. Make the other an example of a closed-loop system.

Objectives

▸▸ **Describe** how certain requirements affect the development of products and systems.

▸▸ **Explain** criteria and constraints, optimization, and maintenance and control.

▸▸ **Identify** tradeoffs and discuss their effect on product development.

▸▸ **Explain** how technology can have positive and negative effects.

Terms to Learn

- **constraint**
- **criteria**
- **optimization**
- **trade-off**

Standards

- Characteristics & Scope of Technology
- Core Concepts of Technology
- Role of Society
- Attributes of Design
- Troubleshooting & Problem Solving
- Use & Maintenance

criteria
standards that a product must fulfill in order to be accepted

Developing a System or Product

▷ When you purchase a new product, what are your primary concerns?

All products are designed and built to meet the needs or wants of the people who will buy them. The most expensive advertising campaign will fail if the product doesn't fulfill wants or needs. Products must be safe, attractive, useful, and reliable. They must sell at a price consumers are willing to pay. See Figure 2-9.

Some new products may be developed before consumers have expressed a desire for them. For example, most people did not want or need a personal video player or an MP3 player before it existed. Public relations and marketing firms can help build consumer interest in these and other products.

Figure 2-9 Would you be interested in buying this product at this price? If the price was more reasonable, would you buy it?

Criteria and Constraints

▷ What are criteria and constraints?

Product designers determine the features products must have. The **criteria** (requirements or specifications) for a winning product might include much more than those relating to usefulness and practicality. Consumers might demand a certain level of performance, special features, a specific designer label, or a low price. Designers look for a winning combination.

For example, to replace the World Trade Center in New York City, architects were challenged to meet certain criteria. Among other things, the design had to memorialize the people who died, create areas for a wide range of uses, improve access and walkways, and blend into lower Manhattan's skyline.

However, there are usually certain **constraints** (limits) on any product design. Limits for the new World Trade Center included the size of the site (16 acres), need for enhanced security, and safer staircase and elevator bank design. The architectural plans also had to meet all the structural requirements that go into the construction of any tall structure. The winning proposal was created by Daniel Libeskind (LEE-behs-kind). See Figure 2-10.

Optimization and Trade-offs

▷ Why should price be only one of many considerations when designing a product?

Your goal as a designer is to create the best system, product, or process by using all of the best tools, materials, and processes available to you. In technology we call this **optimization**. You optimize your product by making the most of its positive features and reducing any negative features. Of course, you must do this within the criteria and constraints.

However, building the best product also involves trade-offs. A **trade-off** is a compromise. You give up one thing in order to gain something else. Many trade-offs involve cost. You choose components that will perform the needed operations at the most reasonable price. You choose the

constraint
a restriction on a product

optimization
creating the most effective and functional product, system, or process

trade-off
a compromise in which you give up one thing in order to gain something else

Figure 2-10 The design for the World Trade Center in New York City includes a building that, with its spire, is 1,776 feet tall.

least expensive materials and the most economical processes. One reason why a perfectly safe car has never been produced is that its cost would eliminate all buyers.

As the designer, you need to ask: Are the materials readily available? Are they really the best choice? What about waste? All these considerations may involve trade-offs.

Maintenance and Control

▷ **How do modern control systems help you maintain performance?**

You probably remember the story of Jack and the beanstalk. The beanstalk grew much larger than normal and got Jack into quite an adventure. The Beanstalk Principle reminds us that systems, processes, and products should not grow beyond an optimal size. If this principle is ignored, you can expect system failures. The larger and more complex a system is, the harder it is to keep it working properly and control the way it functions.

Systems require maintenance to keep them working. To ensure proper maintenance, many products, such as cars and computers, are built with control elements that keep watch over the system and report any problems. Prompt maintenance then prevents system failures.

Other products may not be worth maintaining. It may be less expensive or more convenient just to replace them. For this reason, ours has been called the "throw-away society."

Figure 2-11 This DVD player in a van increases transportation comfort—a positive impact. Fuel burned in vehicles creates air pollution—a negative impact.

Impacts of Technology

▷ **Should we develop a new technology if it could possibly have negative effects?**

As you know, surprises can be good or bad. In the same way, unexpected effects of technology can be positive or negative. See Figure 2-11. There is no doubt that technology has extended our human capabilities. Inventions and innovations in all seven areas of technology seem to constantly improve our lives. (An invention is a new product. An innovation is a change made in an existing product.) However, unexpected effects can hurt the very people that the technology was designed to serve.

For example, the internal combustion engine, the automobile, and our system of roads and highways have given us the ability to travel in personal comfort. Some of these technologies have also polluted our atmosphere and may have helped cause global warming. Technologists continue to work at reducing these negative impacts.

Medical, agricultural, energy and power, information and communication, manufacturing, construction, and transportation technologies are all discussed in this book. As you explore each area, look for the positive as well as the negative effects. To truly become technologically literate, you must learn how to weigh one against the other.

SECTION REVIEW 2C

Recall ▸▸

1. What do we mean by criteria and constraints?
2. What is optimization?
3. Describe how maintenance and control are used to increase the life of some products.
4. What are trade-offs and what effect do they have on product design?

Think ▸▸

5. Why must technologists consider cost when they get ready to manufacture a product?
6. Should we stop developing new technology if we can't ensure that its use won't have unexpected consequences? Why?

Apply ▸▸

7. **Teamwork** Manufacturers often build very similar competing products. Working in teams of three or four, choose a communications, clothing, or personal product that has many competitors. Compare criteria, constraints, optimization, maintenance, and control (where applicable) that the different manufacturers addressed when building their product. Will this knowledge help you decide which product to purchase? Present your findings in a written report, chart, video, or PowerPoint® presentation.

Summary Activity

For each numbered blank, pick the answer from the list on the right that makes the most sense in the entire passage. Write your answer on a separate sheet of paper. No answer will be used more than once.

Most of our __1__ developed because individuals had ideas that they were able to turn into useful devices. The __2__ resources of technology include people, information, tools and machines, capital, time, materials, and energy.

Today, large companies buy all the __3__ needed to create new technology. They hire __4__ with the knowledge and skill needed to use tools and machines to convert materials into __5__ products. The company supplies the energy to run the machines and gives its people the time they need to create new ideas.

Technology has developed many systems and subsystems. A car engine is a(n) __6__ of the automobile. An automobile is a complete system, but it is also a subsystem of our transportation __7__ .

In an open-loop system __8__ is the information, ideas, and activities needed to plan for production. Process is the construction stage, where machines, labor, and materials __9__ the product. Output is what the system has produced. An open-loop system changes into a closed-loop system when __10__ is introduced.

Products are designed to meet the needs and __11__ of consumers. Marketing firms use advertising to create desire for new products. Product designers determine what features will make the products sell, taking into consideration a product's __12__ and constraints. They try to make the most of its positive features while reducing its negative features. Trade-offs may be needed to turn a design into a new product that can be built at a(n) __13__ cost.

To protect people and the environment we need to monitor new technology to determine possible __14__ outcomes. As a technologically __15__ person, you must weigh the positive outcomes against the negative outcomes.

Answer List

- build
- criteria
- feedback
- input
- literate
- negative
- new
- people
- reasonable
- resources
- seven
- subsystem
- system
- technology
- wants

Comprehension Check

1. What is technology?
2. What is a primary tool? Name two examples.
3. What is the difference between a tool and a machine?
4. Is the fabric in a 100 percent cotton shirt a processed material or a manufactured material? Explain your answer.
5. Is a table lamp a closed-loop system or an open-loop system? Explain your answer.

Critical Thinking

1. **Infer.** Why is skill important when creating technology?
2. **Analyze.** How might tools affect the amount of time needed to make a product?
3. **Compare.** The ZAP Company makes batteries that cost only one-third as much as other brands, but they last only two-thirds as long. Is this a good trade-off? Why or why not?
4. **Evaluate.** Describe three positive and three negative impacts of computers.
5. **Design.** Suppose your company is designing a toy truck for children twelve to eighteen months old. Write safety specifications for the toy.

Visual Engineering

Communication devices for vehicles have changed dramatically over the years. First it was radio, then tape and CD players, phones, DVD players, and Global Positioning Systems (GPS). What might be next? Develop your own idea for a new communication device for vehicles. Describe to the class what the device would do and how it would work.

TECHNOLOGY CHALLENGE ACTIVITY

Designing and Building a Game of Skill

Equipment and Materials

- paper
- ruler
- glue
- wood screws
- Masonite® board
- rubber bands
- electric drill press
- speed bores
- woodworking vises
- hand woodworking tools
- large steel ball bearings
- Styrofoam™ plastic sheets and shapes
- markers
- clay
- dowels
- scroll saw
- nails
- wood

⚠ SAFETY

Reminder
Throughout this book you will be using tools and machines to work on various activities. Be sure to always follow appropriate safety procedures and rules. Remember, safety is an attitude that you must develop and maintain at all times.

Background

New products are manufactured to meet many different needs. They include toys, electronic gadgets, cars and trucks, home appliances, clothing, sporting goods, and even business supplies. Manufacturers spend a great deal of money determining what consumers want to buy. They even hire people to test early versions of their products. The suggestions of these product testers are often included into a product's final design.

Goal

The Arcadian Pinball Machine Company has decided to design and market a new game of skill. Your team has been asked to design and build a model of this non-electric game of skill. Figure A shows two very different solutions to meet this challenge.

Criteria and Constraints

▶▶ The game must consist of a board on which a ball or puck will roll, slide, or drop.

▶▶ The game board cannot be larger than 1 × 2 feet.

▶▶ A scoring system must be included that eventually takes the player's ball or puck out of play.

▶▶ The game should have some obstacles that the player must overcome. Extra points should be given if the player is able to reach a more difficult area of the board.

▶▶ The final score achieved by a game player should reflect the player's level of skill.

▶▶ A simple ball roll or knock hockey game isn't a satisfactory solution to this problem. Your game must include at least one simple machine explained in this chapter.

FIGURE A

FIGURE B

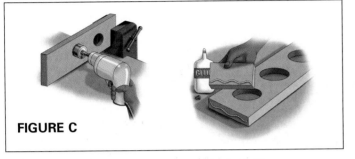

FIGURE C

Design Procedure

❶ Identify the simple machine principles used in the games shown in Figure A.

❷ Discuss ideas for a game board, controllers, obstacles, and method of scoring. See Figure B. Be sure to consider all the requirements listed under Criteria and Constraints.

❸ Develop rough sketches of all promising ideas.

❹ As a team, pick the design that you like best.

❺ List all the materials that you will need to construct the game.

❻ Call in an outside consultant (your teacher) to determine if your design can be produced with equipment and materials that you have on hand.

❼ Select construction materials for your model. This model doesn't have to be built out of the same material as the final project.

❽ Construct your model and test it. See Figure C. Does it hold your attention? Is it a game of skill?

❾ Present your solution to the class.

❿ As a class, pick the best solution or brainstorm how to combine a number of ideas presented into a super game.

⓫ Combine the talents of the entire class and assign parts that you and your classmates will build.

⓬ Construct enough games for each member of the class.

⓭ Decide, as a class, if you want to produce extra games for sale or for donation to charity.

Evaluating the Results

1. What feedback did you and your design team receive during the phases of the project? What did your team do about it?

2. What was the most difficult part of this activity and how could it have been made easier?

3. Did you learn anything that will be of use outside your technology laboratory? Explain.

4. What part of this activity did you enjoy the most?

CHAPTER 3

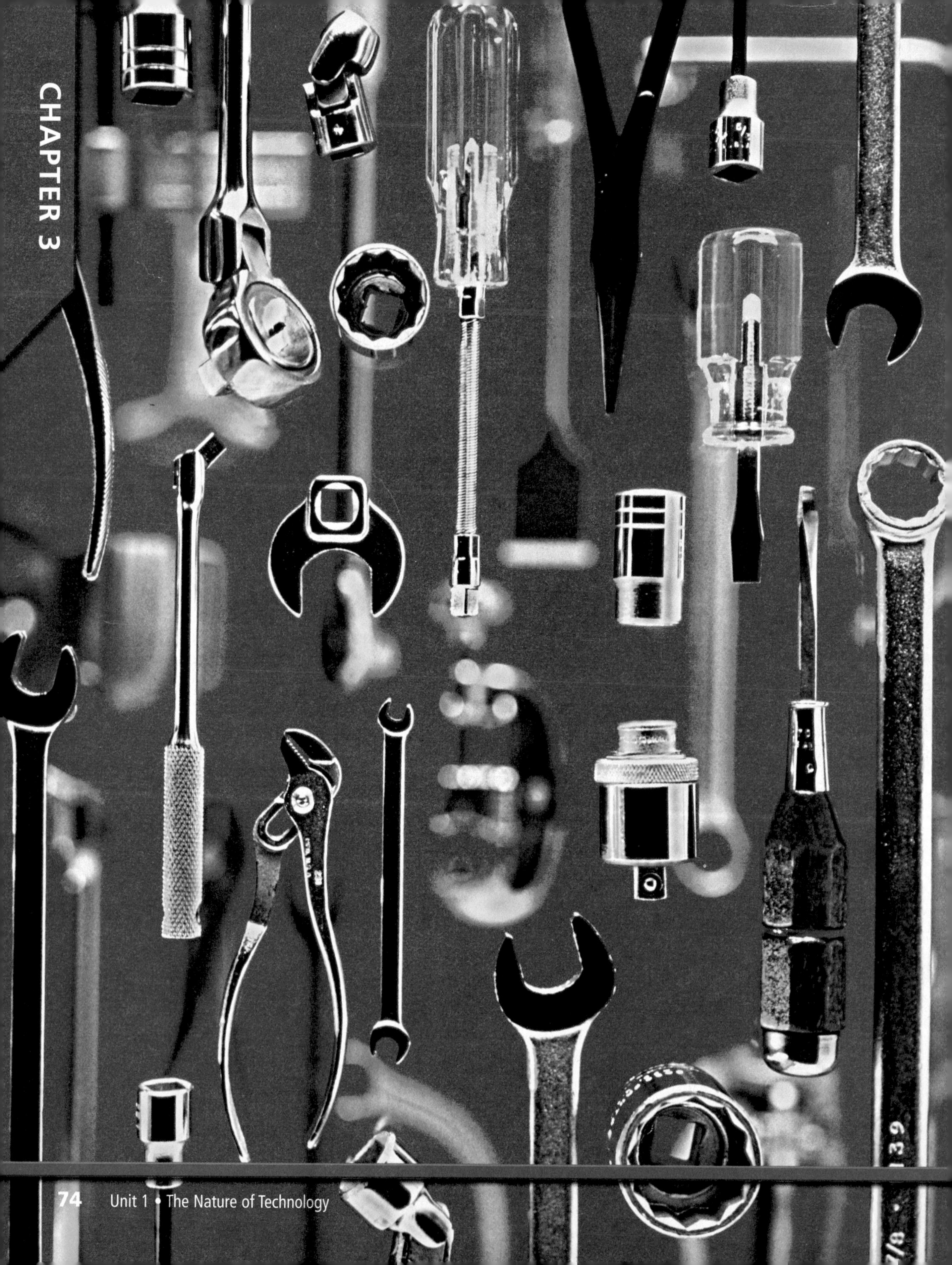

PROCESSES, TOOLS, AND MATERIALS OF TECHNOLOGY

A tool is an instrument that increases your ability to do work. The development of tools, like most technology, moved from simple to complex. The first tools were all hand held and muscle powered. Cave dwellers used them to meet their basic hunting and gathering needs. A tree limb, without much shaping, became a club. A straight branch became a spear. The first machine tool came into existence when an early inventor attached a mechanical power system to a tool.

With tools, early humans could change the materials they found in nature. With axes they could cut down trees and build houses. With simple looms they could weave cloth. As our tools changed, so did some of the materials we use to make products.

In this chapter, you will learn about the tools, materials, and processes found in many school technology labs. In lab activities, you will work with some of them.

Quality tools are important for any project.

Objectives

▶▶ **Describe** separating, forming, combining, conditioning, and finishing processes.

▶▶ **Give examples** of how the different processes are used.

Terms to Learn

- combining
- conditioning
- finishing
- forming
- separating

Standards

- Engineering Design
- Manufacturing Technologies

Technology Processes

▶ **How are products made?**

Whether a product is simple or complex, it has been made using different processes. These steps or operations change materials in some way. See Figure 3-1. For example, to make furniture, wood has to be cut, shaped, fastened, and finished. The same is true for plastic used to make wastebaskets. The plastic has to be cut and shaped, although it may not have to be fastened and finished. Each product is made with processes that are right for the material and the desired result.

Separating

▶ **How are materials separated?**

Separating is removing pieces of a material. One separating process you are probably familiar with is sawing.

separating
removing pieces of a material

Figure 3-1 Whether a product is as simple as a pencil or as complex as a car, the materials used must be processed.

When you saw a board to make it the size you want, you divide it into at least two pieces. Small pieces in the form of sawdust are removed as well. Almost all separating can be done with electrically powered tools. See Figure 3-2. You can:

- drill a hole with an electric drill or drill press.
- grind with a power grinder to remove a small amount of material or sharpen a tool.
- mill with a milling machine that uses a rotating circular cutter to shave off material.
- turn with a lathe to shape metal rods or long items like wooden chair legs.
- use a planer or shaper to slice off thin pieces of material.

Grinding is often used on metals like aluminum and steel. Planing and shaping are done on wood. Drilling is used for wood, plastic, or metal.

Figure 3-2 Different types of separating operations.

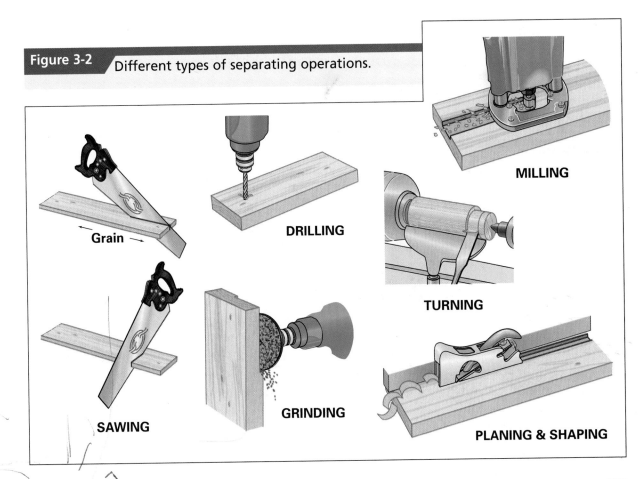

MILLING

DRILLING

TURNING

Grain

SAWING

GRINDING

PLANING & SHAPING

Forming

▷ What are some different forming processes?

Forming changes the shape of materials. If you have ever used your hands to mold clay, then you have formed a material. Forming can be done in several ways.

forming
changing the shape of a material

Bending

▷ How is bending done?

In bending, material is formed by forcing part of it to move into a different position. It is most commonly done with metal. Aluminum is bent into the shape of a gutter for a house roof or steel into the drawers of a file cabinet. Wood can also be bent into different shapes by using heat and moisture. The ribs inside a canoe and the rockers on a rocking chair are sometimes made this way.

Casting

▷ What holds the liquid in casting?

In casting, a liquid material is poured into a mold. The liquid could be a molten metal, glass, plastic, or liquid clay. The liquid takes on the shape of the mold as it hardens. See Figure 3-3. Some glass bottles are cast, as are plastic action figures. Can you think of other examples?

Compression

▷ What is compression?

In compression, a flat material is pressed into a mold by a strong force, and the material takes on the shape of the

Figure 3-3 In casting, the molten metal, glass, clay, or plastic takes on the shape of the mold as it cools.

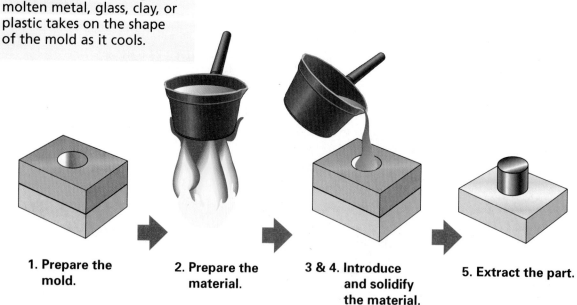

1. Prepare the mold.

2. Prepare the material.

3 & 4. Introduce and solidify the material.

5. Extract the part.

mold. This is commonly done with metal sheets to change them into such things as car doors. Powdered metal and powdered plastic can also be formed by compression. Pressure and heat bond the materials. Mixing different materials results in a solid that could not be formed any other way. Mixing powdered metal and powdered Teflon® to form a bearing is one example. It produces a material that combines the strength of metal with the slipperiness of Teflon.

Forging

▷ **How is forging done?**

The technique of shaping metal by heating it and then hammering it into shape is called forging. Old-time blacksmiths formed horseshoes, door hinges, and other items with forging. Modern forging is done with huge and powerful machines that can be as large as a house. Forging produces a very strong product. Some automobile and truck engine parts are forged.

Extruding

▷ **How is extruding done?**

In extruding, softened material is squeezed through a small opening. You extrude toothpaste from a tube when brushing your teeth. Some common materials formed by extrusion are pipe and wire. Can you think of any others?

Combining

▷ **How do we join materials together?**

It is common to join several parts together to make a finished product. This is called **combining**. For example, a wooden pencil is made from four different materials joined together: a wooden barrel, pencil graphite, a rubber eraser, and a metal ferrule that holds the eraser to the barrel. All four of those items are combined together to form a useful product. Combining is also done when materials are mixed together, as they are for paint or cake mixes.

combining
joining materials together

Mechanical Fastening

▷ **How is fastening done?**

If you have ever nailed two pieces of wood together or tightened a screw, then you have done mechanical fastening. Mechanical fasteners are small pieces of metal or plastic that hold parts together. Examples include nails,

SCREWS

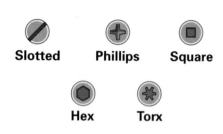

Flat head   Round head   Oval head

Slotted   Phillips   Square

Hex   Torx

Figure 3-4 Screws, nails, nuts, and bolts are all mechanical fasteners. Do you know the term used for a nail's length?

NAILS

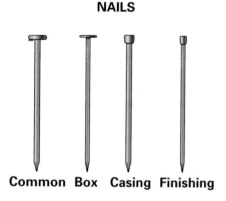

Common   Box   Casing   Finishing

NUTS AND BOLTS

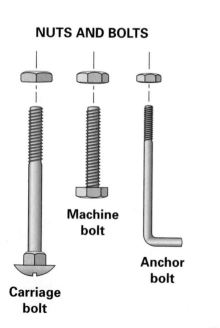

Machine bolt

Anchor bolt

Carriage bolt

staples, wood screws, nuts and bolts, pins, and rivets. See Figure 3-4. Some fasteners are designed to hold parts together permanently, like the nails and wood screws in a piece of furniture. Others allow materials to be taken apart when necessary, like the threaded nuts that hold a bicycle's wheel to the frame.

Snap Joint Fastening A snap joint holds things together without a separate fastener. It is often used with plastic parts like those in disposable cameras and CD cases. A small raised section of one part snaps into a hole in another part.

Heat Fastening When some materials are heated, they soften and flow into each other. This is a common way to combine metal parts and is called welding. A special burning gas or electricity heats the material until the parts flow together. See Figure 3-5. Brazing and soldering are similar. However, the bond is not as strong.

Plastic parts can also be combined using heat. Plastic requires much less heat than metal. If you look carefully at some plastic items, you might be able to see a spot where heat was used to melt one plastic part in order to attach it to another.

Figure 3-5 The tremendous heat generated during welding softens a material until the parts meld together.

Gluing One of the easiest ways to join parts is with glue. Glue is called an adhesive because one part adheres, or sticks, to another. Adhesives form a film on the surfaces being joined. The film adheres to both surfaces, which holds the parts together. White glues are used for wood. Epoxy or polyester resin (REZZ-in) can hold metal and ceramic parts together. Super glues can hold some plastics together, but glue does not work on all plastics.

Conditioning

▷ How does conditioning change a material?

Conditioning is done to change the inner structure of a material. When you bake cookies, for example, you are using a conditioning process. The oven's heat changes the dough from a flexible clay-like mass into a light, crispy cookie.

Materials are conditioned to improve their performance. Heat is used to harden ceramics and some metals. See Figure 3-6. After a steel piece has formed into a certain shape, heat and chemicals can be used to harden its surface. This allows the item to last longer. Other conditioning processes may be used to soften materials. Leather, for example, is softened before it is made into shoes. Still other processes relieve stresses and strains caused by heat and forming methods. Some metals, for example, become brittle and crack if they are not treated.

conditioning
changing the inner structure of a material

Figure 3-6 By heating glass, glass makers can make extraordinary pieces of art.

Finishing

▷ What is finishing?

Finishing is the last step in making a product. Its purpose is to improve the product's appearance. It can be done in several different ways. Finishing can be simply smoothing and polishing the surface of the product. The spoons, forks, and knives you use to eat with were probably finished this way.

Other finishing methods use coatings, such as paints, clear finishes, or plastic. Some clothes hangers are plastic coated, wooden pencils are often painted, and furniture may have an oil finish or clear coating to protect the wood.

3A SECTION REVIEW

Recall ▸▸

1. What is the difference between forging and casting?
2. Name the processes used to shape wire, shape a car door, glue furniture, and paint a bike.
3. Give five examples of mechanical fasteners.

Think ▸▸

4. Why do you think a screw has more holding strength than a nail?

5. Why is using sandpaper considered a separating process?

Apply ▸▸

6. **Design** The original paper clip was patented around 1900. Try to form a better one. Take a standard paper clip, straighten it out, and then form it into another shape. See if your design works better for some applications than others, such as thin sheets of paper or a large number of sheets.

Tools and Machines

▷ What tools and machines are commonly used in the technology lab?

Primary tools and machines are for general-purpose use and are found in all areas of technology, as well as in your technology lab. They are categorized by how they are used to process materials. Some are used for measuring and laying out. Others are used for holding, separating, combining, conditioning, or finishing. Tools and machines are also used to diagnose, adjust, and repair malfunctioning products.

Measuring Tools and Machines

▷ Which tools and machines are used for measuring?

Measuring tools are used to identify size, shape, weight, distance, density, and volume. Some, such as rules, measure materials directly. Others, such as the marking gauge, compass, and calipers, are used to transfer measurements from one place to another. Figure 3-7 shows several different measuring tools.

Objectives

▶▶ **Describe** and tell the purpose of several hand tools.

▶▶ **Describe** and tell the purpose of several portable power tools.

▶▶ **Explain** the importance of safety when using tools and machines.

Terms to Learn

• hand tool
• measuring tool
• portable electric tool

Standards

■ Use & Maintenance
■ Information & Communication Technologies

measuring tool
tool used to identify size, shape, weight, distance, density, or volume

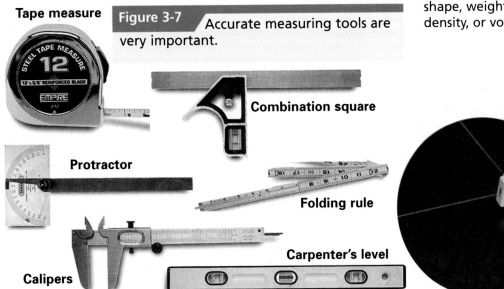

Tape measure

Figure 3-7 Accurate measuring tools are very important.

Combination square

Protractor

Folding rule

Calipers

Carpenter's level

Laser level

Metric Conversions

Although customary measures can be converted into metric measures, the results are not always as good as using metric measuring tools. A conversion table is found in the Appendix of this textbook.

ACTIVITY Measure the length and width of this book directly using customary and metric rules. Then convert the measurements using conversion formulas. Which method gives the best results?

Measuring machines are also used. Those containing lasers measure electronically. Others use infrared beams.

Using Metric Measures

▷ Why are metric measurements easy to use?

Products made in the United States are used all over the world. Products made in other countries are sold here. However, the United States uses the customary measurement system, and many other countries use the metric system, which is also called the International System of Units (SI). The metric system is based on units of ten and is often easier to work with for that reason.

The conversion to metric measurement has been slow to take hold in the United States. Switching can be costly. Machines often have to be rebuilt and expensive tools replaced. Standard sizes have to be changed. One day, however, all the products you buy and all the tools you use may finally be in only metric sizes.

Holding Devices

▷ Why are holding devices needed?

If the material that you are cutting isn't clamped in place, your risk of injury increases. To protect you and your workpiece when cutting, bending, drilling, or hammering, place the workpiece in a vise or clamp. Figure 3-8 shows different types of holding devices.

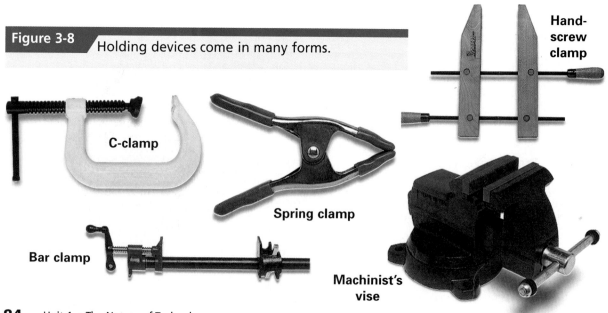

Figure 3-8 Holding devices come in many forms.

Hand-screw clamp

C-clamp

Spring clamp

Bar clamp

Machinist's vise

Hand, Portable Electric, and Machine Tools

▷ What is the main difference between a hand tool and a portable electric tool?

A hand tool or a portable electric tool can often be used for the same process. The difference is that you must supply the muscle power to work a **hand tool** and electricity supplies the power to the **portable electric tool**. Your technology laboratory might also have large machine tools that can perform the same processes. Power equipment is usually faster and more efficient. The following pages show different types of hand, portable electric, and machine tools that are often found in a technology laboratory.

hand tool
tool powered by human muscle

portable electric tool
small, portable tool powered by electricity

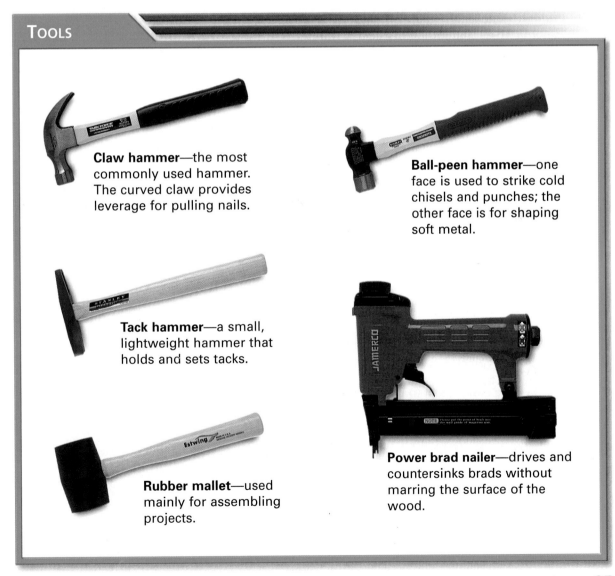

TOOLS

Claw hammer—the most commonly used hammer. The curved claw provides leverage for pulling nails.

Ball-peen hammer—one face is used to strike cold chisels and punches; the other face is for shaping soft metal.

Tack hammer—a small, lightweight hammer that holds and sets tacks.

Rubber mallet—used mainly for assembling projects.

Power brad nailer—drives and countersinks brads without marring the surface of the wood.

Crosscut saw—cuts across the grain of wood.

Band saw—cuts curves and resaws stock to thinner sizes.

Backsaw—is used to make fine cuts for joinery and is often used in a miter box.

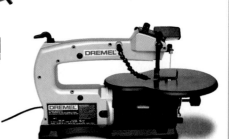

Scroll saw—the best tool for intricate and accurate irregular curves.

Hacksaw—cuts metal.

Circular saw—a portable saw that cuts wood and sometimes other materials.

Jigsaw—makes straight and curved cuts.

Coping saw—used to cut curves, scroll work, and molding as finishing trim.

Table saw—the most commonly used saw. Its size is determined by the diameter of the largest blade it can use.

Brace—bores holes in wood by hand. Special auger bits must be used with the brace.

Twist bit—designed for wood. If you use it with metal, lubricate it with machine oil.

Spade bit—the long point makes it easy to locate the hole exactly where you want it.

Electric drill—comes in three chuck sizes: ¼″, ⅜″, and ½″. Most have a reverse drive and variable speed.

Hole saw—cuts large holes in wood, plastic, and thin metal.

Countersink bit—drills a neat taper for the head of a wood screw.

Utility knife—for safety, the blade can be retracted into the handle.

Aviation snips—easier to use on metal than tin snips; especially designed to make curved and straight cuts in metal.

Screwdriver set—left: Phillips-head and standard stubby screwdriver; middle: standard slotted; right: Phillips-head.

Lineman's pliers—mainly for twisting and cutting wire.

Tin snips—used to make straight cuts in lightweight sheet metal.

Straight-jaw locking pliers—clamps firmly to an object.

Groove-joint pliers—grips objects that are round, square, or hexagonal.

Slip-joint pliers—has small and large teeth to grip objects. The jaw size can be expanded.

Needle-nose pliers—used in fine work such as jewelry-making.

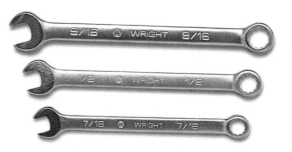

Open-end wrench—has a different-size opening at each end.

Combination wrench—has both box- and open-end heads. Both ends are for the same size bolt.

Adjustable wrench—can be used on a variety of bolts and nuts. It should be used only when a box- or open-end wrench is not available.

Thinking about Safety

▶ Why is safety instruction important?

Before you begin your first "hands-on" activity, stop and *think seriously about safety*. Equipment, tools, materials, and activities determine the dangers of a particular situation. That is why the safety rules around a swimming pool, gymnasium, and technology laboratory are different. Even the rules from one technology lab to the next may be different. That is why your teacher will provide you with a set of safety rules and safety instructions especially for your technology lab.

In general, recognizing hazards is one way to avoid danger. Accidents usually occur because people are not aware of the dangers that exist around them. You can avoid accidents just by paying careful attention to what you are doing. Another way is to develop a safe attitude. That means keeping safety in mind as you work.

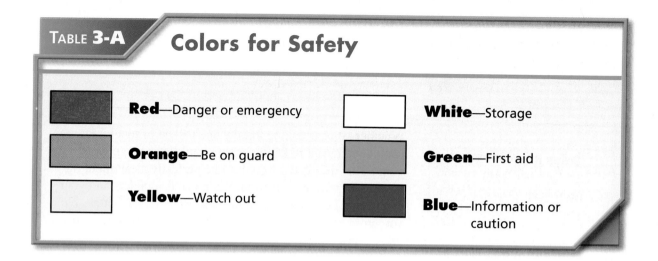

TABLE 3-A Colors for Safety

■	**Red**—Danger or emergency	□	**White**—Storage
■	**Orange**—Be on guard	■	**Green**—First aid
□	**Yellow**—Watch out	■	**Blue**—Information or caution

Six colors are used for signs and labels to indicate danger or other safety factors. See Table 3-A. Be sure you know what they mean as you work in the lab.

The Do's and Don'ts of Lab Safety

▷ What dangers should you avoid to prevent injury?

Using common sense and following some simple safety rules will make your experience in the lab a very enjoyable one. Here are a few basic rules:

- Protect your eyes by wearing proper eye protection.
- Wear a protective apron and roll up your sleeves.
- Never use equipment, tools, and materials unless your teacher has approved their use.
- Never plug in or turn on an electrical device without your teacher's permission.
- Inform your teacher if you are injured.
- Inform your teacher if you find any broken, dull, or damaged tools or equipment.

Most injuries occur because people don't think about what they are doing. Instead, they:

- touch a material that could be hot without wearing heat resistant, non-asbestos gloves.
- touch spinning rollers, which causes their fingers to be pulled into a machine.

- rest their fingers in areas where they can be pinched.
- use chemicals without wearing the proper eye and clothing protection.
- wear loose clothing and jewelry around machines that can grab them.
- use electric tools with broken wires or insulation.
- use tools that should be plugged into a three-prong plug in non-grounded wall outlets or extension cords.

Safe Use of Hand Tools

▷ **What causes most hand-tool accidents?**

Your teacher will show you the correct way to use hand tools as they are needed to complete the activities in this book. Here are some important reminders for using hand tools:

- Never use a tool to perform a job unless the tool was designed to do that job.
- Always cut *away* from yourself. Most accidents with hand tools happen because people cut *toward* themselves.
- Use sharp tools. A dull tool is more dangerous than a sharp one because you often need to force the dull tool, which makes it more likely to slip.
- Never use broken tools or tools without proper handles.

Safe Use of Machines

▷ **What safety precautions should you take when using machines?**

The machines in your technology lab are not toys. They are designed to process materials. They cut, bend, and reshape what goes into them. You don't want to have your hand bent, smashed, or cut.

To avoid any chance of injury when working with machines, keep the following rules in mind at all times:

- Stay out of the safety area that surrounds a machine unless you are the operator.
- Never use any machine until your teacher has shown you how to operate it.
- On a session-by-session basis, never use a machine until your teacher has granted you permission. This will protect you from using a machine that has been damaged since the last time you entered the room.

TECH CONNECT
LANGUAGE ARTS

The People-Friendly Robot

Robots, too, are machine tools used to make products. Most industrial robots are no more than a single arm. ASIMO, however, is a humanoid robot who works at Honda Motor Company's headquarters in Japan.

ACTIVITY Research ASIMO and his job, then write a short news story about the robot for your school paper.

- Make certain that other people are clear of the machine area before you start any machine.
- Work alone.
- Wear safety goggles.
- Watch what you are doing, don't rush, avoid talking, and concentrate.
- Shut down the machine and request your teacher's assistance if you are having any difficulties.
- Never walk away from a machine that is running.

3B SECTION REVIEW

Recall ▸▸

1. Name three tools used for separating.
2. Why is a dull tool more dangerous than a sharp one?
3. Why should you wear safety glasses or goggles when using tools to cut materials?

Think ▸▸

4. Why is a hammer considered a combining tool?

5. Experienced machinists are more likely to get hurt operating their machines than new workers who have received proper instruction. Why do you think this is true?

Apply ▸▸

6. **Safety** Read the instruction manual for a portable power tool to learn about using it safely. Make a presentation to the class about its safe use.

Engineering Materials

▶ **What are engineering materials?**

Objectives

▶▶ **Describe** some basic properties of materials.

▶▶ **Name** and **give examples** of common engineering materials.

Terms to Learn

- alloy
- ceramic
- composite
- inorganic material
- mechanical property
- organic material
- sensory property

Standards

- Influence on History
- Design Process
- Manufacturing Technologies

Materials that are physically turned into products are called engineering materials or production materials. They are the building blocks of our designed world.

Materials must be located and processed before they are used. For example, materials such as oil and natural gas are obtained by drilling. They are then processed in a refinery, where the impurities are removed.

Properties of Materials

▶ **Where do our materials come from?**

Natural materials are classified by how they originated. **Organic materials**, such as wood, leather, and cotton, come from living things. **Inorganic materials**, such as stone, metals, and ceramics, come from mineral deposits. They were never alive.

A material's suitability for a particular project is determined by checking its properties. For example, the covering on an electrical wire must be an insulator. You would look at a material's electrical properties when selecting one to cover electrical wire.

Table 3-B (page 94) shows the **mechanical properties** of materials. They determine how a material reacts to forces. **Sensory properties** are those we register with our senses. Color, texture, temperature, odor, flavor, and sound are all sensory properties. Can you think of materials that have those properties?

Chemical properties control how a material reacts to chemicals. Optical properties have to do with how it reacts to light. Thermal properties control its reaction to heat or cold. Magnetic properties have to do with a material's reaction to magnetism.

Kinds of Materials

▶ **What materials are commonly used to make products?**

Our world is filled with many useful and unusual materials. However, some are more commonly used to make

organic material
material that comes from something that is or was once alive

inorganic material
material that comes from minerals that were never alive

mechanical property
way in which a material reacts to a force

sensory property
property of a material that we register with our senses

TABLE **3-B**

Mechanical Properties of Materials

Strength of a material is determined by how well it will withstand forces like tension, compression, shear, and torsion. Tension is the force that exerts a pull from the center towards the ends of an object. Compression is the force that tends to press a material together. The resistance to these forces in a material is tensile strength, or compressive strength. When materials are pushed in opposite directions, the force is measured by shear strength. Torque, or torsion, is the twisting force on a material.

Elasticity is a material's ability to stretch out of shape and then return to the original shape.

Hardness is a material's ability to withstand scratches, dents, and cuts.

Fatigue is the ability of a material to resist bending and flexing.

products than others. Wood, plastics, metals, ceramics, and composites are discussed in this section.

Wood

▷ What determines if a tree is a softwood or a hardwood?

Each tree type has its own characteristics. However, trees fit into two categories—hardwoods and softwoods. If a tree bears cones and keeps its leaves all year long it is categorized as a softwood tree. If a tree loses its leaves during

CONIFEROUS TREE
(Balsam Fir)

Leaves

Cone

DECIDUOUS TREE
(Black Oak)

Leaf

Fruit
(Acorn)

Figure 3-9 Trees may be classified as either softwood or hardwood. Which type of wood is usually used in fine furniture?

cold or very dry seasons, the tree is categorized as a hardwood tree. See Figure 3-9.

Once harvested, trees are eventually turned into such products as furniture, fuel, construction materials, or paper. Logs are cut into lumber at a sawmill. The lumber is then seasoned to match normal humidity in the air. This is done by air-drying the wood for at least a year or drying it quickly in an oven. See Figure 3-10. The lumber is now ready for additional processing.

Figure 3-10 Wood is put into the kiln for drying.

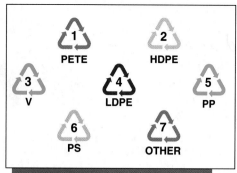

Figure 3-11 These symbols are used to identify recyclable plastics. Generally, only those with numbers 1 or 2 are recycled.

Plastics

▷ What is the difference between a thermoplastic and a thermosetting plastic?

Plastics are synthetic materials, which means they are not found in nature. Many plastics come from petroleum. Some are produced from plants. Chemical processing can create thermoplastics or thermosetting plastics. Thermoplastics are formed into products using heat and pressure. When the original product is no longer needed, the thermoplastic parts can be melted and formed into new products. When coupled with recycling programs, thermoplastics aren't bad for the environment. See Figure 3-11.

A thermosetting plastic can only be heated and formed into a product once. Thermosetting plastics are a problem to recycle. At best, they can be chopped up and mixed with other materials. Otherwise, they remain in landfills unchanged for centuries.

Metals

▷ How are alloys made?

Metals are mined from natural rock deposits. Gold mines, uranium mines, and other mining facilities are set up to separate valuable metals from the rock. A mining operation might need to process tons of rock to produce a small quantity of pure metal. For example, the Mission open-pit copper mine in Arizona processes 2,000 pounds of rock to obtain 13 pounds of copper. See Figure 3-12.

After processing, metals are used directly or mixed with other metals or materials to create **alloys**. For example, by itself copper is used to make copper wire and electronic components. It is also alloyed with tin to make bronze. Iron is alloyed with carbon and other materials to make different types of steel. Other metals commonly used in industry include aluminum, chromium, zinc, and lead.

Metals can be processed into many different shapes. See Figure 3-13. Steel, for example, is made into beams for construction that may be I-shaped, U-shaped, L-shaped, or some other design.

alloy
a material made by mixing a metal with other metals or materials

Figure 3-12 An open-pit copper mine in Arizona.

Figure 3-13 Steel is usually available in these shapes.

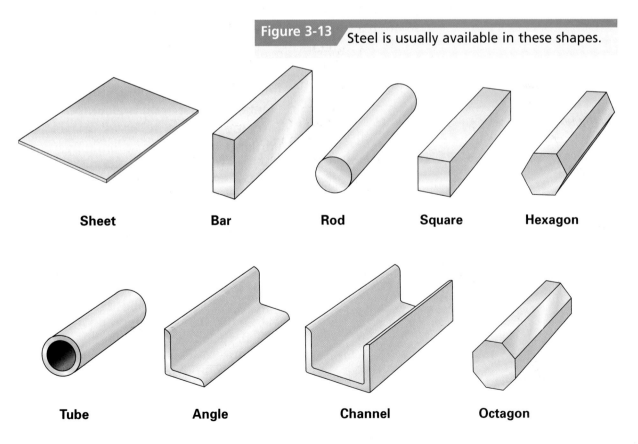

Sheet   **Bar**   **Rod**   **Square**   **Hexagon**

Tube   **Angle**   **Channel**   **Octagon**

Ceramics

▷ What are ceramics?

Ceramics are made from inorganic, mostly nonmetallic, minerals, such as clay, sand, or quartz. High temperatures are used to fuse these minerals into useful products. The firing of clay and sand are ancient technologies that were carried out before scientists could study and explain the process.

The two oldest ceramic products, pottery and glass, are very different. Pottery and almost all other ceramics except glass are thermosetting materials. Once they are heated and formed, they can never be softened again. Glass, however, can be formed again and again as often as needed. Ceramics are used to make sandpaper, pottery, fine dinnerware, bathroom fixtures, bricks, and materials used in spark plugs, furnaces, and space shuttles.

Composites

▷ How are composites better than natural materials?

Composites are created by combining two or more materials to form a new material that is better than the original materials would have been on their own. The composite's ingredients provide the correct physical properties, and a binder holds them together.

Composites are often made with glue, resin, or an epoxy binder to bond layers of wood or wood fibers, glass carbon, Kevlar®, or metal together. See Figure 3-14. Concrete is a composite that is made by mixing sand, gravel, and Portland cement. Most buildings could not be built without concrete.

ceramic
material made from non-metallic minerals that are fused together with heat

composite
material made by combining two or more materials

Figure 3-14 Wood composites from left to right: oriented-strand board (OSB), particleboard, medium-density fiberboard (MDF), hardboard, and tongue-and-groove OSB. Which one is most commonly used for flooring?

LOOK TO THE FUTURE

Futuristic Fibers

Textiles are also engineering materials. Designers are now developing special fabrics that may come in handy in the future. Imagine a shirt smart enough to monitor heart patients, a fabric balloon strong enough to lift a building, or steel fibers that can be knitted into special heat-producing blankets. Soldiers of the future may wear uniforms made of computer-controlled fibers that can transmit an image of the background, making the soldiers invisible.

SECTION REVIEW 3C

Recall ▸▸

1. What are ceramics made from?
2. Name three organic materials that technology turns into products.
3. What is the difference between a thermoplastic and a thermosetting plastic?
4. What is the difference between hardwood and softwood?

Think ▸▸

5. What properties would you look for in a material used to make sandals?

6. Examine this textbook. Name three materials used to construct it.

Apply ▸▸

7. **Design and Construction** Design a composite material that can be made in your technology laboratory. Construct the material and test its properties. Document your work.
8. **Research** Investigate one of the material properties discussed in this chapter. Develop a chart, report, or display that shows what the property is all about.

Summary Activity

For each numbered blank, pick the answer from the list on the right that makes the most sense in the entire passage. Write your answer on a separate sheet of paper.

Whether a product is simple or complex, it has been made using different processes. __1__ is removing pieces of a material, such as in sawing. Forming changes the __2__ of materials. In bending, material is formed by forcing part of it to move into a different position. In casting, a liquid material is poured into a(n) __3__. In __4__, a flat material is pressed into a mold by a strong force, and the material takes on the shape of the mold. The technique of shaping metal by __5__ it and then hammering it into shape is called forging. In __6__, softened material is squeezed through a small opening.

It is common to join several parts together to make a finished product. This is called __7__. __8__ fasteners are small pieces of metal or plastic that hold parts together. Examples include nails, staples, and rivets.

Conditioning is done to change the inner __9__ of a material. Materials are conditioned to improve their performance. __10__ is the last step in making a product. Its purpose is to improve the product's appearance.

Measuring tools are used to identify size, shape, weight, distance, density, and volume. You must supply the muscle power to work a(n) __11__ tool and electricity supplies the power to the __12__ tool. Before you begin your first "hands-on" activity, stop and *think seriously about safety*.

Mechanical properties of materials have to do with how a material reacts to __13__. __14__ properties are those we register with our senses.

Lumber comes from trees. Plastics are __15__ materials, which means they are not found in nature. Metals are used directly or mixed with other metals or materials to create __16__. Ceramics are made from __17__, such as clay. Composites are created by combining two or more materials to form a new material.

Answer List

- alloys
- combining
- compression
- extruding
- finishing
- forces
- hand
- heating
- mechanical
- minerals
- mold
- portable electric
- sensory
- separating
- shape
- structure
- synthetic

Comprehension Check

1. Give at least one example of separating, forming, combining, conditioning, and finishing.

2. How does conditioning change a material?

3. Describe and tell the purpose of a calipers, a C-clamp, a rubber mallet, a circular saw, an electric drill, lineman's pliers, and tin snips.

4. Identify at least three sensory properties and give examples.

5. Name at least three commonly used engineering materials.

Critical Thinking

1. **Infer.** Why do you think it's important that measuring tools be accurate?

2. **Analyze.** Think about accidents you have had in the past. What could you have changed about your behavior that might have prevented the accidents?

3. **Design.** Suppose you wanted to build a cabinet for your home. What material would you use to make it, and what properties of that material influenced your decision?

4. **Connect.** In what ways do you think the development of power tools changed manufacturing and construction?

5. **Decide.** Fine furniture is usually made from wood from hardwood trees that have taken many years to grow in wilderness forests. Some environmentalists object. Do you think cutting old trees to make furniture is a good idea or not? Explain.

Visual Engineering

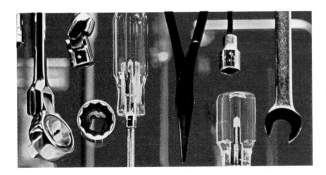

The Best Tool. When buying tools, select the best one you can afford. Buying good quality tools will save you money in the long run. Good tools can last a lifetime and will usually perform better. However, the most expensive tool is not always the best tool. Buyers should do some research, read reviews in trade magazines, and check with professionals who use the tool.

Select a tool shown in this chapter, such as an electric drill. Do some research to find out which type might be the best value for a home workshop. Discuss your findings with the class.

TECHNOLOGY CHALLENGE ACTIVITY

Making a Plaster Casting

Equipment and Materials

- heavy-duty aluminum foil
- scissors
- adhesive tape
- foam shaving cream
- thick, heavy-duty cardboard or lightweight wood larger in area than the image
- plaster of Paris
- bowl, stirring stick, and water for plaster mixture
- paint or dark wax shoe polish

Background Some buildings have cornerstones giving the date of construction or other information. Monuments, too, often have engraved lettering or designs. Copies can be cast of these images by making a mold.

Goal For this activity, you will locate an engraved image that you would like to copy. Then you will make a mold and cast a duplicate.

Criteria and Constraints

▶▶ If you want to copy an image that is part of private property, you must first obtain permission from the owner or manager.

Reminder
In this activity, you will be using tools and materials. Be sure to always follow appropriate safety procedures and rules. Remember, safety is an attitude that you must develop and maintain at all times.

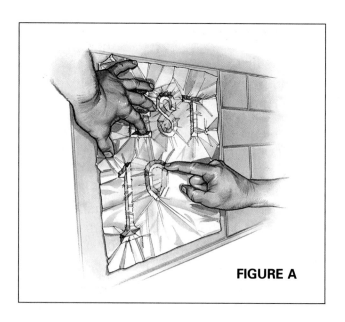

FIGURE A

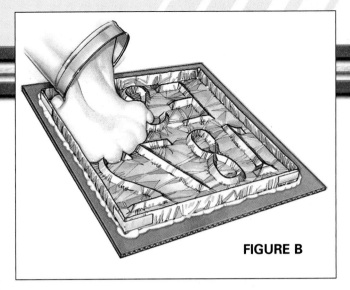

FIGURE B

Design Procedure

1 Smaller castings work better than large ones. Try to find an image that is fairly small, such as 8 inches by 10 inches. You could also consider duplicating only one section of a larger image.

2 Bring the foil, scissors, tape, shaving cream, and cardboard with you to where the image is located. Cut a piece of aluminum foil large enough to give you a border of 3 to 4 inches around the image.

3 Center the foil over the image and carefully push it into the image with your fingers. See Figure A. It is easy to tear the foil, so work slowly and carefully. If you do tear the foil, it is best to start over with a fresh piece.

4 Cover all the foil with shaving cream. Quickly place the heavy-duty cardboard or lightweight wood on the shaving cream. The shaving cream will act as a light glue. Carefully pull back on the cardboard to pull the foil from the image. Place the cardboard on the ground with the aluminum foil facing up.

5 Fold the edges of the foil up, and double them over to form sides. Place tape at the corners and other places that might need strengthening. You have now produced a mold of the engraving.

6 In the lab, mix the plaster of Paris according to the directions. Make enough to cover the foil mold. It hardens in a short time, so work quickly. Pour the plaster into your mold and shake the mold a little as you do it. The shaking helps prevent air bubbles. See Figure B.

7 After the casting hardens, peel away the aluminum foil. (The foil mold can only be used one time.) Your casting will be a brilliant white. Painting or rubbing the background with dark wax shoe polish will give it a more pleasing and readable surface.

8 Display the casting with a brief description of the original image.

Evaluating the Results

1. Did your first casting turn out as well as you expected? Sometimes air gets trapped in the plaster and leaves holes in the casting. Sometimes the casting is not as flat as you expected. If you wish, make another casting to improve your skill.

2. How long did the plaster remain workable? What properties did it have that made it suitable for this activity?

ENERGY AND POWER FOR TECHNOLOGY

Do you have a little brother or sister or know someone who does? When a child runs around, do you think, "This kid sure has a lot of *energy*"? What is this stuff called energy? Is it strength? Is it speed? Is it motion? What do you think energy is?

Nature doesn't usually give us energy in a form we can use directly. It's as if the energy were hiding—waiting for someone to find it. Coal, for example, is a source of energy, but it's just a black rock. It doesn't do anything by itself. You can't see or use the energy. Or can you?

Technology makes it possible to find and release the energy hidden in nature and to put that energy to work. Technologists are constantly looking for new and better ways to make use of nature's energy. For these reasons, learning about energy is an important part of your technology education.

Large windmill farms like this one in southern Minnesota are a source of energy.

Objectives

▸▸ **List** the different forms of energy.

▸▸ **Explain** the differences among renewable, nonrenewable, and unlimited energy resources.

Terms to Learn

- calorie
- energy
- fossil fuel
- geothermal energy
- hydroelectric power
- solar cell
- solar heating system
- wind farm

Standards

- Characteristics & Scope of Technology
- Relationships & Connections
- Environmental Effects
- Design Process
- Assessment
- Energy & Power Technologies

energy
the capacity or ability to do work

Energy Basics
▶ What is energy?

As you go through an average day, you use many different forms of energy. Electrical energy powers the light bulbs when you wake up on a dark morning. Thermal, or heat, energy from your home's furnace keeps you warm. Mechanical energy in the form of bus or car transportation might get you to school. Can you think of other forms of energy that extend human capabilities during an ordinary day?

Energy is the *ability* to do work. Simply having ability doesn't mean, however, that work will get done. A talented young man or woman certainly has ability, but he or she might take a midday nap and not do any work.

Forms of Energy
▶ How many forms of energy are there?

All energy in nature can be grouped into the following six forms:

- Mechanical energy, which is the energy of motion, as in a kicked soccer ball
- Thermal, or heat, energy, as in a hot air balloon
- Electrical energy, as in a bolt of lightning
- Chemical energy, as in a wristwatch battery
- Nuclear energy, as in a power plant
- Light (radiant) energy, as in the solar cells of a calculator

Energy cannot be created or destroyed. However, it can be changed from one form to another. As you read at the beginning of the chapter, coal is a source of energy. To accomplish some work with that energy, however, technology must change its form. For example, to use the coal to produce useful electricity, we would need to change the form of the energy many times. Figure 4-1 shows the path from black rock to electricity and includes the following steps:

1. Burning coal produces heat.

2. Heat changes water to steam.

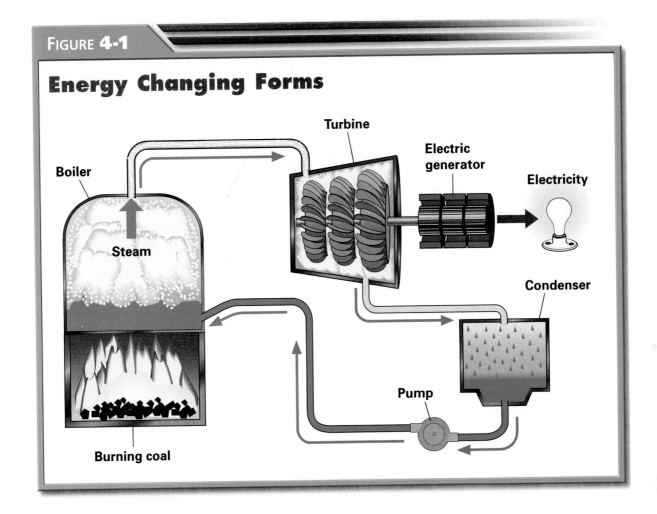

Energy Changing Forms

Boiler

Steam

Burning coal

Turbine

Electric generator

Electricity

Condenser

Pump

3. Steam powers a generator.

4. The generator produces electricity.

Energy Resources

▷ All of today's energy sources are often grouped into what three categories?

Let's take a closer look at today's sources of energy. Because of our concern with the amount of energy available for our increasing needs, these sources are often divided into three groups: renewable, nonrenewable, and unlimited.

Renewable Sources

▷ What is the main advantage of renewable energy?

Renewable energy sources are those that come from plants and animals. They can be replaced or renewed

Chapter 4 • Energy and Power for Technology **107**

Figure 4-2 Swimming uses a lot of energy. From what do we get that energy? With what units do we measure that kind of energy?

when we need more. Two examples include food and alcohol.

Food Your body requires food for the energy you use in walking, blinking your eyes, thinking, and all your other activities. See Figure 4-2. Food energy is measured in **calories**. Some foods contain more energy than others; more food energy means more calories. If you are an active person, your body usually requires between 2,000 and 3,000 calories of energy every day.

Alcohol Alcohol is a liquid made from crops such as corn and sugar cane. It can be used as fuel in special car and truck engines. Those engines can operate with regular gasoline, an alcohol-gasoline mixture, or pure alcohol. They are called flexible-fuel vehicles.

Ethanol and methanol are two types of alcohol that can be mixed with gasoline. *Ethanol* is the technical name for alcohol made from grain. *Methanol* is the technical name for alcohol made from natural gas or biomass. Although they have both been approved for use in gasoline, ethanol is used more often. Methanol is rarely used in gasoline.

Alcohol-fueled cars produce less air pollution than those that use gasoline. Alcohol added to gasoline can extend fuel supplies. A mixture of ten percent alcohol and ninety percent gasoline is sold at some service stations. It can be used in all ordinary car engines.

calorie
the measure of energy in food

Nonrenewable Sources
▷ What are fossil fuels?

Nonrenewable energy sources are those that cannot be replaced once they're gone. Coal, oil, natural gas, and uranium are examples.

HOW FOSSIL FUELS WERE FORMED

The development of fossil fuels began when plants and animals died and their remains formed thick layers on the bottom of swamps.

Eventually, the remains were covered by thick layers of earth.

Pressure and heat over millions of years slowly changed the remains into fossil fuels.

Figure 4-3 Natural gas, coal, and oil are fossil fuels that have formed over millions of years.

Coal, oil, and natural gas are known as **fossil fuels**. A fossil is what remains from a plant or animal that lived long ago. Coal, oil, and natural gas are formed from once-living plants and animals. See Figure 4-3.

fossil fuel
fuel produced from the fossils of long-dead plants and animals

Coal Millions of years ago, as plants died, their remains fell to the ground. Over time and under pressure they formed thick layers. Those layers are now coal seams. A seam is a strip of coal between other rock layers. It can be close to the surface or deep underground. Many seams are only two or three feet thick. See Figure 4-4. Today, most coal is used by power companies to generate electricity.

Figure 4-4 To obtain coal, strip mines are dug in the ground using huge shovels and trucks.

Figure 4-5 An offshore oil rig is used to pump oil from beneath the ocean floor. Workers live and work on the rig for days at a time.

Oil You can thank tiny sea animals that lived millions of years ago for the gasoline used in today's cars. When they died, their remains combined with those from plants to form crude oil. See Figure 4-5. The fuels used by cars, trucks, locomotives, airplanes, and ships come from oil. More of our energy comes from oil than from any other single source. Oil is also turned into useful products, such as plastics, paint, and asphalt for roads.

Natural Gas The slow transformation of plant and animal matter into oil also produces natural gas. As a result, it is often found near oil deposits. Natural gas is a flammable gas.

The United States produces more natural gas than any other country. Many industrial processes use natural gas, but it is mostly used as a fuel for home heating and cooking. What kind of energy is used to heat your home?

Uranium Like fossil fuels, uranium is also a nonrenewable source of energy. Unlike fossil fuels, however, uranium does not come from plants and animals. It's a radioactive, rocklike mineral that is dug from the ground. Uranium is used for fuel in nuclear power plants that produce electricity. It is also used to power some U.S. Navy ships. Most of our supply of uranium comes from New Mexico and Wyoming.

Uranium develops a large amount of heat during a controlled nuclear reaction. In power plants, the heat changes water into steam, and the steam operates generators that produce electricity.

The amount of energy in uranium is amazingly high. One pound of the material can produce as much electricity as 3 million pounds of coal. There are about 400 nuclear reactors in the world. The United States has over 100, more than any other country. They produce about fifteen percent of our country's total amount of electricity.

Unlimited Sources

▷ What are four sources of energy that we will never run out of?

Although we are using up our nonrenewable sources of energy, we have several sources that will never run out. They are unlimited. We will never use up all our solar energy, energy from wind, energy from flowing water, or geothermal energy. They are more plentiful than fossil fuels. However, they are also quite difficult to use.

Solar Energy Solar energy comes from the sun. Unlike coal, oil, wood, or many other sources of energy, solar energy is available all over the world. We use the sun's rays for light, electricity, and heat. Some homes are following an energy-conserving trend of being heated with **solar heating systems.**

An active solar heating system uses mechanical devices to take sun-heated water into the building. The most important parts of the system are large flat panels, called solar collectors, on the roof. Water flows through tubes in the solar collectors and is warmed by the sun. The water continues into the building and is used to heat the interior. Some homes in your community may have black solar collectors on the roof. See if you can find them.

A passive solar heating system uses only windows and walls to take advantage of the sun's warming rays. It is called passive because there are no mechanical devices to collect the solar heat. There are no pumps or flowing water. A passive solar home usually has large windows on the south side. Sunlight streams in and warms the interior. Some of the warmth is absorbed by the walls. The walls radiate the heat back into the room at night.

Solar energy can also produce electricity. This happens when sunlight strikes wafer-thin **solar cells**. The cells are also known as photovoltaic (foh-toh-vohl-TAY-ik) cells, or photocells.

solar heating system
a system in which energy from the sun is used to heat a building

solar cell
device that converts sunlight into electrical energy

Figure 4-6 Solar collectors produce electricity for some flashing traffic signs. These signs are easier to install. Do you know why?

wind farm
a large collection of windmills located in an area that has fairly constant winds

Solar cells are made from specially treated silicon (sand) and other materials. See Figure 4-6. Orbiting satellites obtain their electricity from solar cells. Solar cells are not used as much as they might be because their electricity costs much more to produce than electricity from a power plant.

Wind The motion of air across the earth has filled the sails of ships and turned windmills for centuries. More recently, wind has been used to turn propellers connected to generators that produce electricity. Unfortunately, wind is not a dependable source of energy. Some days it's stronger than others. Some days it doesn't blow at all.

Experiments have been conducted with **wind farms** in California and other places. Wind farms consist of a great many windmills. The wind spins the bladed rotors at about 20 revolutions per minute. The rotors are directly connected to alternators (generators) and controlled by advanced computers. Wind farms usually are located on land once thought to be useless and in areas known to have fairly constant winds. It is not usually practical for one house to have its own windmill, although in earlier

CAREERS

Wind Farm Manager
Wind Energy, Inc., a specialist in wind energy, is looking for people to manage wind farms. You will maintain the windmills and oversee the wind farm. We need people who are responsible and who have a good work ethic. Because we are on the cutting edge of energy technology, we expect to expand rapidly into other areas.

Being Responsible

When you're a responsible employee, others around you will come to trust and depend on you.

When responsible people accept a job, they see that it gets done. They are accountable for their actions and obligations. When you show an employer that you are willing to take on responsibility, you have a better chance for advancement. You can practice responsibility now, at school and at home. How can you show you're ready for responsibility? Here are some tips:

- ✔ Show that you are not afraid of being accountable.
- ✔ Look for chances to do more than you were asked to do. Volunteer for added tasks.
- ✔ When your own work is done, help others with their work.
- ✔ Look for ways to do the job better.

days windmills were often used for pumping water. See Figure 4-7.

Only certain regions of the country provide enough regular wind to make windmills practical. A windmill and other necessary equipment (like large batteries for storing electricity) are quite expensive. The propeller makes noises when it turns. Windmills also interfere with television reception, and some people find them unattractive.

So far, experiments have not shown windmills to be practical for generating large amounts of electricity. Perhaps windmill builders have yet to find the one design that will solve the problems. Do you have any ideas?

Flowing Water We use flowing water to generate **hydroelectric power**. *Hydro* means "water." A controlled amount of water flows through pipes in a dam and into a turbine. A turbine resembles a pinwheel. A spinning turbine, connected to a generator, creates electricity. About ten percent of our electricity comes from hydroelectric dams. See Figure 4-8.

Figure 4-7 Before electricity was readily available, windmills were used to pump water from wells.

hydroelectric power electricity generated by turbines propelled by flowing water

Figure 4-8 Above: Water flowing through an opening in the dam spins a turbine connected to a generator. The generator produces electricity.
Right: The Grand Coulee Dam, which is near Spokane, Washington, is the largest dam in the United States.

Diagram labels: Release gate · Electric generator · Dam · Water into sluice · Turbine activates generator as the water rushes through

geothermal energy
heat energy produced under the earth's crust

Geothermal Energy You may already know that molten rock lies far under the earth's crust. This is where we get **geothermal energy**—heat produced under the earth. Hot water or steam is created when underground water comes in contact with hot materials and surfaces as a spring. The steam can be used to produce electricity. Some geothermal electric power plants are located along the west coast of the United States. Iceland has many hot springs. Almost all the homes in the capital city of Reykjavik (REH-kyah-vik) are heated with this geothermal energy.

4A SECTION REVIEW

Recall ▸▸

1. Name the six forms of energy. Give an example of each form.
2. How are fossil fuels formed?
3. Where are a house's active solar collectors located?
4. How do dams produce electricity?

Think ▸▸

5. Some solar houses have no windows on one side. Why do you think this is so?

6. Suppose your home obtains electricity from only the wind. What would you do if the wind stopped blowing for several hours?

Apply ▸▸

7. **Construct** Build a model of an offshore oil-drilling rig. Put it on display with a description of how it operates.
8. **Construct** Build a model car that runs on solar cells.

Converting Energy to Power

▶ What is the difference between energy and power?

Objectives

▶▶ **Identify** the most common forms of power.

▶▶ **Name** several uses for each of the forms of power.

▶▶ **Explain** how energy and power technologies are selected.

Many people use the words *energy* and *power* to describe the same things. Energy and power are related to each other, but the two words have different meanings.

Terms to Learn

- efficiency
- horsepower
- hydraulic power
- load
- pneumatic power
- power

Power and Work

▷ What is power?

Power is a measure of the work done over a certain period of time when energy is converted from one form to another or transferred from one place to another. Power is a way of rating how quickly the work is done. You can probably cut grass more quickly with a riding lawn mower than with a push-type lawn mower. In both cases, the work is the same: cutting the same area of grass. However, the engine in the riding mower allows you to do it more quickly, so it is the more powerful.

One common measure of power is horsepower. A horse pulling a rope over a pulley can lift 550 pounds one foot in one second. That's the technical definition of one **horsepower**. See Figure 4-9 (page 116). If you're in very good physical condition, you might be able to develop about 0.2 horsepower for several minutes. A three-horsepower lawn mower engine produces as much power as three horses. Horsepower is one way we remember animals and their contribution to technology.

Technologists continually look for new or better ways to use power. Their goal is to improve the overall quality of life for you, your community, your state, our nation, and the world.

Standards

- Core Concepts of Technology
- Relationships & Connections
- Cultural, Social, Economic & Political Effects
- Environmental Effects
- Energy & Power Technologies

power
a measure of work done over a certain period of time when energy is converted from one form to another or transferred from one place to another

horsepower
a measurement of power based on lifting 550 pounds one foot in one second

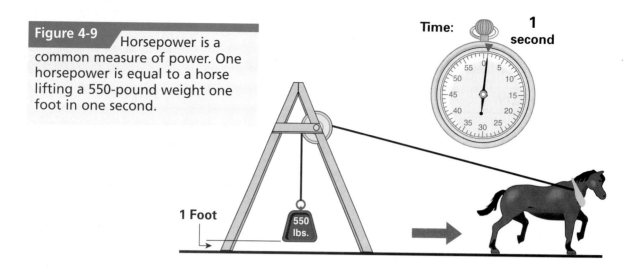

Figure 4-9 Horsepower is a common measure of power. One horsepower is equal to a horse lifting a 550-pound weight one foot in one second.

Time: 1 second

1 Foot

550 lbs.

Forms of Power

▶ What are the most common forms of power?

We commonly use three forms of power: mechanical, electrical, and fluid. See Figure 4-10.

- **Mechanical power.** As you know, mechanical energy is the energy that is involved in motion. When you use a hammer to drive a nail, you are using mechanical energy to do work. You are converting mechanical energy (the motion of the hammer) to mechanical power (the force that pushes against the nail).

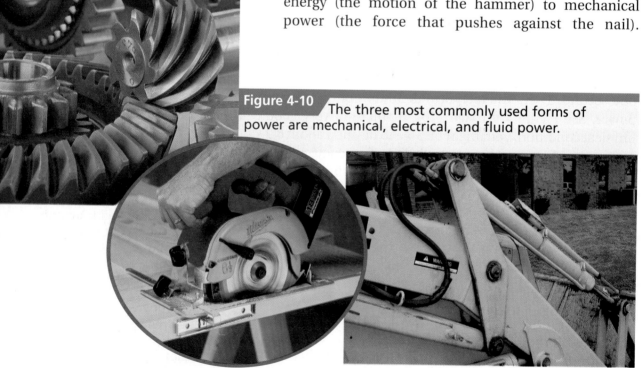

Figure 4-10 The three most commonly used forms of power are mechanical, electrical, and fluid power.

Mechanical power must often be changed or controlled before it can do useful work. Gears, pulleys, and clutches are just a few of the ways we change or control mechanical power.

- **Electrical power.** We convert energy into electrical power in power plants. That power is then used for such things as running machines, lighting our houses, and broadcasting TV shows. Electrical power is discussed in more detail in Chapter 5, "Electricity to Electronics."

- **Fluid power.** When fluids, such as air and water, are put under pressure, they can control and transmit energy. They receive that energy from an outside source, such as a motor. The greater the pressure they are under, the greater the force they exert.

  Fluid power produced using pressurized gases is called **pneumatic power**. Pneumatic (noo-MAT-ik) power operates tools in some automobile repair shops and factories. You might know someone who keeps an air compressor in the garage. Compressors are used to pressurize air for power tools and paint sprayers and to inflate tires. Putting liquids under pressure results in **hydraulic power**. Hydraulic power is popular for heavy construction equipment and factories because it provides more force than pneumatic power.

Power systems are used to drive and provide propulsion (motion or force) to other technological products and systems. Power systems must have a source of energy, a process, and a load. The energy source is part of the input in a power system. The process consists of converting the energy into a form that can do work. The **load** is the output force. The load also describes the amount of resistance the power system must overcome in order to accomplish its task. For example, if you use a truck to carry five hundred pounds of gravel across town, the load is the force the truck must exert to carry that much gravel.

The development of power systems has been important to the development and spread of our culture. The use of airplanes is an example. One hundred years ago it took months to travel around the world. Ideas, customs, and products moved slowly. Today, thanks to machines that fly very fast, every city on our planet is only hours away from every other city. Ideas, customs, and products spread quickly.

> **pneumatic power**
> fluid power produced by putting a gas under pressure

> **hydraulic power**
> fluid power produced by putting a liquid under pressure

> **load**
> the output force of a power system

TECH CONNECT
SOCIAL STUDIES

Changing the World
Henry Ford produced the first car that the average person could afford. By doing so, he changed the world.

ACTIVITY Hold a panel discussion about the ways in which wide use of the automobile has affected other aspects of life, such as health, social interaction, entertainment, the growth of towns, and development of new businesses.

Selecting Energy and Power Technologies

▶ How do you decide which energy type is best for a specific purpose?

You are faced with many choices every day. Some are technical decisions. Selecting a wristwatch is an example. You can buy a digital watch that displays numbers. However, you may prefer an analog model that has hands. Your decision is based on each item's advantages and disadvantages. Selecting specific energy and power technologies is also done by considering their advantages and disadvantages.

What kinds of things must be considered when selecting energy and power technologies? Something engineers consider when designing products and systems is efficiency. **Efficiency** is the ability to achieve a desired result with as little effort and waste as possible. An efficient machine, for example, does a lot of work compared to the amount of energy it uses. A bicycle is a very efficient machine. Pedaling a bike at twelve miles per hour requires only 0.10 horsepower.

However, much of the energy we use is not used efficiently. Have you ever felt the hood of a car that has been running? It probably felt hot. All engines that can perform work release heat energy into the environment. Overall, only about twenty percent of the energy content of gasoline is converted into mechanical power in an ordinary car engine. The rest is wasted.

Other factors that must be considered include cost, availability, ease of use, environmental and social impacts, and time required. Many trade-offs may be involved. Consider the following common situations. What factors and trade-offs do you think play a part?

- Face-to-face speech works quite well when you are near another person. A telephone is better when you are farther apart. At what distance do we change from face-to-face speech to using a phone?

- Most cars use gasoline engines. No ordinary car uses jet engines. Why do you think that's the case?

- Some carpenters use ordinary hammers and nails. Some use pneumatic nailers. See Figure 4-11. What would be some advantages and disadvantages of each method in terms of power?

efficiency
the ability to achieve a desired result with as little effort and waste as possible

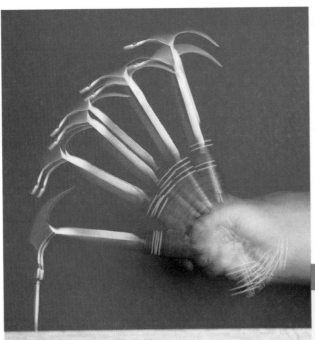

Figure 4-11 A traditional hammer and pneumatic nailer. Most carpenters prefer to use the nailer for jobs requiring many nails. *Which type of simple machine is a hammer?*

SECTION REVIEW 4B

Recall ▸▸

1. Are energy and power the same? Why or why not?
2. What is the measurement of horsepower based on?
3. Name the three forms of power we commonly use.
4. What is efficiency?

Think ▸▸

5. In the following examples, identify the energy source, the process, and the load: a biker riding down the street; a power saw cutting wood.

6. How would pedaling a twelve-speed bike be different from operating one that is less efficient?

Apply ▸▸

7. **Mathematics** Measure your body's power by running up a flight of stairs and timing yourself. Multiply your weight (pounds) times the vertical height of the stairs in feet. Then divide that result by the time in seconds multiplied by 550.

Objectives

▶▶ **Describe** forms of pollution resulting from the use of energy and power technologies.

▶▶ **Discuss** methods to slow depletion of energy resources.

Terms to Learn

- acid rain
- energy conservation
- greenhouse effect
- recycle

Standards

- Relationships & Connections
- Cultural, Social, Economic & Political Effects
- Environmental Effects
- Role of Society

Impacts of Energy and Power Technology

▷ What have been the positive and negative effects of our energy consumption?

The use of technology can have unintended consequences. For example, Americans have good transportation systems, countless electrical devices, excellent housing, and an abundance of food. Our comfortable way of life and successful economy have been made possible in part by advancements in technology. The driving force of technology has been our use of energy. Unfortunately, not all results have been good. Our energy consumption has created some serious problems. We must find ways to balance economic needs with protecting the environment.

Pollution

▷ What are some different causes of pollution?

Pollution results when contaminants—unwanted elements—get into our environment, whether in our air, our water, or our land.

Over ninety percent of our energy comes from burning fossil fuels. You may already know that burning produces a great many pollutants. Burning fills the air with haze and sometimes makes your eyes burn. It can also be a serious health threat. Each fuel produces many pollutants but there is usually one pollutant that is particularly serious in each case.

All coal has a small amount of sulfur. When coal burns, it creates sulfur dioxide (SO_2). The sulfur dioxide combines with water vapor and oxygen in the air to form a weak sulfuric acid. This acid mixes with nitric acid (NO_x), another pollutant, and falls to the earth as **acid rain**. See Figure 4-12. It can kill fish, crops, and trees. Acid rain also damages monuments and statues. There is no complete solution to the problem. However, acid rain can be reduced if power companies and industries use coal containing less sulfur.

▌**acid rain**
a weak sulfuric acid created when sulfur dioxide in the air mixes with rain and oxygen

They can also install special equipment to remove up to ninety percent of the sulfur dioxide from their smoke.

Gasoline forms carbon monoxide when it burns. Carbon monoxide is an odorless, colorless, and poisonous gas. When breathed into your lungs, it reduces the ability of your blood to carry oxygen. If a person already has a lung problem, too much carbon monoxide in the air can make the problem even more serious.

Burning fossil fuels also contributes to the **greenhouse effect**—the heating of the earth's atmosphere. This occurs when too much carbon dioxide builds up in the air. The carbon dioxide prevents heat from escaping. If we produce too much heat and it can't escape, the temperature of the earth's atmosphere may increase.

On the positive side, warmer temperatures might mean longer growing seasons for crops. More food could be produced for the world's increasing population. On the negative side, the polar ice caps could melt slightly. That would raise the ocean level and cause flooding of some seacoast cities.

Waste heat is also a water pollutant. Heat produced by power plants is discharged into lakes, rivers, and oceans. Water plants and animals can be harmed.

greenhouse effect
increase in the temperature of the earth's atmosphere caused by the rise in carbon dioxide

Figure 4-12 Acid rain is created when sulfur dioxide and nitric acid in the air mix with water vapor and fall to the earth.

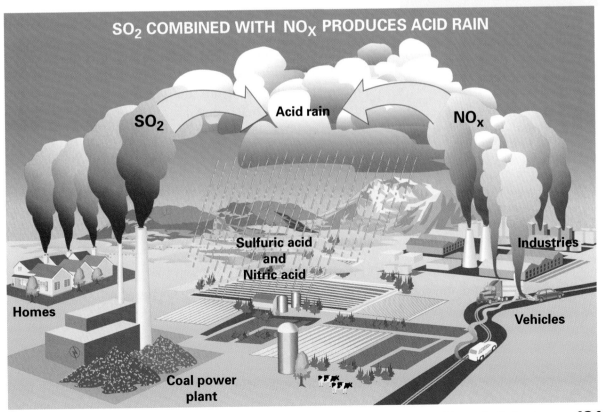

SO$_2$ COMBINED WITH NO$_X$ PRODUCES ACID RAIN

SO$_2$ Acid rain   NO$_X$

Sulfuric acid
and
Nitric acid

Homes

Industries

Vehicles

Coal power
plant

Figure 4-13 Nuclear waste is transferred to sites like this one under very strict safety guidelines.

Nuclear Waste

▷ What is done with nuclear waste?

Nuclear pollution may be the most threatening pollution of all. Nuclear, or radioactive, waste is a solid left over after nuclear fuel is used up. The waste remains dangerous for many years and can cause serious health problems. Proper disposal of this material is difficult and an important social issue.

The waste is placed in special concrete containers. The area is constantly checked for radioactive leakage. See Figure 4-13. In the future, the waste will be buried. One central location being considered by the federal government is at Yucca Mountain, about 100 miles northwest of Las Vegas, Nevada. If approved, the site would be the first underground storage area for nuclear waste.

Some people think that nuclear waste should be sent into space with rockets. What do you think? Future generations may be the ones to figure out how to use this waste in a positive way.

Depletion of Resources

▷ What can we do to maintain a supply of energy for our future needs?

Our earth has a limited supply of some energy sources like fossil fuels. However, we require more and more energy as the world's population grows. Also, each person uses more energy now than in the past, partly because of all the technological conveniences.

We have to be sure that there is enough energy to provide for human needs. One way to do that is to develop alternate sources, such as solar energy. Another is through **energy conservation**. Energy conservation is the management and efficient use of energy sources.

Energy can be conserved if we **recycle** the materials and products we use. To recycle means to use again. See Figure 4-14. Metal, glass, paper, and some plastics can be recycled. Reusing aluminum cans is one of the biggest energy savers. Recycled cans require eighty percent less energy to produce than cans made from raw materials. All glass can be recycled. Paper is usually recycled into bags, paper towels, and packaging materials. Materials that can be recycled have a special triangular symbol on them.

Recycling reduces our use of natural resources. It also reduces the amount of solid waste sent to landfills. Most cities and towns have a recycling center or a civic organization that arranges routine collections. Check to see how recycling is handled in your community.

energy conservation
the management and efficient use of energy sources

recycle
to reuse all or part of substances, such as metal, glass, paper, and plastics

Figure 4-14 These students are taking part in a recycling program. The symbol above indicates materials that can be recycled.

Garbage In, Oil Out

Imagine driving a car that runs on garbage! A new process called thermal depolymerization may help you do just that. The process turns waste products, including animal parts, plastics, medical wastes, and even sewage, into high-quality oil, natural gas, and purified minerals. These products can then be used for fuel, fertilizer, or manufacturing chemicals. No harmful by-products are produced, which makes the process environmentally friendly. It can even be used to process ordinary oil and coal so that they burn cleaner.

TECH CONNECT
MATHEMATICS

Keeping It Cool

Willis Carrier invented air conditioning in 1902, but it took movie theaters to make it popular. Although many homes did not have AC until the 1950s, theaters did, and that's where people went to cool off and learn how nice it was. People who use air conditioning wisely will save money as well as energy.

ACTIVITY A theater has maintained a temperature of 69°. To save energy they dial up to 78°. By what percentage has the temperature increased?

What You Can Do

▷ **How can you help?**

We can't stop using energy just because it causes pollution and some sources are limited. It's also true that we can't ignore these things just because we want energy. We have to find a proper balance between the two.

We can reduce our energy use. We can pass laws and obtain newspaper publicity. This will make everyone aware of the problems and what is being done by our political leaders. We can control how much pollution enters the environment from each energy source. We can never remove all the pollution, but we can always reduce the amount. Little pollution is created when we use solar, wind, water, or geothermal energy. We can develop more and better ways to use these unlimited energy sources.

What can you do? You might think that since you can't vote and don't have much money, your opinion doesn't count. That's not true. Whether or not you know it, you can influence the adults around you. They will listen to you and even follow your advice if they think that you're sincere and that you know what you're talking about.

- Keep informed. Know what's happening in your community, state, country, and world. Read newspapers Web sites, and magazines; watch the news; and communicate with others in your community. Talk to knowl-edgeable adults. Have discussions with your friends.

- Set your home thermostat at or below 65°F in the winter. You will use less energy because the heating system will operate for less time.
- If your home is air-conditioned, set the thermostat at or above 78°F in the summer. You will use less energy because the air conditioner will operate for less time. See Figure 4-15.
- Use less hot water by spending less time in the shower.
- Turn off all unnecessary lights.
- Walk or ride a bike; use buses, trains, or subways instead of automobiles.
- Use renewable or unlimited energy sources whenever possible.

Figure 4-15 By raising or lowering the thermostat by just a few degrees, you can save dramatically on heating and cooling bills.

SECTION REVIEW 4C

Recall ▸▸

1. How do you save energy by setting your home thermostat at 65°F or lower in the winter?
2. What is the greenhouse effect?
3. Why does using low-sulfur coal reduce acid rain?
4. What type of pollution may be the most dangerous?

Think ▸▸

5. What are some ways you can save energy that are not included in this section?

6. Is it a good idea to dispose of nuclear waste by dumping the concrete containers in the ocean? Why or why not?

Apply ▸▸

7. **Experiment** Oil spills waste a precious resource. They also pollute. See how difficult it can be to clean up an oil spill. Place a small amount of cooking oil in a large bowl of water. Try and soak up the oil with paper towels, tissue, or a sponge. Break up the oil spill with a few drops of detergent. What are the results?

Summary Activity

For each numbered blank, pick the answer from the list on the right that makes the most sense in the entire passage. Write your answer on a separate sheet of paper. No answer will be used more than once.

Learning about energy is an important part of your technology education. __1__ is the ability to do work. This is often confused with __2__, which is the measurement of work done. Energy cannot be created or __3__ but can be changed from one form to another.

Our sources of energy are often divided into three groups: renewable, nonrenewable, and unlimited. __4__ energy sources are those that come from plants and animals. These energy sources can be __5__ or renewed when we need more.

Nonrenewable sources of energy cannot be replaced. Four important sources are coal, oil, natural gas, and uranium. All of these sources except __6__ are called fossil fuels because they developed from once-living plants and animals.

The unlimited sources of energy will never run out but are more difficult to use. __7__ energy, which comes from the sun, is used to heat homes. It also produces electricity when sunlight strikes wafer thin cells. __8__ are groups of windmills used to produce electricity. Flowing water is used to generate electricity in hydroelectric dams. __9__ from the ground, known as geothermal energy, produces steam, which can, in turn, produce electricity.

Power is measured when energy is __10__ from one form to another. All power __11__ include a source of energy, a process, and an output load.

Unfortunately, our energy consumption has created serious problems. Burning fuels produces serious threats to our health and environment, including __12__, acid rain, and carbon monoxide.

We can all help control pollution by making everyone aware of the problems. We can __13__ the amount of pollution that enters our environment, and we can develop more and better ways of __14__ energy.

Answer List

- conserving
- converted
- destroyed
- energy
- heat
- power
- reduce
- renewable
- replaced
- solar
- systems
- the greenhouse effect
- uranium
- wind farms

Comprehension Check

1. Explain the differences among renewable, nonrenewable, and unlimited sources of energy.
2. Name several uses for mechanical, electrical, and fluid forms of power.
3. How are energy and power technologies selected?
4. Name at least two possible consequences of the greenhouse effect.
5. What collects the solar energy in a passive solar house?

Critical Thinking

1. **Classify.** Why is a power drill an example of mechanical power?
2. **Classify.** Why is spray-painting a fence a result of fluid power?
3. **Analyze.** How do you think manufacturing technology depends on energy technology? How do you think energy technology depends on manufacturing technology?
4. **Connect.** Give three examples of tasks you did today. How do you think energy and power were involved?
5. **Evaluate.** Which is a more efficient use of energy, walking to the end of the driveway to pick up the newspaper or getting in the car, starting the car, and driving the car to the end of the driveway? Why?

Visual Engineering

Studying Windmills. Harnessing the power of the wind with windmills has been done for hundreds of years. Windmills like the one shown here were used on farms settled by early pioneers. Power from the windmill was used to pump well water. In remote areas, where electricity is not readily available, windmills are still used today.

How do you suppose the turning blades on the windmill power a pump? Make a drawing or build a small model to demonstrate how you think the windmill's rotary motion is changed to the up and down motion of a pump.

TECHNOLOGY CHALLENGE ACTIVITY

Building a Solar Heating System

Equipment and Materials

- two 8 x 12-inch sheets of Styrofoam™ plastic
- one 6 x 10-inch sheet of corrugated cardboard
- one 6 x 10-inch sheet of black (or very dark) paper
- clear plastic wrap
- five feet of small-diameter flexible plastic tubing
- small plastic bottle and bowl
- thin wire
- wire cutters
- drill
- razor knife
- rubber bands
- tape
- red food dye
- thermometer
- clothespin
- wristwatch

Background Most solar collectors are black because dark colors absorb the sun's rays. Light colors reflect the sun's rays. The simplest kind of solar collector has a black heating plate and a glass or plastic cover. Sunlight strikes the plate, and the plate becomes hot. The heat is trapped inside by the glass or plastic cover. The trapped heat is transferred to water flowing through tubing. The warmed water is then sent inside the building to a location where the heat can be used.

Goal For this activity you will make a small solar heating system similar to those used in houses. Then you will test it to see how well it works.

Criteria and Constraints

▸▸ Your solar collector should be made from Styrofoam plastic, plastic tubing, and other materials.

▸▸ Your water supply should be held in a plastic bottle.

▸▸ Your system should be as watertight as possible.

▸▸ When it is finished, you should compare the efficiency of your system with systems built by other students.

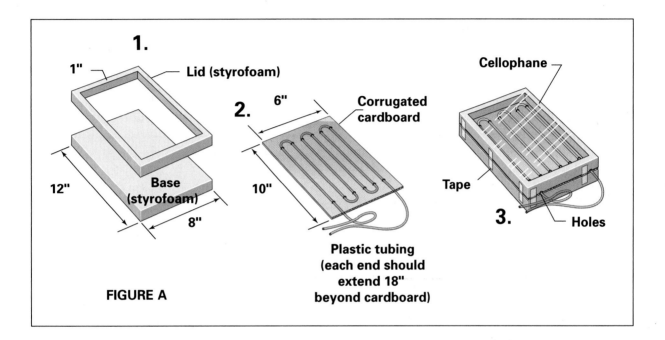

1.

1"

Lid (styrofoam)

12"

Base
(styrofoam)

8"

FIGURE A

2.

6"

10"

Corrugated
cardboard

Plastic tubing
(each end should
extend 18"
beyond cardboard)

Cellophane

Tape

3.

Holes

Design Procedure

Look over the simple plans shown in Figures A and B. Become familiar with the general procedure you will follow.

❶ Use the razor knife to cut the Styrofoam pieces to the sizes shown in Figure A. Save the center cut from the lid.

❷ Make a heating plate by first placing the black paper on top of the corrugated cardboard. Arrange the plastic tubing into S-shaped curves on the black paper. Make as many S curves as will fit. Leave about 18 inches of extra tubing at each end. Fasten the tubing to the black paper and corrugated cardboard with short pieces of wire. Poke a piece of wire through the cardboard and twist its ends together on the underside.

SAFETY

Reminder

In this activity, you will be using tools and machines. Be sure to always follow appropriate safety procedures and rules. Remember, safety is an attitude that you must develop and maintain at all times.

Building a Solar Heating System (Continued)

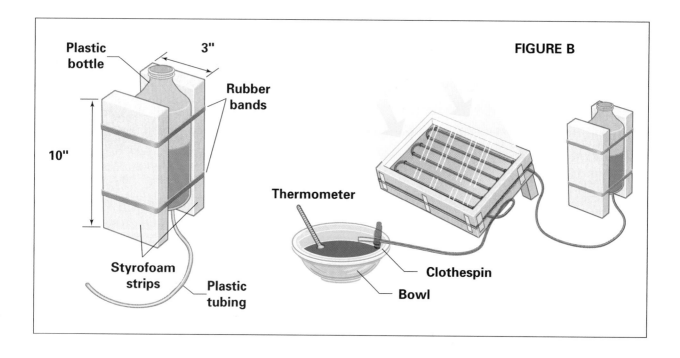

FIGURE B

Plastic bottle — 3" — Rubber bands — 10" — Styrofoam strips — Plastic tubing — Thermometer — Clothespin — Bowl

❸ Use tape to attach the clear plastic wrap over the hole in the Styrofoam lid.

❹ Place the heating plate on the Styrofoam base. Notch holes in the lid for the tubing and place the lid over the base. Tape the assembly together, and your solar collector is finished.

❺ Now complete the solar heating system. See Figure B. Drill a hole in the bottom of a plastic bottle. The hole should be just large enough so that the plastic tubing will fit through. Force one of the free ends of the tubing into the hole. If necessary, seal the connection with waterproof glue.

❻ Make a bottle stand using the piece of Styrofoam cut from the lid center.

❼ Move the solar collector to a place where the sun's rays will strike the heating plate. Placing it at a 45° angle will improve the heating effect. Put the free end of the tubing in the bowl, and clamp it with the clothespin. Fill the bottle with water colored with red food dye. The dye will help you see the water. Place the bottle at least six inches above the collector so that the water can easily flow through the heating plate.

Your system is complete. Now let's test it.

8 Open the clothespin to allow the water to flow through the tubing in the heating plate. Measure its temperature as it flows into the bowl. Close the clothespin.

9 Allow the water to remain in the collector for 2 to 5 minutes. Open the clothespin and measure the water's temperature by flowing water onto the thermometer.

10 Compare your results with the results of others in your class.

SAFETY

Heat Caution!
In this activity you will be working with a solar collector that could become quite hot. Follow all safety precautions for preventing burns. Wear gloves when handling hot items.

Evaluating the Results

1. What was the highest water temperature measured?
2. Cover up half the collector. That's like creating a day with 50 percent cloud cover. Does the water reach only half the maximum temperature?
3. What could you do to improve your solar collector?

CHAPTER 5

ELECTRICITY TO ELECTRONICS

Have you ever walked across a carpet, touched a doorknob, and received a shock? You can see this static electricity jump and hear the crackle of a static discharge. Yet these shocks are only annoying, not dangerous. Another form of static electricity—lightning—can kill you.

Some electrical devices require very little electricity to operate. You can touch the track of an electric train and not get a shock.

Your toaster, television, and stereo use a lot of electricity. Putting your hand inside these devices can kill you.

Every electric circuit contains voltage, amperage, and resistance. Together they determine what electricity can do and when electricity can be dangerous.

In this chapter, you will learn what electricity is and how it is used. You will also learn about the part played by semiconductors, superconductors, and integrated circuits.

SECTIONS

A bolt of lightning carries a tremendous amount of electricity.

Objectives

▶▶ **Explain** what electricity is.

▶▶ **Describe** the three types of electricity and tell the differences among them.

▶▶ **List** uses for direct and alternating current.

Terms to Learn

- AC
- circuit
- DC
- electricity
- frequency
- static electricity

Standards

- Relationships & Connections
- Design Process
- Energy & Power Technologies

Electricity Basics

 What is electricity?

To understand what electricity is, you need to know a little bit about what science has taught us about atoms. Atoms are the building blocks from which all things are made. Atoms contain an equal number of positively charged protons and negatively charged electrons. Most also contain neutrons, which have no charge at all. See Figure 5-1.

The protons and neutrons are located within the atom's nucleus. The electrons circle the nucleus in paths called shells. In some cases, an electron can move or be drawn away from the atom. When an atom loses electrons, it becomes positively charged. Seeking balance, it then tries to pick up electrons from another atom. When an atom has too many electrons, it tries to get rid of them. This movement of electrons is what we call **electricity**. See Figure 5-2.

Atoms with the same electrical charge repel each other. Those with opposite charges attract each other. Atoms obey the same rules of repulsion (pushing away) and

▌**electricity**
the movement of electrons from one atom to another

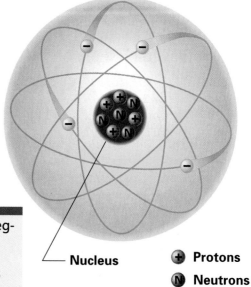

Figure 5-1 In an atom, the negatively charged electrons travel around the nucleus, which contains positively charged protons and neutral neutrons.

— Nucleus

⊕ **Protons**
Ⓝ **Neutrons**
⊖ **Electrons**

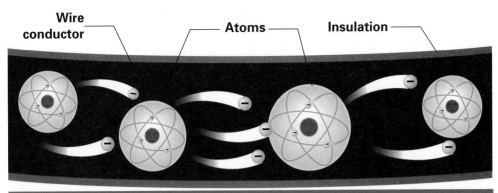

Wire conductor — **Atoms** — **Insulation**

Figure 5-2 Electrons in a circuit move from atom to atom. When an atom loses electrons, it becomes positively charged.

attraction (moving together) as magnets. Under ordinary conditions, atoms will pick up or lose electrons until they are in a neutral state.

Types of Electricity

▷ What are some different types of electricity?

We use electricity for many, many purposes. Is the electricity that comes out of the wall outlets in your home the same as the electricity that comes out of wall outlets in other countries? Are static electricity, lightning, and the power in batteries different from the electricity that powers your home?

Static Electricity

▷ What causes static electricity?

Static electricity occurs when atoms have built up "extra" electrons. These atoms have a negative charge. As soon as they contact an object with a positive charge, they give up the electrons, and a spark occurs. See Figure 5-3. When you receive a shock from ordinary static electricity, you aren't hurt because it doesn't have much power. Static electricity can, however, destroy electronic circuits that need very little electricity to operate. It will also ignite certain gases or flammable liquids.

static electricity
a buildup of electrons that are not in motion and are seeking to discharge

Figure 5-3 When you walk across the carpet, your body builds up a static charge. When you touch a metal doorknob, the charge flows between your hand and the doorknob, and you receive a shock.

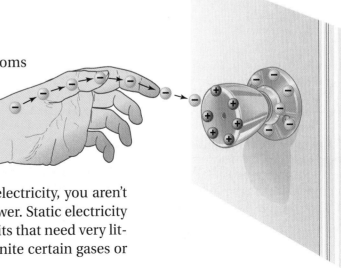

A Shocking Story

You probably have received a minor shock from static electricity. How did it feel?

ACTIVITY Do you ever get "static" from your parents? Write a short paragraph comparing the scientific meaning of static to the meaning used here.

DC
direct current; the one-directional flow of electrons

circuit
the pathway that electricity flows along; often includes parts such as a wire and a device to which the electricity is being delivered

Lightning is a very strong discharge of static electricity. When lightning hits a power line, the electricity surges through the line. Each year surges caused by lightning destroy millions of dollars worth of computers, TVs, and other electronic devices. Did you know that lightning surges can kill people who are using a telephone connected to a phone line that has been hit?

Direct Current

▷ What is direct current?

DC, or direct current, is the flow of electrons in only one direction. At one time, all of the electricity used in the United States was direct current. Today, only one permanent appliance in your home is designed to run on DC electricity—your telephone. A separate DC wiring network connects your telephone to your local phone company.

Devices powered by batteries also use direct current. Your notebook computer, MP3 player, and portable CD player are all equipped with batteries.

Batteries contain stored electricity. You can take them to places that aren't hooked up to a permanent electricity source. When batteries are in use, electrons flow out from the negative terminal of the battery, around the circuit, and then back through the battery's positive terminal. See Figure 5-4. A **circuit** is the pathway electricity takes. Electric circuits are subsystems that carry electricity throughout a device. If a circuit is broken, electricity stops flowing, and the system stops working.

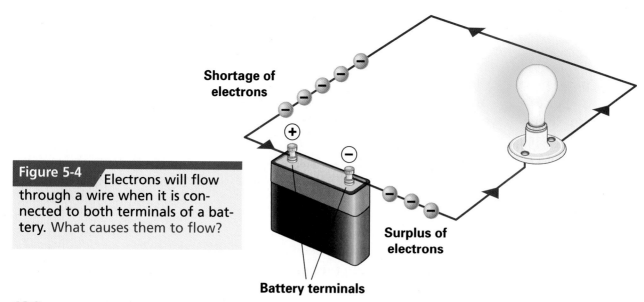

Shortage of electrons

Surplus of electrons

Battery terminals

Figure 5-4 Electrons will flow through a wire when it is connected to both terminals of a battery. What causes them to flow?

Alternating Current

▷ What is the difference between alternating current and direct current?

The wires that come into your home can be traced all the way back to a power-generating station. The electrons that flow through these wires change direction 120 times per second. This constant directional change is the reason that this electricity is called **AC**, or alternating current. Alternating current is easier to transmit and control than direct current.

Each back-and-forth motion is called a *cycle*. The electricity that powers your home is measured at 60 cycles per second. The number of cycles per second is called the **frequency** of the alternating current. (In the metric system the name for cycles per second is *hertz* or Hz.)

AC
alternating current; electrical flow that constantly changes direction

frequency
the number of cycles or changes in direction of alternating current; measured in hertz, or cycles per second

SECTION REVIEW 5A

Recall ▸▸

1. What causes atoms to become positively or negatively charged?
2. Describe the characteristics of static electricity, battery electricity, and household electricity.

Think ▸▸

3. What is occurring when lightning hits a tree?

4. What are the limitations of using batteries to power electrical devices?

Apply ▸▸

5. **Construct** Build a small DC electric motor. (*Note:* Your teacher must approve all experimental plans before you hook up any device to an electric current.)

Measuring Electricity

▶ **How can flowing water help us understand electricity?**

Objectives

▶▶ **Explain** voltage, amperage, and resistance.

▶▶ **Use** Ohm's law to determine measurements in a circuit.

▶▶ **Explain** the purpose of an electric meter.

Terms to Learn

• amperage
• Ohm's law
• resistance
• transformer
• voltage
• wattage

Standards

■ Relationships & Connections

■ Design Process

■ Energy & Power Technologies

voltage
the pressure needed to push electricity through a circuit

transformer
device used to change electricity from one voltage to another

amperage
(AM-purr-age)
electrical flow

resistance
anything that opposes or slows the flow of electrical current

Electricity flowing through a circuit can't be seen. However, its movement through a circuit is similar to that of water through a pipe.

Voltage, Amperage, and Resistance

▶ What are voltage, amperage, and resistance?

You need water pressure to push water through a pipe, and you need electrical pressure to push electricity through a wire. **Voltage** is the name given to the pressure that pushes electricity through an electric circuit. If all other things stay the same, the greater the electrical pressure, the more electricity will be pushed through the circuit. Voltage pressure is measured in units called *volts*.

Alternating current can be changed from one voltage to another with a **transformer**. This is done because some devices require more voltage than others. Also, very high voltages can be sent inexpensively over long distances.

Amperage is the term used for electrical flow. It is measured in units called *amperes*. The rate at which water flows can be measured in gallons per second. Amperage is measured in electrons per second. In cases of electric shock, it is the amount of current (amperage) that makes electricity dangerous enough to kill. However, the exact amount can vary greatly.

In an electrical circuit, anything that opposes the flow of electricity is said to have **resistance**. Resistance is measured in units called *ohms* (OHMZ). Resistance in a water pipe is determined by the diameter, length, bends, and kinks in the pipe. The resistance in an electric circuit is determined by the electric wire's diameter, length, and temperature. See Figure 5-5. For example, the resistance of a wire *decreases* as the wire gets fatter. The resistance of a wire *increases* as the wire gets longer. Adding components (parts) in the circuit, like an electrical appliance, will also increase resistance.

Ohm's Law

▷ **What can you determine by using Ohm's law?**

Building an electric device often means joining many different electrical systems together. These separate systems are all needed to fulfill the requirements of the device. The components must get the proper amount of electricity or the device will burn out, not function properly, or refuse to work. When electric circuits are designed, the power needs of the components must be known.

A mathematical formula based on **Ohm's law** is used to determine the voltage (E), amperage (I), or resistance (R) of an electric circuit. Ohm's law states:

$$Voltage = Amperage \times Resistance$$
$$or$$
$$E = I \times R$$

The law gives the electrical designer a way of determining exactly what the circuit will need to make it work efficiently. See Figure 5-6.

The Electric Meter

▷ **What does an electric meter do?**

The electricity that appliances use is measured in watts. **Wattage** (W) is the voltage of the electric circuit multiplied by the amount of amperage needed to run the appliance:

$$W = E \times I$$

Figure 5-5 Resistance slows down the flow of water in a pipe or the flow of electrons in an electrical wire.

Ohm's law
scientific law expressing the relationship among amperage, voltage, and resistance; it takes one volt to force one ampere of current through a resistance of one ohm

wattage
a measure of electric power; calculated by taking the amperes used times the voltage of the circuit

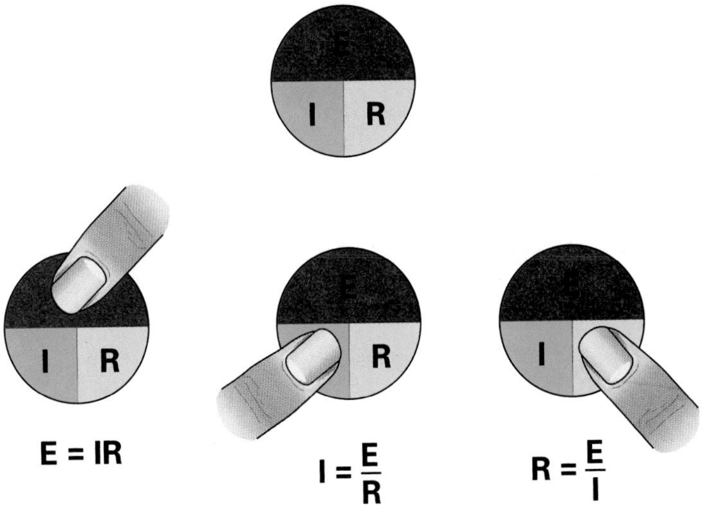

Figure 5-6 To use this Ohm's law diagram, put your finger over the value you want to find. This will give you the correct equation to use. If the voltage is 120 and the resistance is 10, what is the amperage?

E = IR

$$I = \frac{E}{R}$$

$$R = \frac{E}{I}$$

A typical electrical meter for a house. Do you know how to read this kind of meter?

Your electric company meters (keeps track of) the electricity that your family uses. See Figure 5-7. When you turn on an electric appliance in your home, the wheels of your electric meter turn. The moving wheels are measuring the electricity that is being consumed. The more appliances that you run, the faster the wheels turn and the bigger your electric bill. The meter measures kilowatt-hours. A kilowatt-hour is electric power equaling 1,000 watts used for one hour's time.

Most electrical appliances tell you how much electricity will be consumed when they are in use. This information (the wattage) is usually printed on a plate on the bottom of the appliance.

Today, many large appliances are sold with an efficiency rating. The higher this number, the cheaper it is to run the appliance. Electric utility companies encourage people to use less electricity during high-demand periods of the day.

5B SECTION REVIEW

Recall ▸▸

1. What is voltage, and what unit is used to measure it?

2. What is amperage, and what unit is used to measure it?

3. What causes electrical resistance in a wire?

Think ▸▸

4. Based on your understanding of atoms, what do you think causes electricity to flow through a circuit?

5. Under what circumstances might someone need to use the formulas of Ohm's law?

Apply ▸▸

6. Construct Make a model or a chart that shows similarities and differences between water moving through a pipe and electricity moving through a circuit.

Controlling Electrical Flow

How is electrical flow controlled?

SECTION 5C

Objectives

▶▶ **Tell** the difference between a conductor and an insulator.

▶▶ **Name** some uses for semiconductors and superconductors.

▶▶ **Describe** the two basic types of electrical circuits.

Terms to Learn

- conductor
- insulator
- parallel circuit
- semiconductor
- series circuit
- superconductor

Standards

- Core Concepts of Technology
- Relationships & Connections
- Use & Maintenance
- Energy & Power Technologies

Materials that are made up of atoms having a very weak hold on their electrons are called **conductors**. Experiments have shown that electricity will flow through these materials easily. Copper, aluminum, gold, and silver are excellent conductors of electricity. Think of electrical conductors as highways on which electricity travels. Electricity travels along a conductor at a rapid speed.

Other materials contain atoms that have a very tight grip on their electrons. These materials are **insulators**. Rubber, plastic, and ceramics make good insulators. They are used to insulate electrical conductors. This insulation prevents the electrons from leaving their intended path. One way to control electrical flow is to combine conductors and insulators in this way.

If an electrical extension cord has a break in its insulation and you touch the broken area, electricity can flow into your body, giving you a shock. Never use a damaged extension cord.

Semiconductors and Superconductors

What are semiconductors and superconductors?

Some materials can act as either conductors or insulators. These materials are called **semiconductors**. Silicon found in ordinary sand can be used to make a good semiconductor. Adding different ingredients to it and adjusting the temperature affects how well it works. Semiconductors are important to electronic devices like computers.

Superconductors have no measurable resistance to electricity. Superconductors are important because they make it possible to generate more productive and efficient electrical power.

Superconductor technology is in use today in magnetic resonance imaging (MRI) machines in medicine, ship

conductor
material that allows electricity to flow easily through it

insulator
material that resists the flow of electricity

semiconductor
material that can be used as either a conductor or an insulator

superconductor
material that has no measurable resistance to electricity

motors, magnetic levitation (maglev) trains, and lightning-fast electronic switches. Technology developed for one area has been transferred to other areas.

Electrical Circuits

▷ **What does an electrical circuit consist of?**

An electrical circuit begins at a power source and ends back at that power source. The circuit must have at least one device that consumes electricity, such as a buzzer or light bulb. If the circuit isn't complete, electrons will stop flowing, which turns off the electricity. An electric switch is used to open and close the circuit to turn things on and off. See Figure 5-8.

Series Circuits

▷ **How does a series circuit work?**

When you wire batteries and energy-using devices in a line, you create a **series circuit**. See again Figure 5-8. If any item in the circuit breaks down, power is lost to the entire circuit.

When batteries are connected in a series, each one increases the voltage (pressure) in the circuit. To increase power, most battery-powered electric devices have batteries connected in series. When you place three 1.5-volt batteries together in a series, the circuit is powered by 4.5 volts. Do you understand why?

Light bulbs and other electrical devices can also be wired together in a series. In this type of circuit, the electricity must pass through each device on its way to the next. A major disadvantage of series wiring is that if one device burns out, all the devices in the circuit stop working. Are the circuits in your home series circuits? Do all the lights go out when one bulb burns out? Most of the circuits in your home are not series circuits.

series circuit
a single pathway that current flows through to more than one electrical device; if one device in the path stops working, they all stop

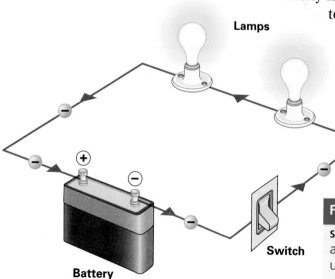

Lamps

Switch

Battery

Figure 5-8 A series circuit provides a single path for current. Can you think of a device in which a series circuit is often used?

Handling Pressure

When you handle pressure well, every job you do will go much more smoothly.

Pressure goes with almost any job. Perhaps it is the deadlines that you must meet or a customer that is dissatisfied with a product or service. Here are a few ways of managing pressure:

Navy Engineering Technician

The U.S. Navy is looking for people to train to be engineering technicians. In this job, you will operate and maintain the world's most advanced electronic equipment. The job also includes inspection, testing, and repair. Applicants must have good mathematics and science skills and be able to handle pressure and stress.

✔ **A big job can often seem overwhelming. Break it down into smaller tasks or steps. Then complete those steps on a planned schedule. The overall task will seem much more achievable. Before you know it, you'll be finished.**

✔ **Have a daily "to do" list. Once you complete a task mark it off the list. As the number of crossed-off items grows, you'll feel a sense of accomplishment.**

✔ **If the job seems like more than you can handle alone, ask for help.**

✔ **Maintain good health by getting enough rest, nutritious food, and daily exercise. A healthy person is able to handle better the effects of pressure.**

Parallel Circuits

▷ How do parallel circuits differ from series circuits?

In a **parallel circuit**, electricity flows along more than one path. A path exists around each energy-using device. See Figure 5-9 (page 144). If one device burns out, the electricity doesn't stop flowing to the other devices in the circuit.

Most of the circuits in your home are parallel circuits. In all parallel circuits, the voltage remains the same throughout the circuit regardless of the number of devices connected. Each device draws the current it needs to operate.

All the electric outlets in your home are connected to a fuse box or circuit breaker panel. See Figure 5-10. These safety fuses or breakers shut off the electricity in case of a power overload. A power overload will happen if energy-using devices pull more electricity than the wire in the circuit can safely carry. If an electric wire is not properly protected by a fuse or a circuit breaker, an overload can superheat the wire and start a fire.

parallel circuit
a circuit having multiple pathways carrying current to individual devices; if one device stops working, the others are not affected

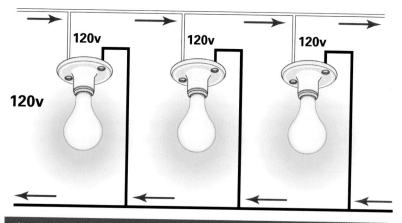

120v 120v 120v

120v

Figure 5-9 A parallel circuit provides at least two different paths for the current.

Figure 5-10 Circuit breakers like these shut off the flow of electricity if the circuit is overloaded. What is sometimes found in older houses instead of circuit breakers?

5C SECTION REVIEW

Recall ▶▶

1. What is the difference between an electrical insulator and an electrical conductor?

2. What is a semiconductor?

3. What are superconductors, and how are they used today?

4. Describe the two basic types of circuits.

Think ▶▶

5. Look up the meaning of the prefix *semi*. Why do you think it is part of the term semiconductor?

6. What kinds of things could go wrong with a circuit that would cause a fuse to blow?

Apply ▶▶

7. **Compare and Contrast** With your teacher's help, wire low-voltage, DC light bulbs together in both a series and a parallel circuit. Remove a bulb from each circuit. What happens?

Electronics

▷ What is an electronic device?

An **electronic device** changes one form of energy, such as sound, into an electrical signal that can be transmitted by a sender through a channel to a receiver. See Figure 5-11. At the receiver, the signal is decoded. Examples of electronic devices include the telephone, the TV, and the radio.

Signal Transmission

▷ How can electronic signals be sent?

Electronic signals can be sent along a wire, through the air on electromagnetic waves, or as pulses of laser light through fiber optic cables.

Figure 5-11 In this electronic device, is the robot the receiver or the transmitter?

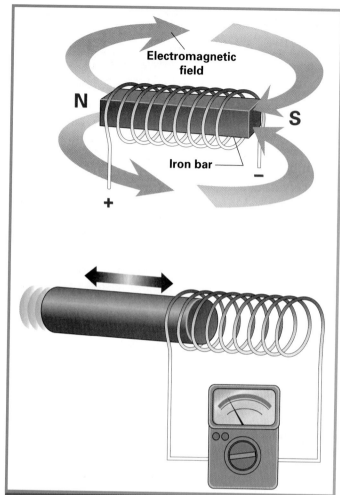

Wire Transmission

▶ How does wire transmission work?

The telegraph, ordinary telephones, and cable TV use wire transmission channels. The electric pulse travels along the wire.

Electricity flowing through a wire creates a magnetic field around the wire. If the wire is wrapped around an iron bar, the bar acts like a magnet when electricity flows through the wire. See Figure 5-12. This type of magnet is called an electromagnet. In 1844, Samuel Morse combined electricity and magnetism to produce the telegraph. An electromagnet at each end helped relay the message.

Alexander Graham Bell's 1876 telephone adjusted the flow of electrons in an electric circuit to replicate the sound of the sender's speech. Varying current traveled over the circuit and caused a varying vibration at the receiving phone.

Telephones of today still use the principle developed by Bell over 100 years ago. Your call is transmitted through a cable made of copper wire. Your phone signal reaches the computer exchange network where it is directed to the proper receiver.

Figure 5-12 Did you know that passing a bar magnet back and forth through a coil of wire will generate electricity? This is the reverse of creating an electromagnet by sending electricity flowing through a coil of wire wrapped around a steel bar.

electromagnetic wave
a wave of electromagnetic energy used to carry an electronic signal through the atmosphere

transmitter
the part of the communication system that changes the sender's message into electrical impulses and sends it through the channel to the receiver

Atmospheric Transmission

▶ How are signals sent through the atmosphere?

The magnetic field created when electricity flows through a wire can also be used to carry a message. By rotating an electromagnet, alternating current is produced. The **electromagnetic waves** created carry radio, TV, and cell phone signals through the air for many miles. These waves move at the speed of light.

The device used to create the waves that carry the signal is called a **transmitter**. You have probably seen tall transmitter towers in your area.

Figure 5-13 Thousands of phone conversations can go through these fiber optic phone lines at the same time.

fiber optic cable
thin, flexible, glass strands that transmit light over great distances

laser
a very powerful, narrow beam of light; all the light rays have the same wavelength

Fiber Optic Transmission

▷ How do fiber optic cables work?

Fiber optic cables are made from very pure glass. The term *optic* is used to describe something related to light. Fiber optic cables transmit light. Each glass strand in a cable can be as thin as half the thickness of a human hair. See Figure 5-13.

The glass strands have an outer coating of glass with different reflective characteristics. This outer layer causes the rays of light traveling at the core to stay inside it, in spite of twists and turns. These cables can carry a signal, without loss of power, for very long distances. When necessary, their power can easily be boosted so that messages will not be lost.

The signals sent along fiber optic cables are powered by lasers. **Lasers** produce very narrow beams of highly focused light. All the light rays in a laser beam have exactly the same wavelength. The beam that is sent by a laser can be visible light or invisible infrared light. See Figure 5-14 (page 148). Lasers that send out visible light are used in communication. Messages are sent as pulses of light along the fiber optic cables.

By transmitting signal beams that are on different wavelengths, it is possible to have many messages travel along the same fiber optic strand at the same time. When they reach their destination, the different wavelengths are separated. This marriage of electronics and optics is called

LOOK TO THE FUTURE

Smaller Is Better

Electronic devices often become handier as their size gets smaller. New wireless sensors no bigger than a matchbox are now being tested for a variety of uses. (Sensors are devices that react to light, heat, and other physical data.) Today, they are monitoring energy use in supermarkets. In the future they may be embedded in roadways to send back traffic reports or used on battlefields to monitor the movement of enemy troops.

Figure 5-14 Lasers are used to read the universal product code on most everything you buy at a grocery store. Can you think of something you buy that doesn't have one?

analog signal
an electrical signal that changes continuously

digital signal
an electrical signal having distinct values

optoelectronics. Every time you watch a DVD, download from the Internet, or use a laser printer, you use optoelectronics. See again Figure 5-14. More than one type of transmission may be used to send a single message.

The Difference between Analog and Digital

▶ What is the difference between analog and digital signals?

Analog signals change continuously, and they are found in all kinds of systems. Alexander Graham Bell learned how to turn electricity into analog signals. The current in his telephone circuit varied with the intensity of the person's speech. If your watch has hands that rotate smoothly, it is an analog watch. See Figure 5-15. The second hand on an analog watch never stops moving.

A **digital signal** is analog information that has been converted into digital information. The changing analog information is sampled frequently and turned into distinct, separate values. Your digital watch represents the seconds as exact numbers. It starts and stops continuously. The time is never "between" one second and the next.

This doesn't mean that one type of signal is more accurate than another. See again Figure 5-15. Because digital signals can be compressed to take up less space in storage, sent faster, and stored longer than analog signals, they are preferred for electronics. For example, when you speak into a phone, your voice, which is analog information, is changed into a digital signal.

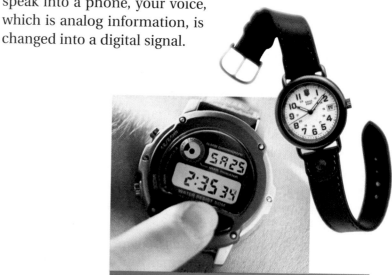

Figure 5-15 A digital and an analog watch. Which do you prefer?

Size and Speed

▷ How small can a cell phone be made?

Each new improvement in electronics is built on the knowledge and accomplishments of earlier inventors. Vacuum tubes were the first devices to work as an electric switch in electronic equipment. They resembled light bulbs, used a lot of electricity, and gave off a lot of heat. Then most vacuum tubes were replaced by the **transistor**. Transistors are small and made of semiconducting materials. The transistor was tiny compared to a vacuum tube, and it didn't produce a lot of heat. However, thousands of hand-soldered connections were needed to join transistors and other components into the complete circuits required by electronic devices.

In the **integrated circuit**, all the circuit components fit on a single chip of silicon. Very tiny passageways in the silicon connect the tiny components together. Today, millions of components can be placed onto a single microchip. See Figure 5-16.

Figure 5-16 Microchips are small when compared to an ant, but they can process huge amounts of information.

transistor
small device made of a semiconducting material and used to control electric current

integrated circuit
a tiny chip made of a semiconducting material that contains many of the components needed to operate an electronic device

SECTION REVIEW 5D

Recall ▸▸

1. Describe the ways in which electronic signals can be transmitted.
2. How did Alexander Graham Bell's original telephone replicate the sound of someone's voice?
3. What are the advantages of an integrated circuit compared to a vacuum tube?

Think ▸▸

4. Jessica is singing a song. Is her song analog or digital information? Explain.

Apply ▸▸

5. **Experiment** Wire a telephone speaker and receiver into a low-voltage direct-current circuit. Use an electric meter to determine the effect that sound has on the electric current that is flowing through the circuit.
6. **Creating Sound** With your teacher's help, use a low-power laser to send sound signals through a fiber optic cable.

Summary Activity

For each numbered blank, pick the answer from the list on the right that makes the most sense in the entire passage. Write your answer on a separate sheet of paper. No answer will be used more than once.

To understand electricity, you need to know a little bit about the atom. Atoms are the __1__ from which all things are made. Within atoms are neutrons, __2__, and electrons. The movement of __3__ from one atom to another is what we call electricity.

Battery electricity is called __4__ current. Your home is powered by a type of electricity called __5__ current.

Electricity is forced through the wire by its voltage. This __6__ pushes the current through the circuit. A circuit wire's thickness, length, and ability to conduct electricity will determine how much opposition it gives to the flow of electricity. Resistance __7__ as a wire gets fatter.

Electrical utility companies sell electricity to people who want to run electrical devices. A(n) __8__ tells the utility company how much electricity has been used. Many new appliances list an efficiency rating. A very efficient machine will use less electricity.

Electrical insulators have a very high resistance to the flow of electricity. They are used to cover electrical __9__. They prevent us from touching the wire and getting shocked. Some materials under special conditions can act as insulators or conductors. New special materials are now being developed that have no electrical resistance. They are called __10__.

In a series electric circuit, electricity must pass through __11__ device. If one device burns out, the electricity will stop flowing. In parallel circuits, the electricity has __12__ pathways to each device in the circuit. If one device burns out, the electricity isn't prevented from reaching the other devices in the circuit.

Electromagnetic __13__ can be used to carry a message from a sender to a receiver. __14__ are also used to carry signals. Since information travels faster this way than it does in a copper wire, computer networks are changing over to fiber optics for the transmission of information.

Answer List

- alternating
- building blocks
- conductors
- decreases
- direct
- each
- electric meter
- electrons
- fiber optic cables
- pressure
- protons
- separate
- superconductors
- waves

Comprehension Check

1. What is electricity?
2. List some uses for direct and alternating current.
3. What is the purpose of an electric meter?
4. What is the primary use of semiconductors?
5. How do fiber optic cables work?

Critical Thinking

1. **Decide.** Joe wants to use the formulas from Ohm's law. Will he be able to obtain his estimate without knowing the voltage?
2. **Evaluate.** Suppose a friend's house has a burglar alarm, a doorbell, and outdoor lights wired in a series circuit. Is this a good arrangement, and why?
3. **Infer.** Based on what you've learned in this chapter, how do surge protectors work?
4. **Compare.** A cable company wants to increase its number of customers and improve wire-transmission quality. Should the company use additional wire cables or fiber optic cables? Explain.
5. **Analyze.** Use the systems model to describe a company that generates electricity. Identify possible inputs, processes, outputs, and feedback.

Visual Engineering

Lasers Today and Tomorrow. Today we use lasers for many things, but it hasn't always been that way. Ask an adult when he or she first noticed the use of lasers. The person might be able to remember how prices had to be manually keyed into a cash register instead of scanned. Start a list of ways lasers are used now. See if you can come up with one or two new applications every day. Research on the Internet and brainstorm with your classmates to find ways lasers might be used in the future.

TECHNOLOGY CHALLENGE ACTIVITY

Building a Simple Alarm System

Equipment and Materials

- clothespin
- 6 inches of #20 bell wire
- wire stripper
- metal punch
- drill
- ⁹⁄₆₄ drill bit
- 3½ x 4½ wood base
- 9-volt battery
- 9-volt battery snap
- piezoelectric buzzer
- two 1-inch ⁶⁄₃₂ machine bolts
- four ⁶⁄₃₂ nuts
- six small washers
- four ½-inch #6 wood screws
- one 1-inch #7 wood screw
- electrical tape
- tongue depressor
- string
- 3½ x ¾ tin plate strip

Electrical Safety
When using tools, machines, and wiring to work on various activities, be sure to always follow appropriate safety procedures and rules.

Background

All alarm systems include a circuit, switch, and some kind of ringer. In an alarm clock the switch is connected to a clock, and the ringer may be a bell or radio that wakes you up in the morning.

The modern burglar alarm system may include sensors that react to physical movement and changes in temperature. The parallel and series circuits that are used in these alarms have been enhanced by computer monitoring that reads the resistance in the alarm's wire. These circuits constantly check for breaks or attempted bypass by burglars.

Goal

For this activity, you will build the simple alarm shown in Figure A so it can be triggered by all kinds of situations. The alarm will sound when the insulation is pulled from the clothespin. You will then find a useful or funny way to use it.

Criteria and Constraints

▶▶ You must build and test your alarm to be sure it works.

▶▶ You must use your creativity to find a use for the alarm. Your use can be serious or funny.

▶▶ You will then demonstrate your alarm to the class. Your demonstration can include visual aids and even a videotape.

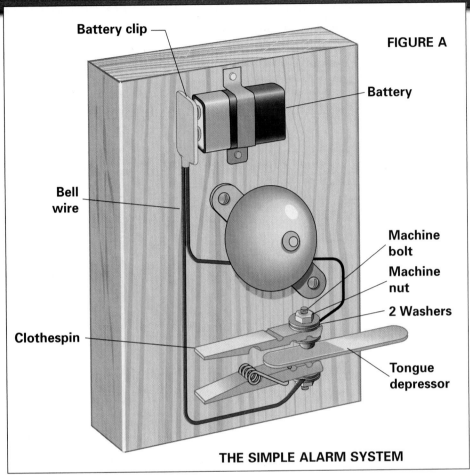

Battery clip

FIGURE A

Battery

Bell wire

Machine bolt

Machine nut

2 Washers

Clothespin

Tongue depressor

THE SIMPLE ALARM SYSTEM

Design Procedure

① Prepare the wood base for your alarm. See Figure B.

② Cut and bend the metal strip that will hold your battery to the base. See Figure C.

③ Use the metal punch to cut the holes in your battery holder. See Figure D.

④ Use two #6 wood screws to attach your battery holder to the base. Do not attach the battery clip to the battery until all work on your alarm is completed.

⑤ Use two #6 wood screws to attach the piezoelectric buzzer to your base.

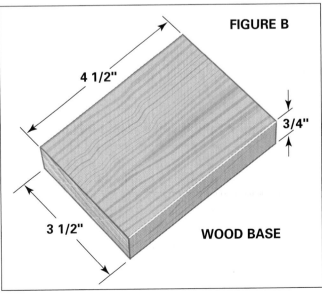

FIGURE B

4 1/2"

3/4"

3 1/2"

WOOD BASE

TECHNOLOGY CHALLENGE ACTIVITY

Building a Simple Alarm System (Continued)

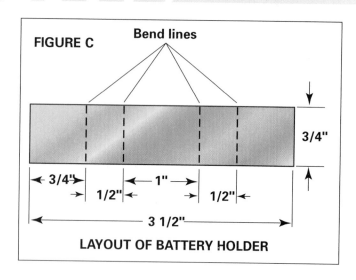

FIGURE C

Bend lines

3/4"

3/4"   1"   1/2"   1/2"

3 1/2"

LAYOUT OF BATTERY HOLDER

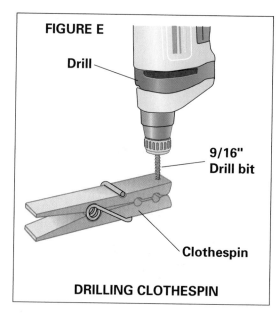

FIGURE E

Drill

9/16" Drill bit

Clothespin

DRILLING CLOTHESPIN

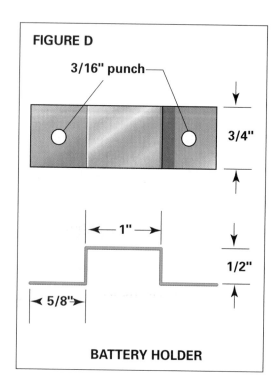

FIGURE D

3/16" punch

3/4"

1"

1/2"

5/8"

BATTERY HOLDER

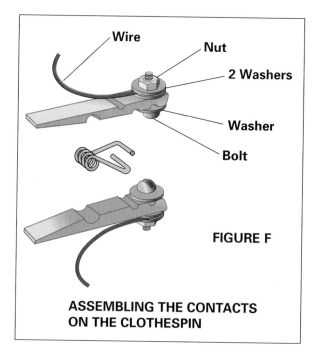

Wire

Nut

2 Washers

Washer

Bolt

FIGURE F

ASSEMBLING THE CONTACTS ON THE CLOTHESPIN

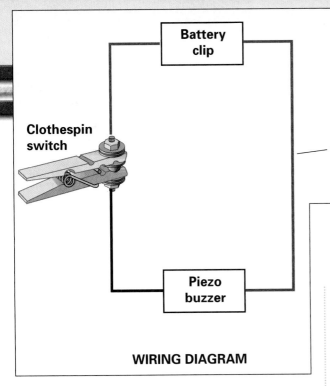

Battery clip

FIGURE G

Clothespin switch

Battery wire spliced to buzzer wire

Piezo buzzer

WIRING DIAGRAM

6. Drill a ⁹⁄₆₄-inch hole through the two closed ends of your clothespin. See Figure E.

7. Through each of these holes, slip a machine bolt and washer. *Note:* Bolt heads will be on the *inside* of the clothespin. See Figure F.

8. Secure each bolt with a nut.

9. Strip ½ inch of insulation from the ends of your wire.

10. Slip two washers onto the ends of each bolt. Hook one end of your bell wire between these washers and secure it in place with a nut.

11. Follow this same procedure to hook the black wire from the piezoelectric buzzer to the other side of the clothespin.

12. Secure the clothespin to the base by screwing the #7 wood screw through the spring.

13. Splice the red wire from the buzzer to the red wire of the battery snap. Tape your connection with electrical tape. See Figure G.

14. Splice the free end of your bell wire to the black wire of the battery snap. Tape your connection.

15. Check that all components have been connected correctly.

16. Place a tongue depressor between clothespin contacts.

17. Connect the battery clip to the battery.

18. Test the alarm by pulling the tongue depressor out from between the contacts.

19. Shorten the depressor, and drill a small hole into its end for a pull string.

20. Adapt your alarm to work where you plan to install it.

21. If desired, create a videotape or other visual aids for your demonstration.

Evaluating the Results

1. Why doesn't the buzzer ring before you pull the tongue depressor?

2. Can a larger model of this alarm protect your home? Why or why not?

3. Is there a relationship between the voltage of the battery and the size of the buzzer? If so, what is it?

Can you think of one technology that doesn't require spoken or written communication? Can you think of one that doesn't require a basic understanding of the principles of science and mathematics? Social studies is also important because society affects ways in which technology is developed and used. Technology brings about social change as well. Foreign language is important because technology has made our world smaller. Today we can fly to other countries in just a little more time than it took our ancestors to walk to a distant neighbor. It used to take months to send a letter to the other side of the world. Today it takes a few seconds to connect by telephone or computer.

What role do mathematics, science, language arts, and social studies play in the development of technology? What role do these subjects play in today's world of work and in your daily use of technological devices? In this chapter, you will begin to learn the answers to these questions.

SECTIONS

6A Science and Mathematics

6B Language Arts and Social Studies

Building and launching your own model rocket can be lots of fun, as well as being educational.

▶▶ **Discuss** ways in which science and technology are connected.

▶▶ **Discuss** ways in which mathematics and technology are connected.

Terms to Learn

- hypothesis
- scientific law
- scientific theory

Standards

- Relationships & Connections
- Influence on History
- Troubleshooting & Problem Solving

hypothesis
an explanation for something that is used as the basis for further investigation

Science and Mathematics

▶ **Why are science and mathematics necessary to technology?**

Primitive people created very simple technologies using crude tools and natural materials. Early hunters tied sharp stones to sturdy sticks to make axes. They did not have rulers for measuring the length of the sticks or scales to weigh the stones. They learned what worked from trying different combinations. Modern technology is quite different. It depends heavily on science and mathematics. Without them, we would not have such devices as computers and space shuttles.

Science

▶ **What is the scientific method?**

Scientists working in the fields of life science, physical science, and earth and space science all apply the same scientific method to solve problems.

1. The scientists make observations and form questions.

2. Next, they gather information about what was observed.

3. They use this information to form a **hypothesis**.

4. Scientists then develop an experiment to test their hypothesis.

5. After completing their experiment, they carefully analyze the results.

6. They then repeat the experiment to see whether they get the same results. Finally, they present their conclusions.

As they seek solutions to problems, scientists use technology as a tool for scientific discovery. Their discoveries often lead to the development of new technologies. See Figure 6-1.

Figure 6-1 Scientists are studying alternative sources of energy so that new technologies can one day be developed.

Scientific conclusions that have been carefully developed through experimentation are called **scientific theories**. Over time, these theories will be tested again and again. Eventually they may be accepted as **scientific laws**. All new technology is based on a clear understanding of the principles, theories, and laws of science.

In time, scientific law can be challenged by new knowledge gained through the use of technology. An example is Isaac Newton's third law of motion.

Newton's third law of motion defines gravity as an attraction force between objects. The greater the object's mass, the greater its attraction force. See Figure 6-2.

scientific theory
scientific conclusion carefully developed through experimentation

scientific law
a theory that has been proven true so often that it is accepted as fact

Figure 6-2 The *Pioneer 10* spacecraft escaped the pull of Earth's gravity and became the first human-made object to leave our solar system.

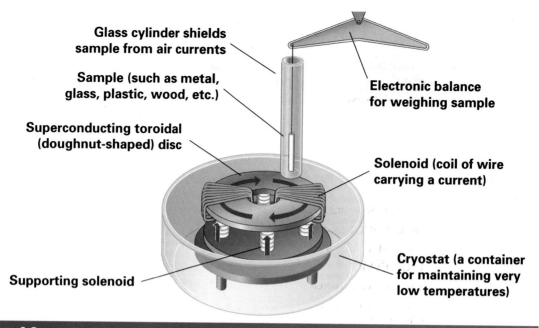

Figure 6-3 This is a simplified drawing of the antigravity machine. While the disc spins, objects can be suspended above it.

Glass cylinder shields sample from air currents

Sample (such as metal, glass, plastic, wood, etc.)

Superconducting toroidal (doughnut-shaped) disc

Electronic balance for weighing sample

Solenoid (coil of wire carrying a current)

Supporting solenoid

Cryostat (a container for maintaining very low temperatures)

TECH CONNECT

MATHEMATICS

Looking for Profit

Mathematics is also needed to calculate the profit from selling a technological device. Profit is the money earned after all expenses are paid.

ACTIVITY Suppose you have invented an automatic bicycle scrubber. Your device costs $112,067 to manufacture, and advertising costs another $34,750. You are selling the device for $39.95. How many devices will you have to sell before your expenses are paid? Round your answer to the nearest whole number.

Newton's law explains why apples fall to the earth and also why the planets in our solar system revolve around the sun. However, telescopes and other instruments showed that light, which has no mass, bends in a strong gravitational field. Newton's law was questioned. Eventually, Albert Einstein's general theory of relativity accounted for the warping of light by the massive objects that make up our universe.

Can new technology change the laws of gravity? Most scientists feel that Newton's law and Einstein's theory can't be reversed. Perhaps they are wrong. A group of NASA researchers is currently using the scientific method to build an anti-gravity machine. See Figure 6-3. Such a machine would revolutionize air and space travel.

Mathematics

▷ How can mathematics help us design technology products?

Few subjects are more important to technology than mathematics. It is used to determine such things as the flight characteristics of military airplanes and the operating speed of computers. Mathematics is often the reason products look the way they do.

For example, a narrow river might require only a simple bridge. A larger river might require a complex bridge with suspension cables. In both cases, mathematical calculations tell technologists which design is better. Let's see how mathematics is used when working with robots and building large ships.

Working with Robots

▷ What can mathematical calculations tell us about robots?

Industrial robots are important machine tools in manufacturing. Most robots have one arm that can be fitted with different tools. They can use electric welders, paint sprayers, grippers, or other devices. Industrial robots are powered by electric motors, hydraulics, or pneumatics. Hydraulic robots use pressurized oil, and pneumatic ones use pressurized air. See Figure 6-4.

Pneumatic robots operate at an air pressure of about 100 psi. Hydraulic robots operate at about 3,000 psi. All other things being equal, a hydraulic robot can lift about 30 times as much because the pressure is 30 times higher. If a pneumatic robot can lift 15 pounds, a hydraulic robot could lift up to 450 pounds (30 × 15 = 450 pounds).

A robot having an arm extension speed of 40 inches per second is not unusual. How long would it take such a robot to move its arm 4.5 feet? The robot could do it in 1.35 seconds! (Convert feet to inches: 4.5 feet × 12 = 54 inches. Divide this result by 40.) Engineers make calculations like these to determine where robots should be used and for which tasks.

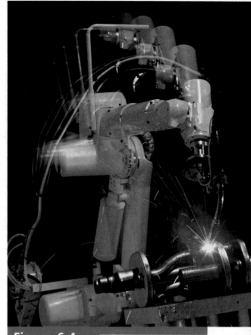

Figure 6-4 This industrial robot is arc welding an exhaust pipe for a new car.

Building Ships

▷ How can mathematics help shipbuilders?

The world's first all-metal steamship made its first ocean voyage in 1845. The iron vessel was designed and built by Isambard Kingdom Brunel, England's greatest nineteenth-century technologist. It was the 252-passenger *SS Great Britain*.

Many people scoffed at Brunel because they knew that most metals sink easily. However, Brunel knew that it's easy to make metal float if it's properly shaped. As long as the density of a rectangular solid is less than that of water, the solid will float. See Figure 6-5. Density is weight divided by volume. The density of water is 0.0361 pounds per cubic inch (pci). Any solid less dense than 0.0361 pci will float.

Figure 6-5 Do you think this cork is more or less dense than water?

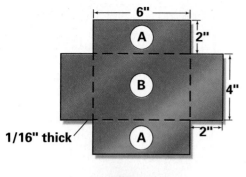

6"

A

2"

B

4"

A

2"

1/16" thick

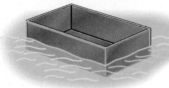

Figure 6-6 In addition to density, what else would be important in keeping this box afloat?

Suppose you wanted to make the flat piece of copper shown in Figure 6-6 into a boat. The piece of copper weighs 1.3 pounds, and its area is 64 square inches. We can find its volume by multiplying its area by its height, or thickness, which is $\frac{1}{16}$ inch. Therefore, the volume of the flat copper piece is 4 cubic inches ($64 \times \frac{1}{16} = 4$). In order to make the copper float, you must increase its volume. Can you do it?

Imagine that you fold the copper into a boxy boat.

- The box's volume is length times width times height.
  Volume = 6 inches × 4 inches × 2 inches
  Volume = 48 cubic inches
- The density of the boxy copper boat is its weight divided by its volume.
  Density = 1.3 pounds ÷ 48 cubic inches
  Density = 0.0271 pci

The density of the copper boat (0.0271 pci) is less than the density of water (0.0361 pci). That means the small boat will float. Brunel used the same logic in constructing the *Great Britain*. It weighed 6.89 million pounds, but its volume was large enough, and it could float.

6A   SECTION REVIEW

Recall ▸▸

1. What is the difference between a hypothesis and a scientific law?

2. Why is calculating a robot's ability to extend its arm important?

3. Why is it possible for a very large, heavy ship to float?

Think ▸▸

4. Suppose you observed that certain metals could be melted at lower temperatures than previously thought. How would you go about establishing your observation as a scientific law?

5. Suppose the copper boat discussed in this section measures 10 × 4 × 2 inches and weighs 3.24 pounds. Will it still float?

Apply ▸▸

6. Science and Mathematics Divide a sheet of paper into two columns. Label one column Science and the other Mathematics. Think of ways in which science and mathematics are connected to technology in your school and at home and write them in the appropriate column.

Language Arts and Social Studies

▶ How are language arts and social studies linked to technology?

Objectives

▶▶ **Discuss** ways in which language arts and technology are connected.

▶▶ **Discuss** ways in which social studies and technology are connected.

Terms to Learn

- **humanities**
- **Information Age**
- **shadowing program**

Standards

■ Information & Communication Technologies

Some subjects that you study in school are referred to as the **humanities**, or liberal arts. They include language arts, social studies, and art. In language arts you study literature, reading, writing, spelling, and grammar to develop the ability to use English correctly. In social studies you explore American and world history, economics, and government. In art you study color, line, shape, form, and other elements that people use to express their creativity. You will learn how art is connected to technology in the next chapter. In this section, we'll look at language arts and social studies.

Language Arts

▶ What is the relationship between language arts and technology?

Verbal and written communication are in many ways the foundation of technology. We need communication in order to develop technology and teach it to others. In turn, technology provides the means for better communication.

Early humans developed language in order to pass along information. See Figure 6-7 (page 164). Of course, a person could demonstrate how to build a fire, for example, but it's easier and faster to use words.

The use of the plow helped change our early ancestors from hunters and gatherers into farmers. With farming came property ownership and the need to keep records. Written and spoken language grew in importance.

In the middle of the fifteenth century, Johannes Gutenberg developed movable metal type. The printed word helped spread the use of technology. In many ways, the manufacture of inexpensive books paved the way for universal education. Before Gutenberg, few people were schooled in reading and writing. As more people became

humanities
subjects having to do with cultural knowledge, such as language and history

TECH CONNECT

LANGUAGE ARTS

The Right Word
If the plans for making a product are poorly written, mistakes might be made during production. Then the product would not work properly.

ACTIVITY Have you ever tried to work from unclear instructions? Write a paragraph explaining what happened.

Figure 6-7 Pictures like this cave painting may have helped early people communicate and develop language.

Information Age
period beginning around 1900 in which human activities focused on the creation, processing, and distribution of information

TECH CONNECT

SOCIAL STUDIES

World Heritage Sites

Countries can select an important structure or cultural site as a United Nations World Heritage Site. Cave paintings in Norway are an example. In the U.S., the Statue of Liberty is a World Heritage Site.

ACTIVITY Use an Internet search engine to find a list of the other sites. Select one and find out more about it. Report your findings to the class.

educated and communication increased, inventions multiplied.

How are the language arts connected to technology today? Suppose someone invented a new kind of computer. He or she would need language to program it and create the plans for producing it. Language would also be needed to create a company, hire employees, keep business records, and communicate with customers. Skill in a foreign language is often needed by people who work with companies overseas. Can you think of other uses for language?

Social Studies

▶ How do technology and social studies work together?

The study of technology deals with more than things. It deals with how people use technology and how it changes their lives. You learn about these interactions in social studies. Most American technical museums are not simply displays of historical items. They show how past inventions have improved people's lives, changed economies, and altered governments.

The Industrial Revolution is a good example. It began about 1750 and resulted in worldwide social changes. Economies had been based on farming. Now, they became based on factory production. People are still researching the effects of the Industrial Revolution.

Today we are in another revolution. Our society has moved beyond factory production into the **Information Age**. We still need factories, of course, but more and more people make their living by creating, processing, and distributing information. See Figure 6-8.

Figure 6-8 The creation and distribution of information are key to most jobs today. This person is creating a Web page.

The Information Age

▶ What is the "Information Age"?

It is difficult to say precisely when the Information Age began. Some say it started in 1844 when Samuel Morse tapped out the first long distance telegraph message between Baltimore and Washington, D.C. Perhaps it began in 1944 when Howard Aiken operated the world's first large-scale digital computer at Harvard University. The Information Age could have started in 1967 when a government group named the Advanced Research Projects Agency (ARPA) established a computer network they called ARPANET. It evolved into the modern Internet. Whenever the Information Age began, it is here to stay.

Linking the world by computers has allowed everyone to share in the specialized knowledge of a few. Medical information, for example, can be immediately sent throughout the world to help save lives.

Global positioning system (GPS) satellites help map the world with great accuracy. This series of twenty-four satellites orbits the earth. From any given spot on the earth, five to eight of the satellites are visible. A small hand-held receiver can pick up data from them. The information is used to precisely determine the receiver's position. See Figure 6-9. GPS information also permits ship captains to plan a long ocean voyage so as to use the least amount of fuel. It allows highway engineers to determine the most practical route for constructing roads in undeveloped regions. Precise satellite mapping helps oil companies locate likely sources of underground petroleum.

The rapid exchange of information has altered the way people visualize the world. Television, computers, fax machines, cell phones, and other products of the Information Age have changed the way schools educate students. Today's young people have almost immediate access to large amounts of information.

We also conduct business differently in the Information Age. In retail stores, universal product code readers rapidly sense the price of food, clothing, and other items. People can get through checkout lines faster. With a credit card and access to a telephone or to an online computer service, people can make travel reservations and buy products. Banking services are also available over the Internet.

Figure 6-9 This forester is using the global positioning system for surveying. Do you know what mathematics principle is used for GPS?

School of the Future

Some researchers think that in the future most education will take place online. Students will have debit cards entitling them to so many hours of learning. Students will enter homework into hand-held computers and then send it directly to another computer for grading. Would you like to be a student in that kind of school?

shadowing program
school program in which students spend time in a work environment

Technology and the Working World

▶ Why are these subjects important to a career?

You study economics because it has to do with the goods and services we produce. It also has to do with how people earn a living. Today's jobs require workers with technology-based abilities. In our Information Age, many jobs are available for people with education and training.

Your future career might require you to communicate and work with people from all over the world. This makes your social studies and foreign language learning experiences as important as those in science and mathematics.

Try this activity challenge. With your teacher's guidance, divide the class into groups of students with similar occupational interests. See Figure 6-10. Each group should research its chosen field to determine what technological devices are used and how language arts, science, mathematics, social studies, and knowledge of a foreign language might be needed.

With your teacher's help, establish a **shadowing program** in which you and your classmates spend a day in the work environment of your chosen field.

Figure 6-10 Occupations can be divided into job clusters, like these, which have been established by the U.S. Office of Education.

Agribusiness & Natural Resources

Business & Office

Communication & Media

Construction

Family & Consumer Services

Environment

Fine Arts & Humanities

Health

Hospitality & Recreation

Manufacturing

Marine Science

Marketing & Distribution

Personal Service

Public Service

Transportation

Organizational Skills

Good organizational skills are important on every job.

Teachers must have many skills if they hope to educate their students. The ability to organize is one of the most important. This is especially true for a technology teacher. Can you imagine a teacher directing the activities of twenty-four middle school students without a good organizational plan? It would be impossible! Maybe your career goal is not to be a teacher, but whatever you choose, the ability to organize will be a big help. Here are some guidelines on how to become better organized.

- ✔ **Keep good records.**
- ✔ **Budget your time.**
- ✔ **Keep everything stored in its place.**
- ✔ **Make neatness a priority.**
- ✔ **Work on tasks in order of their importance.**
- ✔ **Throw away or delete outdated files and papers.**

Technology Teacher
Must be creative, organized, and licensed. Must have knowledge of the National Standards for Technological Literacy (STL) and technology education degree. Must be able to integrate technology with other academic subjects and help solve problems and build experiences. Will be expected to start a TSA chapter in the school.

SECTION REVIEW 6B

Recall ▸▸

1. Name at least three subjects included in the humanities.
2. How did Johannes Gutenberg influence both language arts and technology?
3. What was ARPANET?

Think ▸▸

4. How do you think the global positioning system could be used to help find lost children?

Apply ▸▸

5. **Research** Do some research to find the difference between the Internet and the World Wide Web. Write a paragraph explaining your findings.

Summary Activity

For each numbered blank, pick the answer from the list on the right that makes the most sense in the entire passage. Write your answer on a separate sheet of paper. No answer will be used more than once.

Modern __1__ depends heavily on spoken and written communication and a basic understanding of the principles of science and mathematics.

Scientists all apply the same __2__ to solve problems. Scientific conclusions that have been carefully developed through experimentation are called scientific __3__. Over time they may be accepted as scientific __4__.

Few subjects are more important to technology than mathematics. For example, a narrow river might require only a simple bridge. A larger river might require a complex bridge with suspension cables. In both cases, mathematical __5__ tell technologists which design is better.

In language arts you study literature, reading, writing, spelling, and __6__ to develop the ability to use English correctly. Verbal and written __7__ is in many ways the foundation of technology. We need it in order to develop technology and teach it to others.

In social studies you learn that the study of technology deals with more than things. It deals with how people __8__ technology and how it changes their lives. Most American technical museums show how past __9__ have improved people's lives, changed economies, and altered governments.

Our society has moved beyond factory production into the __10__ Age. We still need factories, of course, but more and more people make their living by creating, processing, and distributing information. Linking the world by __11__ has allowed everyone to share in the specialized knowledge of a few.

Today's jobs require workers with technology-based abilities. With your teacher's help, you could establish a(n) __12__ program in which you and your classmates spend a day in the work environment of your chosen field.

Answer List

- calculations
- communication
- computers
- grammar
- Information
- inventions
- laws
- scientific method
- shadowing
- technology
- theories
- use

Comprehension Check

1. What is the first step in the scientific method?
2. What is a scientific theory?
3. What force does Newton's third law of motion involve?
4. When did the Information Age begin?
5. What is a shadowing program?

Critical Thinking

1. **Decide.** Has this chapter affected how you regard one of the subjects discussed in it? If so, how?
2. **Plan.** Could you improve your work in one of the subjects discussed in this chapter? If so, what do you plan to do about it?
3. **Connect.** Think of a subject not discussed in this chapter. Then think of at least two ways in which that subject is linked to technology.
4. **Relate.** Describe at least two ways in which your own life is part of the Information Age.
5. **Infer.** Think of a job you might like to have some day. How do you think technology might be a part of that job?

Visual Engineering

Rocketry has been in existence for many years, but only in the past fifty years has it been used to explore outer space. Science, engineering, mathematics, language arts, and even social studies have been important to the space program in the United States. NASA has several excellent Web sites filled with information about space exploration. Use a search engine to find these sites and learn some fascinating facts. For example, how does a space shuttle stay in orbit? How fast does a shuttle move? How long did it take to go to the moon and back? What is the future of space exploration? Share your findings with the class.

TECHNOLOGY CHALLENGE ACTIVITY

Producing a Documentary Video about Technology Connections

Equipment and Materials

- video camera and videotapes
- tripod
- art materials for signs and title cards
- paper and pencils
- props for your video

FIGURE A

FIGURE B

Background

Have you ever seen a documentary film or TV show? A documentary usually gives facts and information rather than telling a story. However, some documentaries are about events that unfold like a story. The best documentaries show action, or something happening.

Goal

You and your team will produce a documentary video. The subject of your video will be ways in which a particular technology (or an event involving that technology) is connected to at least three other subjects taught in school, such as history, mathematics, science, language arts, economics, and art.

Criteria and Constraints

▶▶ Your documentary must examine connections to at least three different school subjects. Your documentary may involve the way the technology was developed or a way in which it is used. An example is given in the Design Procedure. Your teacher must approve your choices before you begin.

▶▶ The connections to school subjects must be pointed out in your documentary, either by a narrator or with signs and labels.

▶▶ Your documentary must be at least ten minutes long, but no longer than twenty minutes.

▶▶ You must produce a script for either the narrator or any actors in your documentary.

FIGURE C

FIGURE D

Design Procedure

❶ Brainstorm with your teammates to select a technology in which you are interested. See Figure A. What would you like to investigate about that technology? Then think of at least three subjects you can connect to your technology. Be creative and have fun with it. For example, if you are interested in transportation technology, you might want to create a documentary about the sinking of the *Titanic*. The *Titanic* was a huge passenger ship that was supposedly unsinkable. However, it did sink on its very first voyage in 1912. Your documentary might be about the cause of this terrible event. You might make a model or use a plastic boat to demonstrate what happened. If you cover such things as the ship's weight and density, you would have connections to both science and mathematics. What effect did the disaster have on the travel industry at the time? That would be a connection to social studies. A connection to language arts could be shown in the advertisements for the voyage or in newspaper stories and books written about the disaster.

❷ Create a plan and a script for your documentary. See Figure B. What will you show and how will you show it? How much time will you spend on each aspect of your subject? Keep in mind the audience, purpose, and nature of your message. Will you use a narrator or actors? Who will be "on stage" and who will be behind the camera?

❸ Gather or build your props and set up any experiments or demonstrations. See Figure C. If you're doing an experiment, you may want to tape only the interesting parts and the results and have your narrator sum up the rest. If you're doing a demonstration, run through it once or twice to be sure it works before you tape it. Create any signs or other graphics.

❹ Rehearse your production.

❺ Videotape your documentary. See Figure D.

Evaluating the Results

1. Was it easy or difficult to find connections to other subjects? Explain.

2. How long did it take to plan, write, and produce your documentary? What percentage of that time was spent on scenes actually videotaped?

3. Did your video camera have an editing function? Did you use it? If so, did it help or hinder your efforts? Explain.

Engineering
Problem Solving Leads the Way

Many engineering achievements began as an attempt to solve a problem. Some of the solutions led to other technologies down the line.

1

Did you know that if the Greek inventor Hero of Alexandria had not invented the pump to bring water up from wells, the automobile might never have come into being? Pumps were also used to drain water from flooded mines, but the pumps were not very powerful. Then, in England **in 1698**, Thomas Savery invented the steam engine to add power to the pumps and drain the mines more quickly.

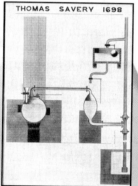

2

Around 1978, Nicholas Joseph Cugnot of France attached a steam engine to a "horse-less" carriage, and the automobile was born. **Investigate:** *Steam did not prove very workable in automobiles. Find out what transportation methods did use steam effectively.*

For many years people had dreamed of digging a canal across the Isthmus of Panama in Central America to shorten the route ships took between the Atlantic and Pacific Oceans. Several attempts were made, **beginning in 1881**, but construction faced many problems. The most serious problem was workers becoming ill and dying from yellow fever.

Then, in 1901, researchers proved that yellow fever was transmitted to humans by mosquitoes. In 1904, the U.S. government asked William Gorgas, an army surgeon, to rid the zone around the proposed canal of yellow fever. He designed a system for sanitation that destroyed the mosquitoes. In two years, yellow fever was eliminated from the canal region. Construction on the canal continued, and countless lives were saved.

Investigate: *The water in the two oceans is at two different levels. Find out what device is used to ease ships from one level to another.*

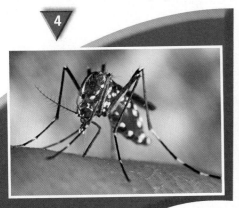

In 1904, electricity was a hot new product, and everyone wanted to use it. Demand for electrical equipment, like these street lights, soared and some materials were in short supply. Leo Baekeland decided to develop a synthetic material that could be used to insulate electrical coils.

Baekeland conducted dozens of experiments that failed. Finally, by heating the chemicals phenol and formaldehyde he produced a gummy goo that when heated again became hard and translucent.

Baekeland called his substance Bakelite. Soon it was being used to make dozens of products. Today, we call it plastic. Without it, our lives would be very different!

Investigate: *Find out who discovered celluloid and what material he was trying to replace.*

Design influences how well things work. One remote control device might have buttons of different sizes and in handy locations. It is probably easier to use than one with buttons of all the same size that are hard to find. Some Internet Web sites are easier to navigate than others are. Finding things takes less time.

Another aspect of design is appearance. Railroad boxcars are an example. Boxcars must meet transportation needs. They do not have to be beautiful. However, do you think that a house should look like a boxcar? For some things, appearance is very important.

When you design something, you arrange parts in a well-planned and skillful way. Design is a creative process that leads to useful products and systems. It often involves problem solving. The steps in problem solving help designers arrive at a design that does the best job. In this chapter, you will learn how design and problem solving are often done together.

Teamwork is a big part of any design or problem-solving activity.

Objectives

▶▶ **Discuss** the roles of creativity, engineering, and appearance in design.

▶▶ **Discuss** ways in which designs are evaluated.

Terms to Learn

- balance
- human factors engineering
- innovation
- invention
- proportion
- unity

Standards

■ Characteristics & Scope of Technology

■ Troubleshooting & Problem Solving

The Design Process
▶ Where does good design begin?

Big or small, expensive or inexpensive, metal or plastic, all products are designed using the same basic procedures. Some you learned about in Chapter 2, "Concepts of Technology." Others will be discussed in this section.

The Role of Creativity

▶ How does creativity influence design?

The design of all new products and processes begins with an idea in someone's mind. John Deere was the first person to use a steel plow in farming. Previously, farmers had used only wooden or iron plows. Deere's 1837 steel plow seems simple today, but no one had thought of it before. His design became the foundation for one of the largest agricultural equipment manufacturing companies in the world. John Deere must have done some creative thinking, which is now considered to be the first step in the design process. Creativity often results in ideas that are original, imaginative, and uncommon.

LOOK TO THE FUTURE

Magic with Mirrors

Thomas DeFanti, the inventor of CAVE technology, got his idea while being measured for a new suit in front of a three-way mirror. He wondered if computers and projectors could create three-dimensional likenesses, and Cave Automatic Virtual Environments were born. In a CAVE, visitors are surrounded by a virtual world that they can see from any angle. Today, scientists and engineers use them to analyze data. In the near future, they will be commonly used by universities, architects, drilling companies, museums, and game designers. Would you like to explore a CAVE?

Figure 7-1 For Thomas Edison's earliest phonograph, music was recorded on wax cylinders.

Thomas A. Edison held 1,093 patents—the most ever granted to one person by the U.S. Patent Office. He often said that his favorite invention was the phonograph. See Figure 7-1. No one before him had ever attempted to patent a similar recording device. **Invention** is turning ideas and imagination into new devices or systems.

If an existing product can be modified to improve it, that's called **innovation**. Bill Gates and Paul Allen were introduced to computers when they were in high school. Their school computers were difficult to operate and were used mostly for solving problems in science and mathematics. Gates and Allen thought that the computers' operating systems could be improved to make them more useful. As a result, they developed their innovative MS-DOS (Microsoft Disc Operating System).

The personal characteristics of designers and engineers may influence their ability to achieve their goals. In addition to being creative, they are often resourceful. They can visualize how a product will look or work. They can also

invention
turning ideas and imagination into new devices and systems

innovation
(in-noh-VAY-shun)
the process of modifying an existing product or system to improve it

ETHICS in ACTION — Can Designs Be Stolen?

As you know, designers and inventors often build on the work of others. What is the difference between building on someone else's work and stealing it? When you steal an idea, you use it without getting permission from the designer or giving him or her credit. You pass the idea off as your own. When you build on someone's idea, you might use it for inspiration, improve on it, or add to it in some way. For example, suppose Joe has designed a new kind of automobile tire that never wears out. Marcela likes Joe's idea. By changing how the tires are made and adding two new ingredients, she designs tires that resist punctures as well. She has built on Joe's design. If she's ethical, she gives him credit for the original idea. Rudy, however, takes Joe's idea, puts his own name on it, and sells it to a manufacturer. Joe never benefits in any way. Rudy is clearly stealing.

ACTIVITY Patents issued by the government can protect inventors. Find out how to apply for a patent and what kind of protection is offered.

think abstractly. In other words, they can think in terms of principles and theories as well as in terms of concrete objects.

The Role of Engineering

▶ Is engineering different from design?

It is one thing to come up with a new idea and another to be sure that the idea will work. Engineering uses mathematics and science to help calculate strength and other important characteristics of a design. Engineering makes sure the product works well, is durable and reliable, and is easy to maintain. For example, some automobile body parts can be made out of plastic instead of metal. They spring back from small impacts, don't scratch easily, and last a long time. When it cannot carry much structural load, plastic can be an ideal choice for some automobile body parts. See Figure 7-2.

Engineering also considers how a product relates to the human body. Clothes are more comfortable, playgrounds are safer, and tools are easier to use when their designers

Respecting Differences

We all want and need respect.

People from many different backgrounds live and work in the United States. You don't have to be close friends with coworkers to respect their differences and work well with them. If they have come from different cultures, American customs may seem strange at first. It may take them a while to get used to new things. They may want to retain some customs from their former country that feel comfortable or that represent their heritage. You can develop respect for people's differences by:

✔ **Giving people time to fit in.**
✔ **Getting to know them before judging them.**
✔ **Trying to see things from their point of view.**

Product Design Engineer
W. Coulter Company, an innovative manufacturer of indoor and outdoor lighting, has three openings for product design engineers. Candidates must like working with their hands, be able to visualize designs, and be able to respect cultural differences among coworkers. CAD experience helpful.

Figure 7-2 The rear bumper of this new car is all plastic.

think about how they will fit the people who use them. This is called **human factors engineering**, or ergonomics.

Engineering can also determine a product's flexibility. The design may have to be flexible enough to meet the needs of many potential users. Some refrigerator doors can be hinged on either the left side or right side. See Figure 7-3. Clothing marked "medium" must fit a range of sizes. Some software works with more than one computer operating system.

The Role of Appearance

▶ What makes a product attractive?

The appearance of a product is often important. Dinnerware, for example, is designed to be attractive, as are clothes, cars, and DVD players. See Figure 7-4. The function of all of those products is enhanced when they

human factors engineering
the design of equipment and environments to promote human safety, health, and well-being; also called *ergonomics*

Figure 7-3 This refrigerator is flexible. The doors can be hinged on either the left or right side.

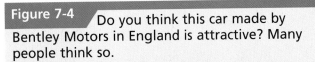

Figure 7-4 Do you think this car made by Bentley Motors in England is attractive? Many people think so.

balance
a state of steadiness or stability

proportion
the correct relationships among sizes and quantities in a design

unity
a state in which all parts of a design work together

Figure 7-5 A chair's design would be evaluated for comfort.

are attractive. Although we can eat just as easily from a tin plate as from attractive china, the china makes the meal more enjoyable and festive. Although an unattractive DVD player might work just as well as one with a sleek, high-tech design, the attractive one looks better in the room and gives an impression of quality.

We all have different ideas about what we find attractive. That's one reason why products often come in varying designs. However, designers follow certain principles that usually produce attractive and successful results. Some of these principles include balance, proportion, contrast, and unity.

- A sense of **balance** is achieved when different elements in the design are placed in such a way as to seem stable and steady. For example, you probably wouldn't place all the furniture on one side of your living room. The arrangement would seem off balance.

- Different parts of a design should be in proportion to one another. **Proportion** usually has to do with size or quantity. Putting huge tractor wheels on a tiny car would create a problem of proportion.

- Sometimes, however, differences in size or materials can deliberately call attention to one part of a design. This is called contrast. For example, a black shirt might be given a red collar for contrast.

- How do you know when everything in a design works together? The design has **unity**. Function and form have been carefully planned and are in harmony with one another.

Evaluating Design

▶ How are designs evaluated?

All during the design process, the design needs to be checked and critiqued. See Figure 7-5. Sometimes basic ideas about the design are changed. For example, someone designing a TV set may suddenly learn about a new technology that will make the original idea out of date before the TV is even built. Ideas are always being refined and improved. Sometimes experiments are made in order to test ideas and products.

After a new design has been developed, engineers often construct a model or prototype on which to test their ideas. When an airplane company designs a new aircraft, they build one for flight testing. A pilot follows a carefully planned flight procedure. Those tests are evaluated and the airplane is changed as necessary before manufacturing begins.

Today, many product designs can be tested using special computer programs. These programs analyze different aspects of the design, such as safety factors. See Figure 7-6.

Figure 7-6 Computer programs can be used to test various aircraft designs.

SECTION REVIEW 7A

Recall »

1. Why is creativity considered the first step in the design process?
2. What influence does engineering have on a design?
3. Name at least three principles designers follow when considering appearance.
4. Name three ways in which designs may be evaluated.

Think »

5. Name some items with which you are familiar and that you think are well designed. Explain. Name some you think are poorly designed. Explain.

Apply »

6. **Experiment** Obtain one or two glider airplanes made of balsa wood. Fly one of them several times to observe typical flight characteristics. Make a list of criteria and constraints that apply. Make some design changes by trimming a little wood from both ends of the wing and see how well it flies. Do the same for the top of the vertical fin at the rear and the ends of the horizontal stabilizer at the rear. Keep trimming and flying, and then draw some conclusions.

Objectives

▶▶ **List** the six steps in problem solving.

▶▶ **Write** an effective problem statement.

Terms to Learn

- brainstorming
- problem statement

Standards

- Core Concepts of Technology
- Relationships & Connections
- Attributes of Design
- Engineering Design
- Troubleshooting & Problem Solving
- Design Process
- Assessment

problem statement
a statement that clearly defines a problem to be solved

Problem Solving

▷ How can problem solving be applied to design?

When you work on a problem, you must first be aware of what you already know and what you need to find out. People sometimes overlook information that could save them time and energy.

Problem solving involves a set of steps. These steps can be performed in sequence, or in a different sequence, and repeated as often as needed. The steps include defining the problem, collecting information, developing possible solutions, selecting one solution, putting the solution to work, and evaluating the solution. See Figure 7-7.

Define the Problem

▷ What does it mean to define the problem?

Before you can solve a problem, you must know what the problem is. Design problems are seldom presented in a clearly defined form. Sometimes defining them is easy to do, but even when it's easy, it's important.

The Problem Statement

▷ How do manufacturers identify a problem?

Engineers and product designers often create a problem statement. A **problem statement** clearly identifies the problem that the product will solve. Suppose you have been asked to design a bookcase for a friend. In order to write a problem statement, you would need to know what problems the bookcase is to solve. For example, where will it be placed, in the living room, bedroom, family room, or workshop? That will help determine its appearance and the materials used to make it. How many books must it hold, a few or many? That, too, will help determine the materials used and also its size. Cost is another factor. Is there a limit as to how much can be spent on materials? Are the materials readily available? All these criteria and constraints will be part of the problem statement.

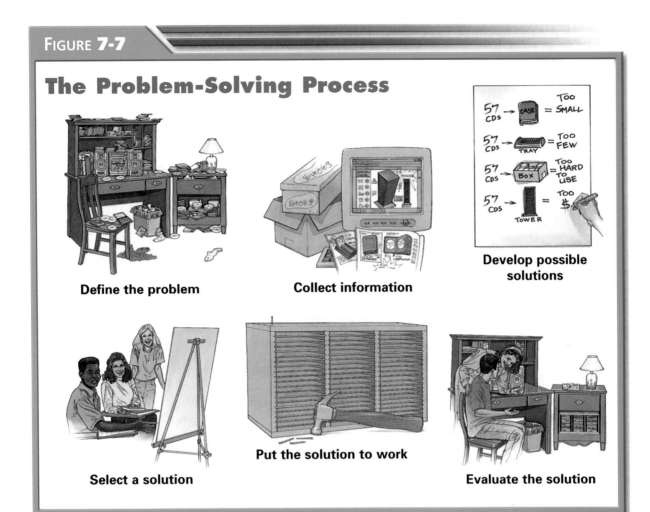

The Problem-Solving Process

Define the problem

Collect information

Develop possible solutions

Select a solution

Put the solution to work

Evaluate the solution

A problem statement for the bookcase might look like this:

Design a bookcase with two shelves that measures 28 inches tall, 33 inches wide, and 12 inches deep so it will fit beside the fireplace. It must be made from wood finished to match the living room furniture, which is walnut. Cost of materials must not be more than $75.

As you can see, a problem statement helps focus your thinking. A poorly worded problem statement results in wasted effort. *Design an improved animal carrier* is not a very effective problem statement. Does it mean a carrier for a cat, a bird, a goldfish, or perhaps some other animal? Maybe it means a trailer for horses. Do you see the difficulty here? Each of those animals would require a completely different type of carrier. See Figure 7-8 (page 184).

Road Warriors

City traffic patterns are very difficult to design. Traffic must move quickly with as few stops as possible. Safety is important.

ACTIVITY Identify a half-mile stretch of road in your community that might be improved with innovations in traffic control design. It should have at least three intersections with traffic lights and/or four-way stop signs. Make a scale drawing of the road. Use problem solving to decide how the traffic could be improved and make a scale drawing of your solution.

Figure 7-8 Be very clear when wording your problem statement.

The statement does not provide enough information for you to get started in the right direction.

An improved problem statement for the animal carrier might be: *Design a plastic and wire-screen carrier for cats weighing up to ten pounds. It must have a latching door hinged at the side and a top-mounted handle.* That statement is more precise. Of course, there is no guarantee that the design will be perfect for all cats. There is no perfect design, and there never has been. Products must be continually re-evaluated to meet the needs of the customers.

Criteria and Constraints

▷ When are criteria and constraints established?

Criteria are usually identified when the problem is defined. They may be included in the problem statement. These requirements and specifications help establish what designers must achieve. The constraints, or limits, on the design must also be determined at the beginning, or designers might waste time working on a design that can't be used. Sometimes, criteria and constraints may change as the design evolves. For example, designers might learn during testing that cats are allergic to a material they planned to use in the animal carrier. A new limit must be added.

Collect Information

▷ What are the advantages of finding out about past solutions to problems?

You don't want to rediscover the wheel. If other people have already solved the same problem or a similar problem, look at what they have done. This could save you a great deal of time and lead to a better solution to your problem.

You also need to learn more about all the factors associated with the problem. For example, you might want to test other animal carriers to learn which are most appealing to cats. You might want to research the anatomy of cats. How do their joints and muscles coordinate?

Develop Possible Solutions

▷ Does every problem have one perfect solution?

As you generate ideas, you will probably realize that most problems have more than one possible solution. However, some may work better than others. Some may be more practical, less costly, more efficient, or easier to produce. As you know, there are trade-offs to every solution. Now is the time to develop the solutions that seem to have the most potential. Always try to explore at least two.

Sometimes people generate ideas by "sleeping" on a problem. Overnight, solutions may present themselves. **Brainstorming** is a group technique that helps you develop possible solutions to a problem very quickly. A number of people freely call out possible solutions. Someone writes all the ideas down, no matter how silly they may seem. Silly ideas often trigger other ideas that will work. Brainstorming can provide sparks of inspiration. See Figure 7-9.

To use this procedure, your class must break up into small groups. One member of the group is chosen to be the recording secretary, and another member is chosen to be the group leader.

TECH CONNECT
MATHEMATICS

Flakes, Nuggets, or Little Os?
Different people like different breakfast cereals. That's why so many kinds are available.

ACTIVITY Suppose you've been asked to create a new cereal. Design and carry out a survey to collect information about cereal preferences. Graph your results.

brainstorming
a group problem-solving technique in which members call out possible solutions

Figure 7-9 Brainstorming can be used in all kinds of problem-solving situations. No ideas are too silly to be considered because they may spark other, more useful ideas.

The group leader states the problem, and each member of the group calls out possible solutions. No one stops to criticize or explain a suggested solution. It's important to consider all ideas, and brainstorming helps that process.

Select One Solution

▷ What must be considered when selecting a solution?

The most important factors to consider when selecting a solution are the criteria and constraints set down when defining the problem. If those are not met, the solution should not be developed, no matter how well it otherwise seems to work.

Next, each solution must be evaluated in terms of its advantages and disadvantages. These may be related to the criteria and constraints, or they may be something different. For example, suppose two kinds of plastic have been proposed for your animal carrier. Both meet all the requirements, but one is more attractive than the other. The appearance of its carrier may mean nothing to a cat, but it might mean something to the cat's owner.

Put the Solution to Work

▷ What is the best way to test a solution?

The best way to test a solution is to put it to work. If the solution is a process, you might carry it out. For example, suppose you wanted to mass-produce animal carriers. You would set up work stations and make a test run to be sure the work flows properly.

If the solution is a product, you would probably want to make a prototype and test it. See Figure 7-10. The best kind of test involves actual use. If you made a prototype cat carrier, you would want to test it with a real cat.

Evaluate the Solution

▷ Should you always stick with your first idea?

Did your product or process work the way you thought it would? Did it meet all the criteria and constraints and any other specifica-

Figure 7-10 Some prototype automobiles don't make it to market. Why do you think that could happen?

tions? Did the cat easily damage the animal carrier? Did potential customers find it unattractive? Every solution must be evaluated and most designs must be refined.

Occasionally, the source of problems is not easily identified. Then troubleshooting methods may be used. Troubleshooting helps identify the cause of a malfunction in a technological system. It is usually done by going through a process step by step.

Sometimes problems can be easily fixed. Sometimes they cannot. Learn from your mistakes and begin again. Be on the lookout for better solutions and use new knowledge gained from experimentation to create new and better solutions. Don't feel that you must stick to your first solution. You will probably see ways to improve it as you work on your project.

Any results or processes must then be communicated to team members, coworkers, or customers. If a new solution must be chosen, many problem-solving steps must be repeated.

SECTION REVIEW 7B

Recall »

1. Name the six steps in problem solving.
2. Describe what goes on during a brainstorming session.
3. During which step in problem solving are criteria and constraints often identified?
4. What is troubleshooting?

Think »

5. What do you think is the most important step in problem solving? Why?
6. You learned about kinds of systems in Chapter 2. If you test and adjust your solution, are you using an open-loop or closed-loop system?

Apply »

7. **Writing** Select a product that you use frequently, such as a hairbrush. Write the problem statement that you think might have been used when the product was designed.
8. **Brainstorm** As part of a small group, brainstorm ideas for a product that would make a bicycle more useful.

Summary Activity

For each numbered blank, pick the answer from the list on the right that makes the most sense in the entire passage. Write your answer on a separate sheet of paper. No answer will be used more than once.

The design of all new products and processes begins with __1__ and an idea in someone's mind. __2__ is turning ideas and imagination into new devices or systems. If an existing product can be modified to improve it, that's called innovation. __3__ makes sure the product works well, is durable and reliable, and is easy to maintain. Personal characteristics of designers and engineers may influence their ideas. In addition to being creative, they are often resourceful and can also think abstractly.

We all have different ideas about what we find attractive. That's one reason why products often come in varying designs. Designers follow certain principles that usually produce attractive and successful results. Some of these principles include balance, __4__, contrast, and unity. After a new design has been developed, engineers often construct a(n) __5__ or prototype on which to test their ideas.

Problem solving involves a set of steps. These steps can be performed in sequence, or in a different sequence, and repeated as often as needed. The steps include __6__ the problem, collecting __7__, developing possible solutions, selecting one solution, putting the solution to work, and evaluating the solution. A problem __8__ clearly identifies the problem that a product will solve.

Sometimes people generate ideas by "sleeping" on a problem. Overnight, solutions present themselves. __9__ is a group technique that helps you develop possible solutions to a problem very quickly. The most important factors to consider when selecting a solution are the criteria and __10__ set down when defining the problem. If the solution is a product, you would probably want to make a model and __11__ it. The best kind of test involves actual use.

Occasionally, the source of problems is not easily identified. Then __12__ methods may be used. They help identify the cause of a malfunction in a technological system by going through a process step by step.

Answer List

- brainstorming
- constraints
- creativity
- defining
- engineering
- information
- invention
- model
- proportion
- statement
- test
- troubleshooting

Comprehension Check

1. What is human factors engineering?
2. What is a problem statement?
3. List several personal characteristics of creative designers and engineers.
4. How is an invention different from an innovation?
5. What is the best way to test a solution?

Critical Thinking

1. **Arrange.** Solve the problem of a river flooding annually by putting the following steps in order: a) Investigate the river's source and its geography. b) Build a model dam. c) Create several dam designs. d) Write a problem statement. e) Test the model. f) Select the most promising dam design.

2. **Connect.** Identify at least three criteria and constraints that might apply to the problem in number 1 above.

3. **Judge.** Which of the steps in problem solving seems least important to you? Why?

4. **Infer.** You know that creativity is important to design. Why is it also important to problem solving?

5. **Design.** Suppose you designed an animal carrier like the one described in this chapter. Write safety specifications for the carrier.

Visual Engineering

Brainstorming. One of the best tools for problem solving is brainstorming. As you know, brainstorming is often done by groups of people. Did you know that it can also be used by one person alone? Think of a problem in your daily life that you would like to solve. Take a few minutes to write down on a piece of paper as many ways as you can think of to solve your problem. Write down even silly ideas. After several minutes, stop and review what you have written. Do some of the ideas look like they're worth exploring? If you are not successful the first time, try it again later. You'll be surprised by how many new ideas will occur to you.

TECHNOLOGY CHALLENGE ACTIVITY

Designing a Product to Overcome a Disability

Equipment and Materials

- pencil and paper
- ruler
- poster board
- permanent markers

Background

In the past, people with physical disabilities were often very limited in what they could accomplish. Living with a disability is still not easy. Today, however, technology is helping more and more people work around these limitations by means of new devices.

Goal

You and your team will identify a disability and produce the design for a product that will help solve a problem for someone with that disability.

Criteria and Constraints

▶▶ Your product must safely aid a person with a physical disability.

▶▶ Your product can be a completely new invention or an improvement on an existing device.

▶▶ You must produce a poster or simple model of your product. A poster must include a drawing of the product.

▶▶ You must turn in a problem statement and a written explanation of your solution.

FIGURE A

FIGURE B

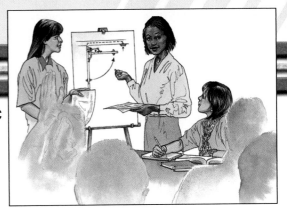

FIGURE C

Design Procedure

1. With your team, select a colorful name, such as the *Eagles* or the *Super Sports*.

2. Discuss some of the difficulties of daily living that people with disabilities might encounter. Consider those who are wheelchair bound or who have visual or hearing impairments. Do some research. If possible, interview someone with a disability and learn about the obstacles that person must overcome each day. As an alternative, simulate a disability by wearing a blindfold or restraining one arm and then try to do a simple task like making a sandwich. What was most difficult about the task and why?

3. Identify one problem to solve. Here are some suggestions. Try to think of others.

   - Device to help a wheelchair-bound person pick up small items from the floor.

   - Device to help a wheelchair-bound person open wide, outward swinging doors. See Figure A.

   - System to assist a visually impaired person in locating the entrance to a public building.

   - System to assist a visually impaired person in determining the color of clothes and whether they match.

   - System to wake a hearing-impaired person in the morning.

   - Device to help someone whose hands are impaired open doors with knobs.

4. Write a problem statement.

5. Use brainstorming to arrive at a potential solution to the problem. Be sure that all group members participate. See Figure B.

6. Select two or three potential solutions and develop them with sketches and written descriptions.

7. Review all the sketches and descriptions. As a team, select one idea that meets all the criteria and constraints in your problem statement. Make any adjustments that are necessary to refine it.

8. Create a poster or simple model that shows how your solution will work. Your poster or model must include the problem statement and a brief explanation of the solution. A poster must also include a simple drawing of the device that will solve the problem.

9. Present your solution to the class. See Figure C.

Evaluating the Results

1. Why did your team decide to work on this particular design?

2. How did the problem statement help you as you worked?

3. How did each team member contribute during the brainstorming process?

SUPERSTARS OF ENGINEERING

INVESTIGATION

Ole Evinrude developed the first American consumer product to be recognized as a National Historic Mechanical Engineering Landmark in 1981. Do some research to find out what that product was.

Leonardo da Vinci
(1452-1519)

Many people know that Leonardo da Vinci was the famous Italian artist who painted the Mona Lisa in 1504. However, many do not know that he was also an engineer. He worked for the Duke of Milan as a designer of dozens of new devices. He built machinery to cut screw threads and nails and invented a water-powered saw. He also built a pump to lift water from a nearby stream into the duke's castle. Fascinated with flying, he created designs for both an airplane and a helicopter centuries before flight was finally achieved by the Wright brothers.

Norbert Rillieux
(1806-1894)

During the nineteenth century, the sugar used for cooking was dark brown rather than white. This was because a lot of heat was needed to evaporate the water. Sugar production was dangerous because people had to hand-pour the boiling juice. Norbert Rillieux made production safer with his invention of a low-pressure evaporator, and the sugar no longer got brown. His invention revolutionized the sugar industry.

Grace Murray Hopper
(1906-1992)

One of the first computer program designers was Grace Murray Hopper. During World War II, Hopper was a naval officer who worked on the United States' first computer, the Mark I. Her knowledge of mathematics helped her develop Flow-matic, one of the first computer languages. It was later used as the basis for several other computer languages. Hopper retired as a rear admiral in 1986, the first woman to hold that rank. She received the National Medal of Technology in 1991.

Robert Goddard
(1882-1945)

Whenever a rocket lifts off anywhere in the world, it uses devices that were originally developed by Robert Goddard. He invented the liquid-fueled rocket engine, ways to control its flight, and many other devices. Goddard was a professor of physics at Clark University in Massachusetts. His rocket experiments were so noisy that unhappy neighbors complained, forcing him to move to New Mexico where he finally did most of his research. NASA's Goddard Space Flight Center in Maryland is named in his honor.

William Hewlett (1913-2001)
David Packard (1912-1996)

William Hewlett and David Packard met while they were engineering students at Stanford University in California. Interested in manufacturing electronic testing equipment, they formed a partnership in 1938 with just $538. The Hewlett-Packard Corporation was one of the first to be established in what is now called Silicon Valley, an area just south of San Francisco. Their first product was a device used with sound equipment in motion picture studios. Walt Disney bought several while making his 1940 movie *Fantasia*. Hewlett-Packard produced the world's first pocket calculator, and its first computer came out in 1966.

Percy Spencer (1894-1970)

The next time you make popcorn in a microwave oven, think of Percy Spencer, the engineer who invented that speedy device. In the 1940s, Spencer was not a biotechnologist. He was developing components for radar, the use of radio waves to detect hidden objects, such as submarines. One day he left a half-eaten candy bar next to a piece of equipment, and the candy melted. Curious, Spencer investigated, and soon the microwave oven was born. The first oven weighed 750 pounds and needed plumbing to prevent overheating. Twenty-six years passed before Amana introduced the first household model.

FROM DRAWINGS TO PROTOTYPES

Can you remember ever saying, "I wish I had a picture to show you what I'm talking about"? You say this when an object or event just can't be described fully with words alone. This is why we have the expression "A picture is worth a thousand words."

This is also why pictures in the form of drawings are used to construct the machines, buildings, highways, products, and systems of our technological world. These drawings can be created by freehand sketching, technical drawing, and computer-aided design and drafting.

In this chapter you will learn about technical drawing, modeling, prototype construction, and the use of computer-aided design and drafting. You will see how the graphic language is the foundation of our designed world.

In a commercial for M&M candies, the TOPIX animation team used SOFTIMAGE/XSI software to create a playful take on the old Godzilla movies. Computer software is also used to create drawings of the products you buy.

Objectives

▶▶ **Explain** the difference between freehand sketching and technical drawing.

▶▶ **Describe** the alphabet of lines and drawing to scale.

▶▶ **Tell** the purpose of each kind of technical drawing.

Terms to Learn

- CAD
- dimension
- drafting
- isometric drawing
- multiview drawing
- oblique drawing
- perspective drawing
- pictorial drawing
- scale drawing
- section drawing
- technical drawing

Standards

- Relationships & Connections
- Attributes of Design
- Use & Maintenance
- Information & Communication Technologies

drafting
the technique used to make drawings that describe the size, shape, and structure of objects

The Graphic Language

▷ **What kinds of drawings are used for technology?**

Drawings are used by architects, designers, engineers, technicians, tradespeople, and many others to describe the size, shape, and structure of objects. The technique used to make these kinds of drawings is referred to as **drafting**. Drafting is done to help people make things. It is done in science and engineering to help people understand data. It may also be used to create images that will become part of a video game or movie.

The different areas of technology sometimes have different rules for how drafting is done. However, they are based on the same general principles.

Freehand Sketching

▷ How are freehand sketches used for technology?

Do you or your friends like to do freehand drawings of cartoon characters or cars? These freehand sketches express the size and shape of objects. Freehand sketching can be the first stage in the development of a drawing that will be used to construct a building, bridge, or automobile. Designers often make many freehand sketches of possible designs to get their ideas down on paper. They may try to be creative. Then they use the sketches to evaluate their designs and pick the best one.

Freehand sketches are usually drawn using only pencil and paper. To make your first sketching experience easier, try drawing on graph paper. The ruled lines can guide you in drawing straight and angled lines. See Figure 8-1.

In the technology laboratory, you will often use freehand sketching to show the size and shape of a project you plan to construct. Drawing is one of the simplest ways of communicating ideas to other people.

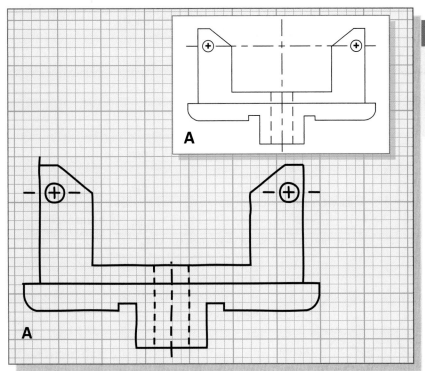

Technical Drawing

▶ How does a technical drawing differ from a freehand sketch?

Can you sketch an object accurately enough to build it? Many simple projects can be made from freehand sketches. As the project becomes more complex, it becomes necessary to have more accurate plans. If the project consists of a number of parts that will be made by different people, the plans must be extremely accurate.

Technical drawing is using mechanical or electronic tools to accurately show the size and shape of objects. It is also sometimes called mechanical drawing. When you produce a technical drawing, you may use pencils, pens, paper, a drawing board, rulers, triangles, compasses, a computer with special software, and other tools and machines. All these tools give the drafter—the person doing the drawing—the ability to draw perfectly straight, round, or curved lines.

technical drawing
accurately representing the size, shape, and structure of objects; also called *mechanical drawing*

Drawing Techniques

What techniques do drafters use?

If all your friends looked exactly alike, you would have trouble identifying them in a crowd. If all the lines of a

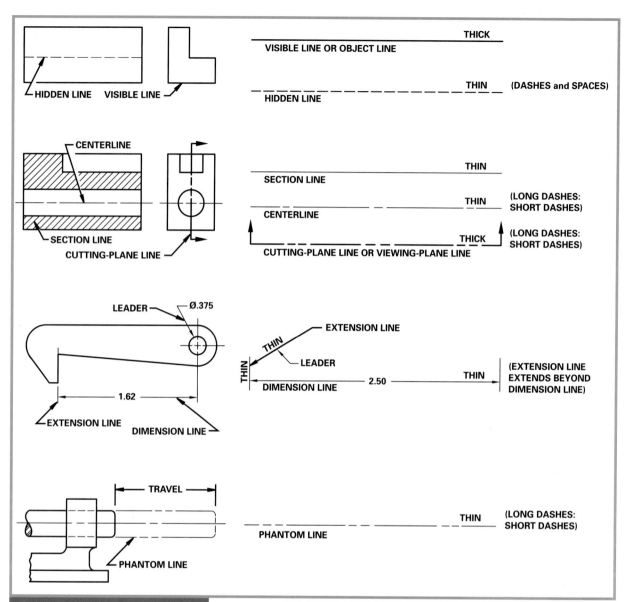

Figure 8-2 Lines in a drawing vary depending on their purpose. Can you tell a phantom line from a visible line?

scale drawing

drawing of an object that is not true size but that is in the correct proportions

technical drawing were exactly the same, it would be confusing. Drafting therefore has rules about how each type of line should be drawn. Figure 8-2 shows a drawing with the different line thicknesses labeled. These lines are often referred to as the alphabet of lines.

Can you imagine walking around with a set of house plans that are as big as a house? Drawings are very often created smaller or larger than the objects that they represent. They are then easy to carry and use. For example, a set of house plans might be drawn so that one-quarter inch equals one foot. This is called a **scale drawing**.

Positive Attitude

A positive attitude is a valuable tool in the workplace.

Did you know that your attitude has a lot to do with how successful you are at a job? It determines how you react to others and how others react to you. Employers like workers with a positive attitude because they get along well with other employees and with customers. Positive thinkers also handle problems more effectively, because they approach a problem believing it can be solved. To build a positive attitude, try the following:

✔ **Instead of just complaining about a problem, take some action to resolve it.**

✔ **When a situation looks bad, try to find some good things about it, no matter how small. You'll feel better, and you'll probably handle the situation better too.**

✔ **Positive thinking can be contagious, so hang out with positive thinkers. (Keep in mind that negative thinking can be contagious, too. Don't spend a lot of time with negative thinkers.)**

✔ **If some people think you are too optimistic, don't let them discourage you. Your positive attitude will help you move forward, and that's all the proof you need that it works.**

Although the object is shown smaller or larger, the proportions are correct.

When technical drawings are prepared, drafters size their drawings to fit the paper they are drawing on. When drawings are prepared on a computer, drafters draw using full-size dimensions. The computer automatically scales their drawing to fit on the computer screen. Let's take a closer look at the different types of technical drawings.

Multiview Drawings

▶ What are multiview drawings?

Think of an object placed inside a square box that is made out of glass. If you took a picture through each side of the box, you would have six pictures to show each individual side of the object. You can also draw an object the

multiview drawing
a drawing that shows an
object from several different
views

same way, showing the different sides. This is called a **multiview drawing**.

Although six drawings can be produced, most objects can be fully explained by drawing the front, top, and right side. Figure 8-3 shows a multiview drawing. A drawing needs only as many views as necessary to provide the full-shape description. When necessary, the drafter creates

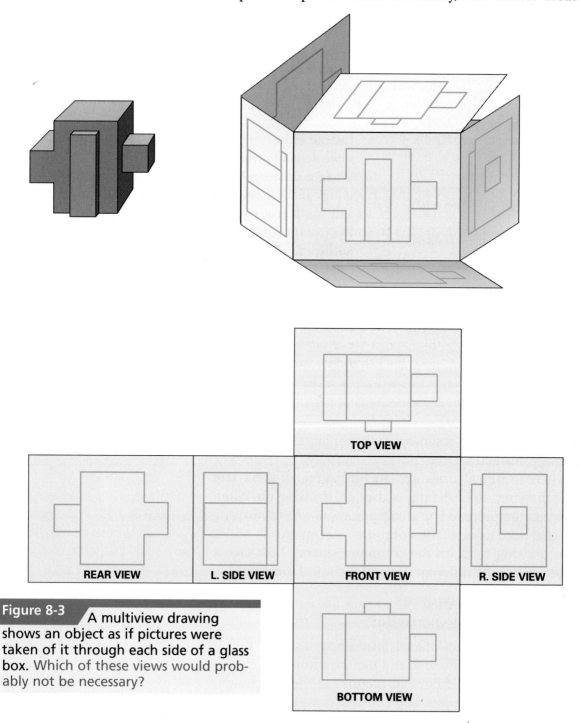

Figure 8-3 A multiview drawing shows an object as if pictures were taken of it through each side of a glass box. Which of these views would probably not be necessary?

additional views of a part so that every detail needed for construction is shown.

It is easy to place dimensions on a multiview drawing. **Dimensions** give the size of the object and the exact size and location of holes, cutouts, and other features. When a small feature must be shown very clearly, a detail drawing of it may be made.

When the drawings are finished, many copies are made. Some sets are given to people who will check for design problems. Other sets are filed with government officials when licenses, patents, or permits must be obtained. Additional copies may be needed at the construction site or factory where the finished product is made.

Pictorial Drawings

▷ What is a pictorial drawing?

It is sometimes hard to visualize what the object actually looks like in a multiview drawing. For this reason, when drafters show an object as a multiview drawing, they often include another drawing that shows it more realistically. A **pictorial drawing** is used to create a realistic looking picture.

Isometric Drawings In an **isometric drawing**, the object being drawn is rotated thirty degrees and tilted forward thirty degrees so that three sides are shown. The edges of the box in Figure 8-4 (page 202) that now face toward you all form the same angle. An isometric drawing is the only pictorial drawing that can be measured along all three axes.

Oblique Drawings In **oblique drawings**, you see a perfect, undistorted front view of the object. See again Figure 8-4. If the object has circular parts on one surface, it is easier to make an oblique drawing than an isometric drawing. Oblique drawings do not always present a clear picture of an entire object. Therefore, they are the least drawn pictorials. They are most often used in the furniture industry.

Perspective Drawings In a **perspective drawing**, the object appears as it would in real life. See again Figure 8-4. This means that the parts of the object that are farther from your eyes will appear smaller. Parallel lines appear to vanish into the distance, which is commonly called the vanishing point.

dimension
a size or location of object parts indicated on a drawing

pictorial drawing
a drawing that shows a three-dimensional object realistically

isometric drawing
a pictorial drawing that shows three sides of an object which has been rotated thirty degrees and tilted forward thirty degrees

oblique drawing
a pictorial drawing that shows one side of the object face on without any distortion and the other sides at an angle

perspective drawing
a realistic pictorial drawing in which receding parallel lines come together at a vanishing point

TECH CONNECT
MATHEMATICS

Mainly Metrics
Drawings and models made for the international market must be made using the metric system of measurement.

ACTIVITY Select a drawing or model you have made for this course and convert the measurements to metrics. If you used CAD, this can be done automatically.

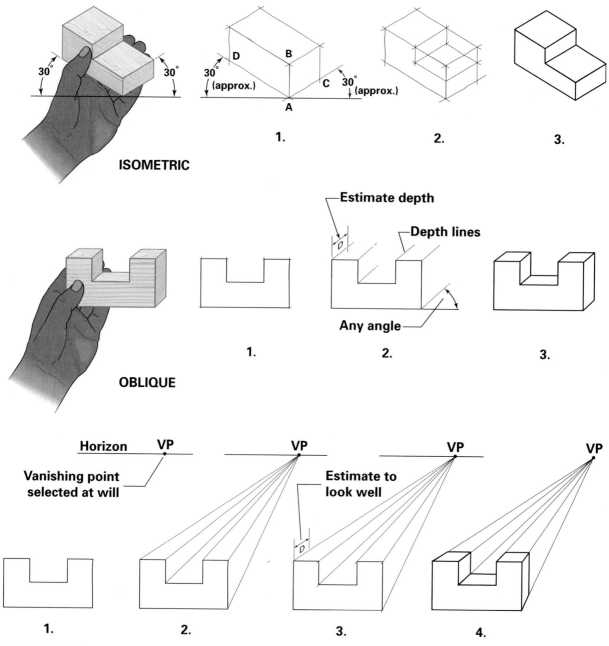

ISOMETRIC

1. 2. 3.

OBLIQUE

Estimate depth

Depth lines

Any angle

1. 2. 3.

Horizon   VP   VP   VP   VP

Vanishing point selected at will

Estimate to look well

1. 2. 3. 4.

PERSPECTIVE

Figure 8-4 Three kinds of pictorial drawings: isometric, oblique, and perspective.

This is the hardest of the pictorial drawings to do, yet with certain objects it is the easiest to understand when viewing. See Figure 8-5.

Renderings are usually perspective drawings to which shading, color, and other details are added. They are usually used for presentations.

Can you see how your positioning of the object before it is drawn affects your final drawing? Can you see why isometric, oblique, or perspective drawings are better with some objects than they are with other objects? Study the

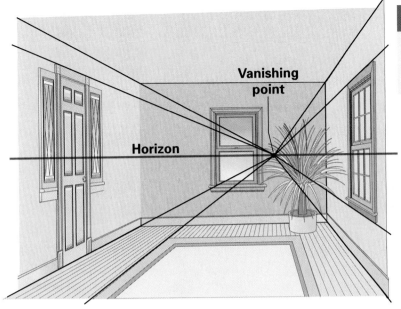

Figure 8-5 A perspective drawing of a room. Color can be
added to make the drawing more
realistic.

Vanishing
point

Horizon

section drawing
a drawing that shows the
interior of an object

examples to see the relationships
among the object, the drawing,
and the observer.

Section Drawings

▷ What is the purpose of section drawings?

If you wanted to see what was
inside a candy bar, what would
you do? You would cut open the
candy bar. In the same way, a
drafter can show the inside of
objects by cutting them open and
drawing a view of the interior.
Drawings that show the inside of
an object are called **section drawings**. See Figure 8-6.
These drawings are used when the insides of objects are
complex.

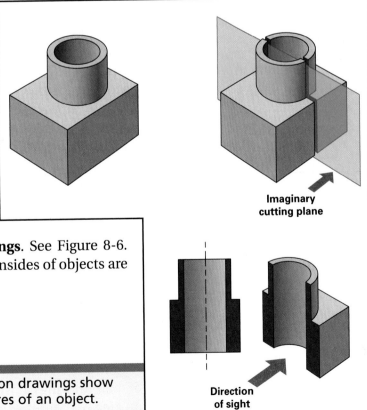

Imaginary
cutting plane

Direction
of sight

Figure 8-6 Section drawings show
the internal features of an object.

Figure 8-7

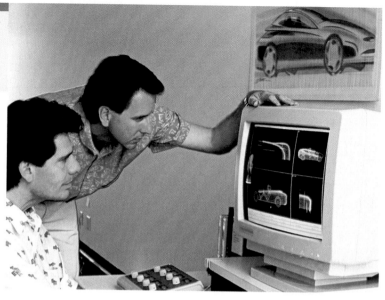

Computer-Aided Drafting and Design

▷ How does CAD differ from traditional design and drafting?

CAD
computer-aided drafting (or computer-aided design); the use of a computer system in place of mechanical drawing tools to create technical drawings and/or design an object

Designers often use a computer to develop and test their designs. Many drafters use a computer instead of a drawing board, paper, T-square, and triangles to make the drawing. See Figure 8-7. The use of computers to do design and drafting is known as **CAD**.

The computer serves as a very powerful tool because it lets you change and even test your design while you are still drawing it. For example, an automotive design can be tested for wind resistance right on the computer. Even an expensive computer, however, can't turn poor designs or drawings into good ones.

With CAD, the drafter selects the type of lines to be drawn from an assortment of drawing tools. The drafter enters the location of the line into the computer by indicating its starting and ending points. The computer program also contains many pre-drawn symbols that the drafter might want to include in the drawing. The drafter picks what is needed and then cuts and pastes it into the drawing.

Computer-aided drafters increase or decrease the size of their drawings at will. For fine detail work, they magnify an area of their drawing, add the details, and then reduce the area back down to size.

When the drawing is finished as in Figure 8-8, the computer will add the dimensions, draw the pictorial, rotate the object so that you can see what the other side looks like, and even send the drawing to the printer. Many copies can easily be made. Large plotters and printers transfer the drawings to paper using ink pens, lasers, inkjets, or electrostatic methods. The completed plans are now ready for other workers to use to make bridges, buildings, or products. Computers are so helpful that today the majority of drafters and designers use them.

Modeling, Rendering, and Animation

▶ How can drafters make models lifelike?

Using a computer, a drafter can produce a three-dimensional model. The model is built by combining various shapes, lines, and arcs to create a solid surface or wire-framed structure. A wire-framed object appears to be built out of wire mesh rather than solid surfaces. See Figure 8-9 (page 206).

Rendering gives a drawing depth. It produces the reflections that are created when light hits an object. A powerful computer and rendering software can make three-dimensional models look lifelike.

| Figure 8-8 | Designers can use CAD to view their drawings from many different angles and even in 3D. CAD can also figure the dimensions. |

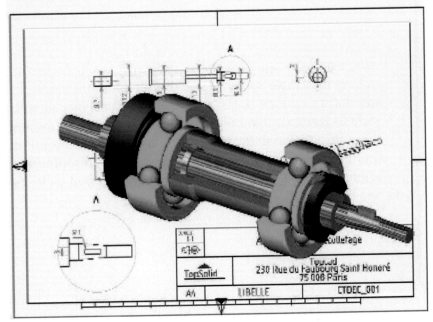

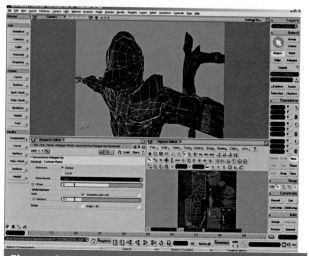

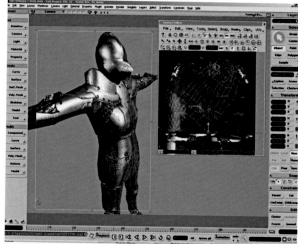

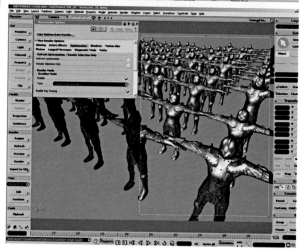

Figure 8-9 After a wire-frame object is drawn, it can be changed in many ways. It can also be the starting point for animation.

To make a drawing move, the drafter adds animation. By simulating moving parts, engineers can use special software to test a design before it is ever built. By combining three-dimensional drawing, rendering, and animation, drafters can create images for architectural presentations, scientific visualization, video games, and motion pictures. See Figure 8-10.

Figure 8-10 With the kinds of drawing software available today, you can draw almost anything you can imagine.

SECTION REVIEW 8A

Recall »

1. Give three reasons why people might make freehand sketches.
2. What is the purpose of a multiview drawing?
3. Name and describe the three types of pictorial drawings.

Think »

4. Explain in your own words what is meant by the saying: "A picture is worth a thousand words."
5. Describe objects that can best be represented by isometric, oblique, perspective, or section drawings.

6. Name some objects that would need to be drawn to a different scale. Would they be drawn larger or smaller?

Apply »

7. **Sketching** On a separate sheet of graph paper, freehand sketch an object in the classroom or complete a multiview drawing of a simple shape, such as a box of tissues. Be sure to use accurate dimensions. What scale will you use?
8. **CAD** Use a computer drawing program to create the drawing you made for the first activity.

Objectives

▶▶ **Give** examples of drafting applications.

▶▶ **Tell** the purpose of working drawings and schematic diagrams.

Terms to Learn

- assembly drawing
- schematic diagram
- working drawing

Standards

- Core Concepts of Technology
- Relationships & Connections
- Attributes of Design
- Information & Communication Technologies
- Manufacturing Technologies

working drawing
a drawing or set of drawings that gives all information needed to make a product or structure

assembly drawing
drawing that shows how to put parts together to make an item

Drafting Applications
▷ How is drafting used in the "real" world?

Architects, engineers, technicians, scientists, and tradespeople need drawings that have been created to meet their own specific needs. For example, drawings that provide all the information needed to build a house look very different from drawings used to build a TV set. Drafters become specialists in such fields as architecture, maps, engineering, manufacturing, and electronics. Let's see what some different drafting applications are all about.

Drafting for Manufacturing

▷ How does industry use drawings?

Without technical drawing, modern industry would never have developed. The machines of industry and all the products that are made by these machines started out on the drawing boards of drafters.

Manufactured objects begin as **working drawings** that give all the information necessary to build them without any further instructions. This information includes complete shape and size description, as well as additional instructions in notes that are written right next to the drawing. Working drawings are usually multiview drawings.

The complete working drawing package consists of multiview drawings, pictorial drawings, and assembly drawings. **Assembly drawings** show workers or consumers how to put parts together to make the object. See Figure 8-11. In some manufacturing systems, drawings are sent from the drafter's computer to the machine that makes the product.

Figure 8-11 An assembly drawing shows how an object should be put together.

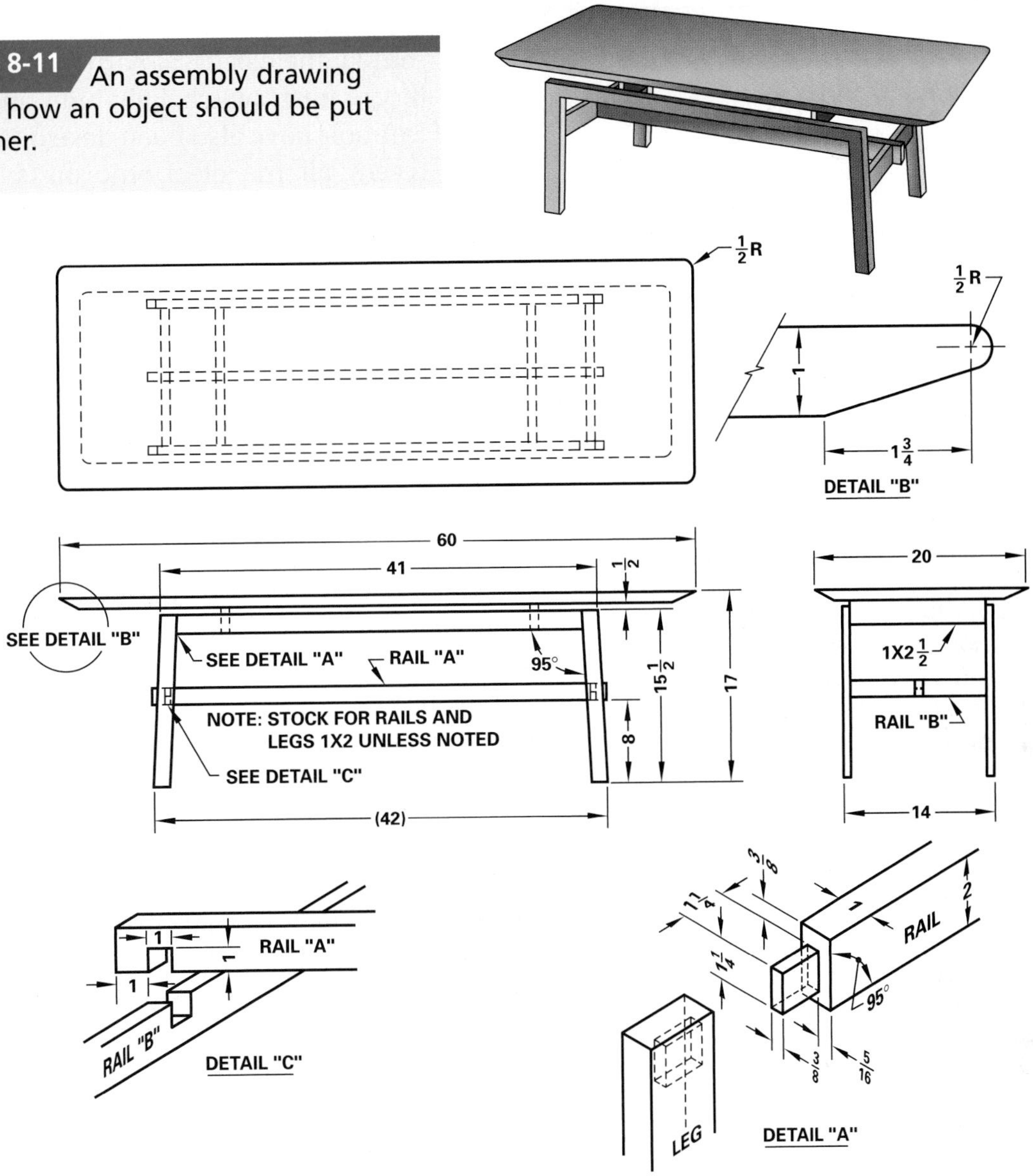

½R
½R
1
1¾
DETAIL "B"
60
41
½
20
SEE DETAIL "B"
SEE DETAIL "A"
RAIL "A"
95°
15½
17
1X2½
8
NOTE: STOCK FOR RAILS AND
LEGS 1X2 UNLESS NOTED
SEE DETAIL "C"
RAIL "B"
(42)
14
1
1
1
RAIL "A"
RAIL "B"
DETAIL "C"
3/8
1¼
1¼
1
2
RAIL
95°
3/8
5/16
LEG
DETAIL "A"

Drafting for Electricity and Electronics

Why are electrical drawings different from other kinds of drawings?

Electrical items have circuits, electric components, and electronic components. Just as you have your own way of expressing things, drafters working in the electric/electronics field phrase things in their own way. Drawings that show how electric and electronic components are connected together are called **schematic diagrams**. See Figure 8-12 (page 210).

schematic diagram drawing that shows the circuits and components of electrical and electronic systems

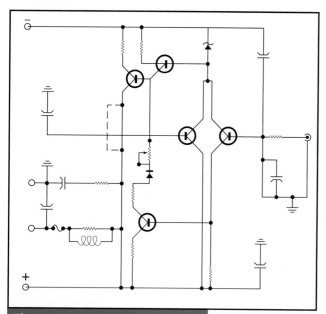

Figure 8-12 A schematic drawing uses lines and symbols to show electrical circuits and components, such as resistors and batteries.

Symbols have been developed to represent all the electrical parts, such as a battery or plug, switch, light bulb, and socket. Symbols have also been developed to represent all the electronic parts, such as transistors and integrated circuits.

Drafting for Engineering and Construction

▶ **What kind of information is needed in working drawings for a construction project?**

Architects and engineers use design and construction knowledge that has been developed over centuries. The first designers probably outlined plans for their structures in the dirt or on cave walls so that helpers could work together.

Today a set of construction drawings includes a complete shape and size description of the foundation, the inside and outside of the structure, the electrical system, and the plumbing. See Figure 8-13. Pictorials are included that show how the structure should look when finished. Depending on the type of structure, the drawings might also include driveways, parking lots, landscaping, and connecting roads.

Designing a building involves more than just deciding what it will look like. Certain areas are more susceptible to earthquakes, termites, floods, and ground slides. When building in these areas, it is necessary to plan and build a structure that can survive the elements.

Designing a home should be done from the inside out. First you determine the activities that will go on in the house. Then you decide on the types and sizes of rooms that will be needed. Must every room be enormous? Remember that a larger house costs more—and larger doesn't necessarily make a house better. The architect's goal is always to design a house big enough to meet the client's needs and small enough to fit the client's pocketbook. Designers must consider all criteria and constraints when developing a design.

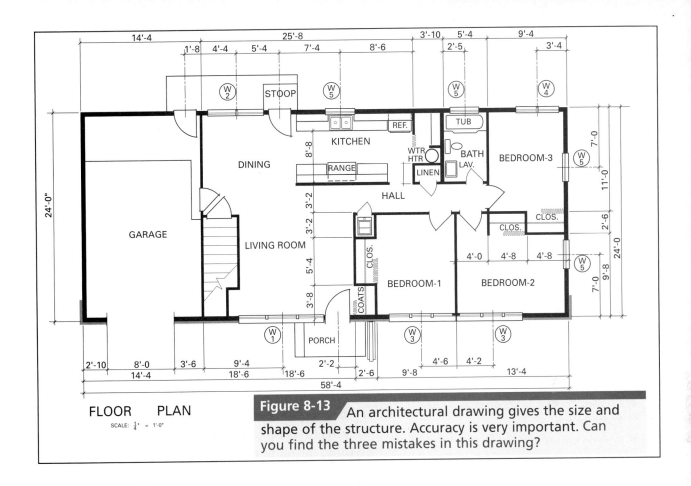

FLOOR PLAN

SCALE: $\frac{1}{4}$" = 1'-0"

Figure 8-13 An architectural drawing gives the size and shape of the structure. Accuracy is very important. Can you find the three mistakes in this drawing?

SECTION REVIEW 8B

Recall »

1. What drawings are included in a set of working drawings?
2. What are schematic diagrams and what are they used for?

Think »

3. Do you think the builders of the Pyramids in Egypt used construction plans? Explain.

4. What do you think is meant by the expression "back to the drawing board"?

Apply »

5. **Communication** Create a list of instructions for a simple task, such as brushing your teeth or making a peanut butter sandwich. Then make a series of sketches that describe the same steps. Your goal is to make the process clear to someone who speaks another language.

Objectives

▸▸ **Tell** the purpose of models and prototypes.

▸▸ **Describe** rapid prototyping.

▸▸ **Explain** the purpose of scientific and engineering

Terms to Learn

• model
• prototype
• rapid prototyping
• visualization software

Standards

■ Characteristics & Scope of Technology

■ Relationships & Connections

■ Engineering Design

■ Design Process

■ Use & Maintenance

model
replica of a proposed product that looks real but doesn't work

prototype
working model of a proposed product

Building Models and Prototypes

⏵What are models and prototypes and how are they used?

Many projects require construction of models and prototypes. A **model** doesn't actually work. It is created to determine how a product, building, or system will look. A **prototype** is usually a model that works. Models and prototypes can be crafted by hand, by machine, or with a computer system.

If a prototype is an accurate representation, it can be used to test such things as product reliability and safety. These tests all take place before the product is mass-produced. The drawings that will be used to build the product might be changed many times to reflect improvements discovered during testing and inspection.

At one time, airplane manufacturers had no choice but to build wood models of their airplane designs. They would fly these models in wind tunnels. If the model had good flying characteristics, the next step was to build a prototype that actually worked.

Some designers like to build models out of clay. See Figure 8-14. However, computers have made virtual three-dimensional models and prototypes possible. For example, NASA uses both the oldest and newest technologies to mold the shape of future flight. See Figure 8-15. On a tour of NASA, you would see wood models, high-tech computer work stations, and machines that can make prototypes.

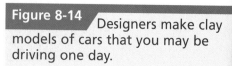
Figure 8-14 Designers make clay models of cars that you may be driving one day.

Figure 8-15 This is an artist's idea of an RLV—a Reusable Launch Vehicle. Do you think we will be using them to travel in space by 2025?

Rapid Prototyping

▷ **What is rapid prototyping?**

Rapid prototyping can turn drawings into three-dimensional objects. Figure 8-16 shows some objects that have been molded using a rapid prototyping (RP) machine. The process uses a laser and a vat of special light-sensitive plastic.

After a CAD drawing is made, the information is entered into the prototyping machine. The machine makes very thin layers out of plastic. These thin layers are then formed on top of each other to build the object from the bottom up.

rapid prototyping
using CAD and a special machine to make a three-dimensional model of an object

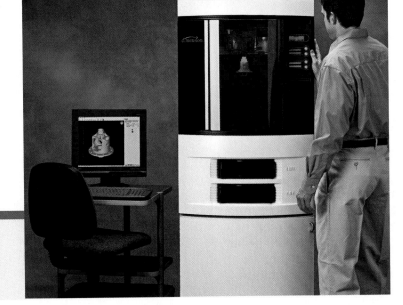

Figure 8-16 A rapid prototyping machine allows a designer to create a prototype of a product in just a few minutes.

Scientific and Engineering Visualization

▷ How do computer models help scientists and engineers understand data?

In the past, to understand tiny structures scientists often turned to LEGO® blocks and Tinker Toys® to make models. **Visualization software** creates virtual models using CAD.

For example, scientists used CAD to create a model of a heart. Medical equipment gathered the data on the size and shape of the heart's muscles, veins, arteries, valves, and chambers. This information was used to create a 3D drawing. This drawing was then rendered and animated. Scientists are using this heart model to test new medicines and explore possible cures for heart disease. See Figure 8-17.

Engineers use the same tools to develop models, gather data, and perform data analysis. Huge structures like dams can be tested for strength during simulated earthquakes. Models of skyscrapers can be tested for wind resistance.

visualization software
CAD software used to create virtual models

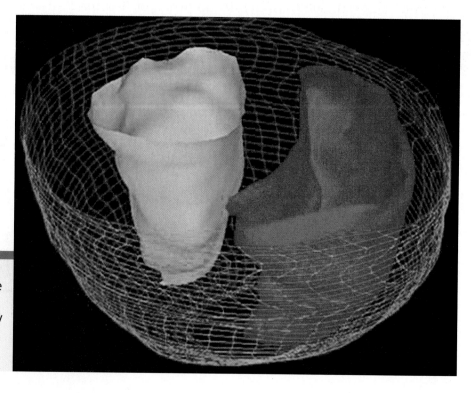

Figure 8-17 This image of a rabbit's heart can be studied in detail for research purposes. It may lead to future cures for human heart problems.

Almost Like *Star Trek*

On *Star Trek*, "replicators" make real things, not just models. Laser Engineered Net Shaping (LENS) will soon allow manufacturers to work in a similar way. Powdered tool steel, titanium, tungsten, or other powdered materials will be fused into solids with lasers. The finished parts and products will be produced directly from CAD drawings.

SECTION REVIEW 8C

Recall ▸▸

1. What is rapid prototyping?
2. Why are models and prototypes created?
3. What is the purpose of scientific and engineering visualization?

Think ▸▸

4. Name several manufactured products that you think might have required a model or prototype.
5. Why do you think the appearance of a product is important?

Apply ▸▸

6. **Construction** Build a model airplane of your own design and test it in a wind tunnel of your own design.
7. **Data Analysis** Use computer software to graph the data from the model plane activity. What conclusions can you draw about your model?

Summary Activity

For each numbered blank, pick the answer from the list on the right that makes the most sense in the entire passage. Write your answer on a separate sheet of paper. No answer will be used more than once.

___1___ are often used by scientists, architects, engineers, designers, and illustrators to show what objects look like. When a freehand sketch is made to describe the ___2___ of the object, the drafter works without mechanical tools. It is very difficult to draw a(n) ___3___ picture without the use of rulers, compasses, and other mechanical drawing tools.

A(n) ___4___ shows the front, top, and right views of the object. Extra views are drawn if they are needed. When the internal structure must be seen, it is drawn as if part of the object were cut away. This is called a(n) ___5___.

It is often difficult to picture the object from multiview drawings. Therefore, the drafter often makes a(n) ___6___ of the object. Several sides of the object are shown at once and represent what the object will look like after it is made.

Drafters represent an object by using different kinds of ___7___. Thickness and length are determined by the rules of drafting. ___8___ give the size and location of the object's features.

In CAD, the designer creates and tests his or her designs directly on ___9___. Printers and ___10___ transfer the drawing to paper.

Almost every object that is made must first be drawn. Drawings made larger or smaller than the real object are said to be drawn ___11___. The parts, however, are in the correct proportions.

A(n) ___12___ is a replica of the product; however, it doesn't work. A(n) ___13___ does work and may even be tested for reliability or safety.

___14___ makes replicas from plastic. The replicas are first drawn with CAD. CAD is also used to make models for scientific and engineering ___15___. The models are used for research and testing.

Answer List

- accurate
- dimensions
- drawings
- lines
- model
- multiview drawing
- pictorial drawing
- plotters
- prototype
- rapid prototyping
- section drawing
- shape and size
- the computer
- to scale
- visualization

Comprehension Check

1. What is the difference between a freehand sketch and a technical drawing?

2. What is the alphabet of lines?

3. Name three main drafting applications discussed in this chapter.

4. Describe drawing to scale.

5. What is the name for an electrical drawing?

Critical Thinking

1. **Classify.** What type of drawing(s) would you use for each of the following? a) Designing a TV remote, b) a kit for making a model plane, c) the outside of a finished house.

2. **Decide.** Which would you rather work from, a multiview drawing of an unidentified product or a very detailed written description? Why?

3. **Plan.** Suppose you have been asked to create a children's toy that has a hidden compartment. Which drawings would you probably use and why? Which parts would you make first and why?

4. **Infer.** Do you think a certain amount of drawing skill might be helpful when drawing with CAD or not? Give your reasons.

5. **Compare.** How do you think drawings done for manufacturing are similar to drawings done for construction? How are they different?

Visual Engineering

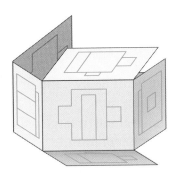

From Another Angle. It is very important for a designer or engineer to be able to visualize ideas. One of the easiest ways for doing that is to create a multiview drawing. Imagine a design for a simple bookcase. Now, draw six views of the bookcase that are laid out as shown in this illustration. Fold your paper into a cube and tape it. Can you visualize the object better now?

TECHNOLOGY CHALLENGE ACTIVITY

Making a Back-Massaging Vehicle

Equipment and Materials

- drawing board
- drawing tools
- computer
- computer printer
- drawing software
- toy factory software
- modeling clay
- 2 x 4 lumber
- ¾-inch plywood
- woodworking tools
- thumbtacks
- sandpaper
- drill set
- drill press
- wood glue
- paper
- pencils
- dowels
- scroll saw
- wood vises

SAFETY

Reminder

In this activity, you will be using tools and machines. Be sure to always follow appropriate safety procedures and rules. Remember, safety is an attitude that you must develop and maintain at all times.

Background

When you prepare a technical drawing, you "break down" an object into different views. You must also learn how to determine the appearance of an object with only the information supplied by those views.

The van shown in the drawing has hexagonal wheels. It is sold in novelty stores as a back massager. When you roll it on a person's back it gives a massage.

Goal

Your goal is to design your own back-massaging car or truck. You will then build it using your own technical drawings.

▶▶ You must prepare a multiview drawing that gives

Criteria and Constraints

the size and shape of your vehicle and any other information needed to build it.

▶▶ You must prepare a pictorial drawing of your vehicle.

▶▶ You must prepare a bill of materials needed to construct the vehicle.

▶▶ Your drawings may be done by hand or using a computer drawing program.

▶▶ Your vehicle can have small or large wheels; they can be of any shape *except* round.

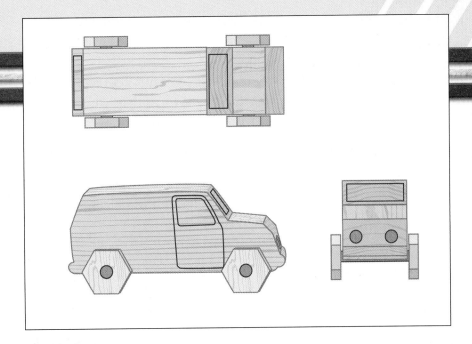

Design Procedure

1. Prepare sketches of different car or truck shapes, or design shapes using modeling clay. Consider the requirements placed on your design.
2. Select the design that meets the requirements best. Prepare a full-size front, top, and right side view of your final design. These views should give you a complete description of your vehicle's shape.
3. Add the needed dimensions to give size description.
4. Prepare a bill of materials that lists all the parts that you will need to construct your vehicle.
5. Prepare a pictorial drawing of your vehicle.
6. Make the body of your vehicle out of 2 x 4 lumber.
7. Make your wheels out of plywood.
8. Cut the axles for your wheels from the dowels.
9. Locate and drill the holes for the axles.
10. Locate and drill the wheel axle joints.
11. Sand all parts.
12. Slip the axles through the body, and slip the wheels onto the ends of the axles.
13. The holes in the body must be large enough for the axles to freely turn.
14. Glue the axle ends into the wheels.
15. Be certain that no glue gets into the body of your vehicle, or your wheels won't be able to turn.

Evaluating the Results

1. Could you build your vehicle using only the size and shape description provided by your drawing?
2. How could you have improved your multiview drawing?
3. If another person was to build your vehicle from your plans, would he or she end up with exactly the same finished product? Explain.

Computers
From Beads to Chips

Today almost every form of communication involves some form of computer technology. Here are some key inventions that made this technology possible.

In 1642, the French mathematician Blaise Pascal built a wooden adding machine for his father, a tax collector. Pascal's machine added and subtracted numbers through the movement of mechanical wheels.

1

One of the earliest mechanical computers was the abacus. It was created in Babylonia about **5,000 years ago.** Calculations were made by moving beads from side to side.

In the 1830s, the English mathematician Charles Babbage designed a machine that could perform complicated calculations by following a set of instructions on punched cards. Babbage's "analytical engine" was never built, but computer designers of the twentieth century adapted some of his ideas.

Investigate: *What did early computers have in common with the Jacquard loom?*

3

The Mark I was the first computer powered by electricity. It was designed by American professor Howard Aiken and completed **in 1944**. The Mark I had many switches that were opened and closed by electricity. The opening and closing of switches formed a type of code that the machine could follow.
Investigate: *How much did the Mark I weigh?*

In 1946, two Americans, J. Presper Eckert, Jr., and John Mauchly, completed the ENIAC. It used vacuum tubes instead of the mechanical switches of the Mark I. The ENIAC could perform 5,000 mathematical operations per second. However, it had to be rewired for each new job, and that took days.
Investigate: *What did the letters ENIAC stand for?*

In 1947, three scientists working at ATT&T's Bell Labs invented the transistor. It was much smaller, used less power, worked faster, and lasted longer. The first computer designed to use transistors was built in 1956. By 1960, all computers were using transistors instead of vacuum tubes.

In the 1960s, engineers developed a way to put dozens of transistors onto a single chip called an integrated circuit. The first chips were used in calculators.

In 1969, the Intel Development Corporation produced the first programmable computer chip. This technology made possible the design of personal computers. During the 1970s and 1980s, personal computers became more powerful, less expensive, and easier to operate. Today, almost half of all U.S. households have at least one computer. **Investigate:** *Research and report on early PCs, such as Honeywell's "Kitchen Computer."*

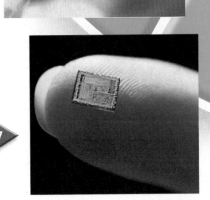

Communication is sending, receiving, and responding to messages. If you hurt yourself and scream out in pain, you are sending a message. Your message contains information. However, if no one hears your call for help, no communication is taking place. Do you see why?

If you are heard by someone who can't understand what you are trying to say, communication still isn't taking place. Your message must be sent, and your message must be received, understood, and sometimes answered for the communication process to be completed.

Suppose someone heard your cry and came to see what was wrong. Does that mean your message was received? Try defining communication right now. Don't be afraid to put things in your own words. Then try to think of other examples of communication that take place every day, including the one you're a part of right now!

It is very hard to get lost if you have a GPS mapping system like this in your car.

SECTION 9A

Objectives

- **Apply** the systems model to communication.
- **Identify** communication subsystems.
- **Discuss** different forms of communication.

Terms to Learn

- **communication technology**
- **electromagnetic carrier wave**
- **graphic communication**
- **sound waves**
- **telecommunication**

Standards

- Core Concepts of Technology
- Information & Communication Technologies

Introducing Communication Technology

What is the difference between communication and communication technology?

Imagine that you are standing in the hallway of your school talking to your friends. Are you using communication? Yes. Are you using communication technology? No. When you are talking face-to-face, you are using communication but not communication technology.

If, on the other hand, you communicate using a public address system, a microphone, or a written note, then you are using communication technology. See Figure 9-1. **Communication technology** is the transfer of messages (information) among people and/or machines through the use of technology. This information is then used to help humans make decisions and solve problems.

Communication technology includes all the knowledge, skills, and tools that led to the development of all of the communication devices and methods ever invented.

communication technology the transfer of messages among people and/or machines through the use of technology

Figure 9-1 When these teenagers talk face-to-face, they are using communication. When they use their cell phones, they are using communication technology.

The Systems Model

▷ How does communication fit the systems model?

As you know, systems can be charted on a diagram. The diagram breaks systems into input, process, output, and feedback.

Communication systems include all the inputs, processes, outputs, and feedback associated with sending and receiving messages (information). The message is the input, how the message is moved is the process, and the reception of the message at the other end is the output. Feedback may involve such things as information about static or other problems in clarity.

Let's use your school newspaper as an example. See Figure 9-2. Suppose you write an article for the paper about the computer lab. Your words, any pictures, the time you spent, and the computer you used are all inputs.

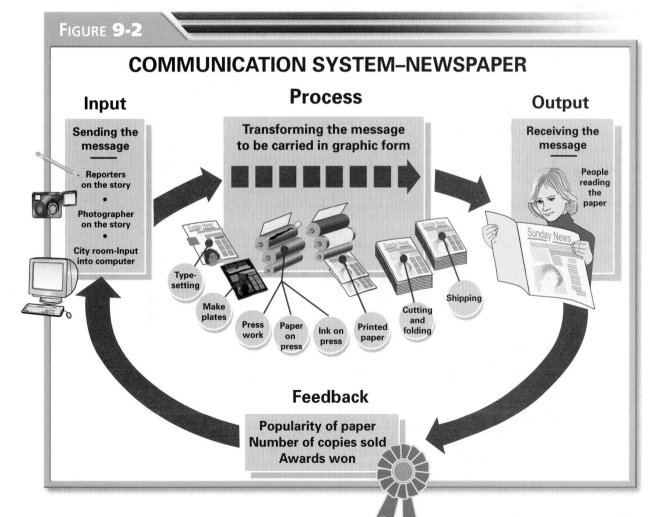

FIGURE 9-2

COMMUNICATION SYSTEM–NEWSPAPER

Input
Sending the message

- Reporters on the story
- Photographer on the story
- City room-Input into computer

Process
Transforming the message to be carried in graphic form

Type-setting
Make plates
Press work
Paper on press
Ink on press
Printed paper
Cutting and folding
Shipping

Output
Receiving the message

People reading the paper

Sunday News

Feedback
Popularity of paper
Number of copies sold
Awards won

Putting the newspaper together and printing it are parts of the process. The primary output is of course the newspaper itself.

When you read a book or magazine, you are at the output end of the communication system, receiving the message. When you use a telephone to talk to a friend, use a computer, or shoot a movie with your video camera, you are controlling both the input and output. What parts of a system are you involved with when you play a video game, watch television, or build a telegraph system?

Your radio and television are the output devices of their communication systems. Computers, CD recorders, and video recorders are communication systems that often contain the input, process, and output devices all in one unit.

Communication Subsystems

▷ What subsystems may be found in a communication system?

Communication systems usually include several subsystems that help transmit information. These subsystems are made up of a source, an encoder, a transmitter, a receiver, a decoder, and a destination. See Figure 9-3. The source is the sender and it could be a person or a machine with a message to send.

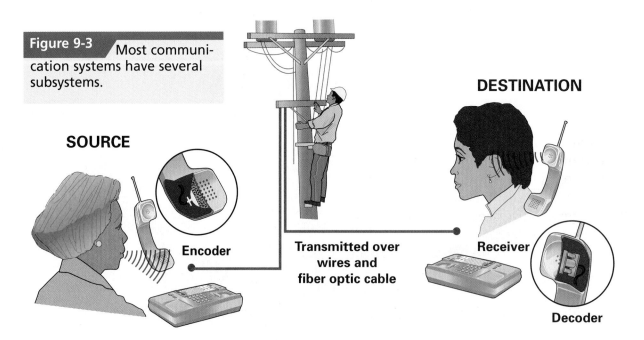

Figure 9-3 Most communication systems have several subsystems.

SOURCE

Encoder

Transmitted over wires and fiber optic cable

DESTINATION

Receiver

Decoder

The encoder changes the message so that it can be transmitted. When you write a note, type on a computer keyboard, or talk into a telephone, you are encoding your message. Your message could be written as words in Spanish or be sent as an electronic signal through a wire.

The receiver of the message at the final destination could be a person or a machine. The message is then decoded, which means symbols on paper must be read or electronic impulses must be turned into information that a person or machine can understand.

Some communication systems have two subsystems designed to store and retrieve information. A telephone answering machine, an MP3 player, and a CD burner are all parts of communication systems that have the ability to store and retrieve information for later use. See Figure 9-4.

When you play a video game, the controller pad encodes and transmits your moves to the machine's central processing unit. The machine determines its own moves and encodes this information so it can be displayed on the screen. You and the machine take turns being the source of information and the final destination for information until one of you wins.

When you are playing a video game you are actually using the tools of communication to process information, solve problems, and make informed decisions. The more you play the games, the better you become at interpreting and using the information on the screen. What other communication devices help you make informed decisions?

TECH CONNECT
SOCIAL STUDIES

Party Line
Until the late 1950s, many people shared telephone lines with complete strangers. Only one party could use the phone at a time. When you picked up your phone, you might hear the other party talking, and vice versa.

ACTIVITY Write a paragraph comparing a telephone party line with a computer chat room.

Message Design

⊳ What factors are important when designing a message?

If you wrote an article for the school paper, you would probably think about the people you were writing it for. You would want to write about things they'd be interested in. The design of a message is influenced by the intended audience.

The medium used—in this case a school paper—is another factor. A story written for a school paper would be different from a story written for a national news program on TV.

Figure 9-4 This telephone has a built-in answering machine. Information is stored for later use.

The nature of the message would also affect how it was designed. A story about the school computer lab would be set up differently than instructions for building the computer lab.

Also important to designing a message is the language used to express the ideas. The language you're writing or speaking in, such as English, is one example. Symbols, measurements, and drawings are other kinds of "languages" people can use to communicate clearly.

For example, technological knowledge and processes are communicated using symbols, measurements, icons, graphic images, and special vocabularies. Letters, characters, icons, and signs can represent ideas, quantities, elements, and operations. The information may be expressed as things we see, hear, or touch. For example, instructions to workers for building the computer lab would include drawings (graphic images), measurements, and construction terms the workers would understand. See Figure 9-5. Ideas would be expressed in a common language that promotes clear communication.

The Purpose of a Message

▷ What purposes do messages have?

Every message has a purpose. That purpose may be to inform, persuade, entertain, control, manage, or educate.

Your article about the school computer lab would probably inform and educate your readers. Perhaps you think more students should use the computer lab after school. In that case you might also use your article to persuade. If the lab teacher asked you to warn students not to play computer games in the lab, your article would be used to control or manage. If it included a funny story about a homework-eating computer, it could be meant to entertain.

All the different ways to communicate information, including graphic and electronic means, can be used for these purposes.

Forms of Communication

▷ **What is one way of grouping communication systems?**

Although the forms overlap, communication systems can be grouped by the way they carry messages. Let's look at the different ways you can transmit a message.

Biological Communication

▷ **What type of communication occurs without technology?**

Ordinary biological communication isn't part of technology. You will study it in depth in science. It is important for you to realize that communication exists outside of technology and that most living things communicate.

Biological communication includes all forms of communication that use natural methods, such as the voice, ears, arms, and hands, to transmit and receive messages. Examples of this form of communication include speaking (language), facial expressions, and hand signals. See Figure 9-6.

Figure 9-6 American Sign Language (ASL) helps people who are deaf or hearing impaired communicate. Do you know any signs?

Listening Skills

Communication isn't just about speaking. Listening is just as important.

Even if you are not a sign language interpreter, developing good listening skills is to your advantage. They will make you more successful in school, and your friends will appreciate you more. In the future, good listening skills will make you a better employee. Here are some guidelines for active listening:

✔ **Listen for the speaker's purpose. What is he or she trying to accomplish?**

✔ **Identify the speaker's main ideas. Your listening will be more accurate.**

✔ **Learn to tell the difference between fact and opinion. Everyone is entitled to an opinion, but it's not the same thing as a fact.**

✔ **Identify what the speaker's voice and body language tell you about what he or she is trying to say.**

✔ **When you're listening to something you need to remember, like instructions, take notes.**

Graphic Communication

▷ What things are used to make graphic communication possible?

graphic communication the methods of sending and receiving messages using visual images and printed words or symbols

Graphic communication includes all forms of communication that send and receive messages visually through the use of drawn or printed pictures and symbols. Printing is the most common example. Magazines, newspapers, messages on clothing, billboards, road signs, and computer images are all forms of graphic communication. See Figure 9-7. In these cases, people send and receive information through reading, writing, drawing, and painting.

Wave Communication

▷ How do waves transfer messages?

Wave communication refers to all forms of communication that move through air, water, outer space, or some

other medium in waves. These forms make use of some of the newest and oldest technological inventions. See Figure 9-8.

Our early ancestors used hollow logs as drums to send coded messages. Banging on the logs caused the air to vibrate with **sound waves**. The sound waves reached people far away. All musical instruments and face-to-face communications also depend on sound waves that travel through the air.

> **sound waves**
> vibrations traveling through air, water, or some other medium that can be perceived by the human ear

Figure 9-8 The waves created by a stone hitting a pool of water are visible. The waves created by hitting a drum are invisible, but your ears hear them.

Figure 9-9 Satellite dishes and tower antennas are necessary for this TV/radio station.

electromagnetic carrier waves
waves of electromagnetic energy that are used to carry signals through the atmosphere

telecommunication
communicating over a distance

Our radio and TV programs are converted into electrical signals. These signals can then be carried by **electromagnetic carrier waves** through the atmosphere. Large antennas and dishes receive the signals. See Figure 9-9. The radio wave and the audio part of the TV signal are converted into sound waves, which you hear when they leave the speakers.

If your telephone company uses fiber optic cables, the signal is transmitted through the cable as waves of light. When you use a camera to take a picture, light waves from your subject bring the image to your eyes and to the camera lens.

Telecommunication

▶ What are some devices commonly used in telecommunication?

Communication over a distance is called **telecommunication**. Today most telecommunication systems use electronic or optoelectronic devices. Did your ever use a telecommunication machine? You most certainly did if you ever used a phone, television, or radio. They are all examples of telecommunication devices.

Satellites are also telecommunication devices. Satellites placed 22,300 miles above the earth and traveling at the same speed as the earth spins are in a geosynchronous orbit. This means that the satellite always stays above the same part of the earth. Its lack of movement in relation to the ground could almost give the impression that it was attached to the end of a very long pole. When a satellite is in a geosynchronous orbit, how do you think it appears to move in relationship to objects that are on the ground?

TECH CONNECT
SCIENCE

Now Hear This!
You have your own built-in sound receivers— your ears. Sound waves striking your eardrums make them vibrate.

ACTIVITY Do some research and then make a drawing showing the main parts of the ear. Label the parts as to their function.

Satellites can help produce maps or observe other parts of the world. See Figure 9-10. The United States had many spy satellites looking down at the Middle East during Operation Iraqi Freedom in 2003. During that war, *Predator* drones and satellites were sending real-time video pictures of enemy positions to commanders on the ground. Satellite-guided targeting systems were so precise that bombs could be sent to military targets without damaging nearby civilian homes. Some of these spy satellites were taking detailed photos using equipment similar to that used on the *Hubble Space Telescope*. It was possible to see an object as small as a grapefruit sitting on the desert sand.

Figure 9-10 This photo was taken by the QuickBird satellite nearly 300 miles above the earth. Can you identify the object in the photo?

SECTION REVIEW 9A

Recall ▸▸

1. What factors are involved in message design?
2. Name the six purposes messages can have.
3. Name the subsystems found in communication systems.
4. What is telecommunication?

Think ▸▸

5. Why would smoke signals be classified as a form of telecommunication?

6. Identify the purposes of the following communications: a radio ad for a political candidate, a movie about American soldiers, a road map, a list of ingredients on a cereal box.

Apply ▸▸

7. **Diagram** Prepare a systems diagram for the communication system of your choice. Use magazine pictures, drawings, and other graphics to show the input, process, output, and feedback of your system.

Objectives

▶▶ **Describe** the different modes of communication.

▶▶ **Explain** how communication technology has made more modes of communication possible.

Terms to Learn

- machine-to-machine communication
- mode

Standards

- Characteristics & Scope of Technology
- Core Concepts of Technology
- Design Process
- Information & Communication Technologies

mode
a way of doing something

Modes of Communication

▶ Why are new modes of communication necessary?

Technology has given us new **modes** of communication. Have you used communication technology to talk with a nonhuman lately?

People to People

▶ What progress have people made in their ability to communicate with each other?

People-to-people communication wasn't always as it is today. At one time people had to communicate with each other within the limits of their own physical makeup (biological communication).

Do you have a baby brother or sister or know someone who does? Have you noticed how the baby communicates its needs? It points, cries, grunts, stamps its feet, grabs what it wants, or speaks just enough "baby talk" to be understood. Our earliest ancestors probably communicated in a similar way. They used simple sounds and gestures.

Over time, people have learned to create new and more powerful methods of communication. They then passed this knowledge on to their children and other members of their clan, or family. People gained the knowledge and skill needed to build complex communication devices. They created all kinds of graphic communication systems to carry their messages through the printed word. Finally, they developed communication based on electrical signals.

People to Machines

▶ How can a person communicate with a machine?

Until the development of electronic communication devices, people were talking only to other people. The machines that they had built could only carry the mes-

Figure 9-11 Communication happens each time you play a computer game. Why would this be considered people-to-machine communication?

sage, and this message could only be understood by other people. Today, however, we have people communicating with the machines that they have created. Examples of people communicating with machines include someone setting an electric timer to turn on a lamp, a computer programmer typing a program into a computer, and a person using a keyboard and joystick while playing a computer game. See Figure 9-11.

Machines to People

▷ How does a machine send a message to a person?

Today machines also send messages to people. Examples of machines delivering their own messages include a whistling teapot that tells you the water is boiling, an alarm system that tells you someone has entered a protected store or home, and a smoke detector that senses a fire and warns you of the danger. See Figure 9-12. Can you think of other examples of machines communicating with people?

LOOK TO THE FUTURE

Designer Computers

How would you like to wear your computer someday? Researchers at NASA are working on a Wearable Augmented Reality Prototype (WARP) that astronauts can wear, leaving their hands free for other tasks. The computer will be a lightweight, voice-activated box with headset and eyepiece. When the user looks into the eyepiece, he or she will see a screen that appears as if it's a few feet away.

Figure 9-12 This smoke detector is a machine that communicates with people.

Machines to Machines

▷ What is an example of machine-to-machine communication?

machine-to-machine communication
the transfer of messages from one machine, usually a computer, to another

Machine-to-machine communication is quite common today. Your computer, for example, gives instructions to your printer, telling it to print your report. In an automated factory, computers attached to sensors control the flow of raw materials and the operation of the machines. The assembly and finishing processes that need to be performed on the product are also controlled by machines communicating with other machines. Finally, the packaging of the finished product may be handled by machines under the direction of—you guessed it—still other machines.

9B SECTION REVIEW

Recall ▸▸

1. How does a baby who cannot talk communicate?
2. How have electronic devices changed modes of communication?

Think ▸▸

3. Identify the mode of communication being used in these examples: a computer program tells a robot how to paint a car; your computer spell-checker finds errors in your report and highlights them; a friend passes a note to you.
4. What is happening if, instead of printing your report, your printer prints pages of nonsense?

Apply ▸▸

5. **Design** Use your artistic talents, a computer graphics program, and magazine pictures to create a large poster. The theme of the poster should be: Communication happens between people and machines.
6. **Construction** Build a model of a machine that can send messages to people. Possibilities include:
   - a warning system to alert a family that their basement is flooding
   - a system to tell you that someone has entered your room or opened your drawer or closet
   - a device that tells you that the toast is done or the water is boiling

   Be prepared to explain your model to the class.

Impacts of Communication Technology

▷ **What impacts have resulted from communication technology?**

Objectives

▶▶ **Assess** the positive impacts of communication technology.

▶▶ **Assess** the negative impacts of communication technology.

Terms to Learn

- **biometrics**
- **personal privacy**
- **tolerance**

Standards

- Cultural, Social, Economic & Political Effects
- Environmental Effects

When people talk about our shrinking world, they don't mean that our planet is really getting smaller. They mean that our technology has made it possible for us to communicate almost instantly with any person located at any spot on our planet.

Although by itself communication technology is neither good nor bad, decisions about the use of its products and systems can have both good and bad consequences. Political, social and cultural, economic, and environmental issues are influenced by the development and use of communication technology.

Political Impacts

▷ **How has communication technology affected our political process?**

Today you can receive information on what is happening all over the world at the same time the events are taking place. See Figure 9-13. When world leaders speak, you hear their statements instantly through satellite communication. Mass communication systems report events and help form world opinion.

Your knowledge of world events will help you decide who can best lead our country. You may not be old enough to vote, but you probably have much more political information than your grandparents had when they were of voting age.

Figure 9-13 Television news reporters can report live from almost anywhere in the world.

Do You Believe Everything You Read?

Mistakes, deliberate falsehoods, and personal opinion make their way into newspapers, books, and magazines everyday. Both ethical and unethical people use communication media. How can you protect yourself from misinformation? The best way is to stay informed using a wide range of sources. During an election, for example, read or listen to what both sides have to say about the candidates, not just one side. Try to gather all the facts.

ACTIVITY Read stories in two different newspapers about the same event. Compare the information. Are all the facts the same? Does one include information the other doesn't? What conclusions can you draw from this?

Social and Cultural Impacts

▷ How has our society been changed by communication technology?

The use of technology can affect our choices and attitudes about the technology's development and use. Communication technology has had important impacts on education and the spread of technology. It has also affected personal privacy.

Education

▷ What effect has communication technology had on education?

All knowledge was at one time passed on from one person to another by word of mouth. As new things were learned, there was more to teach to the next generation. Our early ancestors shared their knowledge only with their own people. People who lived in other areas had to form their own knowledge. The key to the advancement of education and our technology was the development of communication. It helped us to learn from the achievements of people who lived far away or from people who lived and died long before we were born.

Before the invention of printing, only churches, royalty, and the very wealthy owned books. Printing made it possible for more people to own books. See Figure 9-14. Can you see why the growth of the printing industry was so important to education and the spread of technology?

Figure 9-14 The Gutenberg press is considered the first printing press to use movable metal type. What printing process is used?

With the development of printing and increased literacy, more and more people could learn from and about others. Today, so much information is available that ours is called the Information Age.

Personal Privacy

▷ In what ways can your personal privacy be threatened?

We in the United States have tried to make sure that communication technology can never be used by people to watch or control our lives. However, this is becoming difficult, and **personal privacy** is threatened. Banking, stock market transactions, and credit card purchases are all monitored by communication systems. It is even possible to create a full profile of a person by using the Internet. This particular impact of communication technology may be good—as long as it is controlled by people who believe in freedom of expression and democracy. However, in some countries it is used to spy on people and to control what they do.

Biometrics is the science of measuring a person's unique features, such as fingerprints, facial features, voice, and retina of the eye. For example, a biometric reader can convert your fingerprints or facial features into a mathematical image. It may then compare that image with all the other images it has on file in order to identify you. Soon retina scans and voiceprints will be as common as fingerprints for identifying people. They may even replace keys and locks. Although biometric security systems will make it harder for thieves to take money from your bank account or use your credit card, it will also be harder for you to retain privacy.

personal privacy
the right of individuals to keep certain information away from public view

biometrics
the science of measuring a person's unique features

Tolerance

▷ What good might come from communication technology's effect on our understanding of different cultures?

Our communication systems have made us more aware of the different cultures that share our planet. Through motion pictures and television you have been invited into the homes of families of different cultures, religions, and nationalities. They, in turn, have been learning about us.

It is hoped that the awareness of the customs and traditions of others will lead to more **tolerance** and a greater acceptance of others.

tolerance
respect for others

Economic Impacts

▷ **How has communication technology affected the world economy?**

Because communication technology is so fast today, news travels around the world at lightning speed. What happens to our stock market affects the stock markets in Great Britain, Japan, and other countries within hours. See Figure 9-15.

Communication technology has also changed how business is conducted. Most companies today have Web sites that advertise or sell their products. Many manufacturers depend on computerized machines to make products.

Figure 9-15 Communication takes place on the trading floor of the New York Stock Exchange.

Environmental Impacts

▷ **How has communication technology affected our environment?**

You have already learned that systems can be broken down into input, process, output, and feedback. Unfortunately, many of our processes not only produce what we want, but also what we *don't* want. They can harm our environment.

Many communication systems use paper as if it grew on trees. The problem is that paper doesn't grow on trees—it is made out of trees. See Figure 9-16. At one time trees were cut down to feed the ever-hungry paper mills without any concern about future needs.

The computer, copying machine, and fax machine have increased our demand for paper. Do you use a computer for your school reports? How many paper printouts go into the garbage before you get your final report printed?

Today, paper manufacturers have replanting programs to guarantee trees for the future. Papermakers also must take great care not to contaminate nearby rivers and streams with the waste products of the manufacturing process.

Many chemicals, metals, and plastics are used to manufacture communication equipment. Still more chemicals and paper are used to make the books, magazines, and newspapers that you read. Our environment is harmed when these materials are not disposed of properly. What's more, workers can be injured when they work with some of these materials.

Figure 9-16 Logs from trees are often floated down rivers to the paper mills.

The power lines and transmitters that carry our communication signals affect the appearance of our communities. Some medical studies indicate that the signals can physically affect our bodies.

When we think about environmental impact, we usually think about what is wrong with technology. In developing new systems, cost and profit are often most important to the manufacturers and distributors. Environmental and economic concerns often compete with one another. Sometimes people aren't aware of the dangers of the new system until after the product is being used. Government agencies and consumer protection groups try to protect us from these dangers.

SECTION REVIEW 9C

Recall ▸▸

1. Before the development of printing, who owned books?
2. How has communication technology made our world smaller?
3. What are some of the negative and positive impacts of communication technology?

Think ▸▸

4. World events appear on your television screen as they happen. How does this instant communication affect our world?
5. In recent years, certain people have been prosecuted for making copyrighted music available for free over the Internet. Why do you think this has been made illegal?

Apply ▸▸

6. **Design and Construction** Design and construct a special container that could be used for the collection of recycled paper products.
7. **Design and Research** Design an environmental impact study to determine what your community is doing about recycling paper. Interview school, town, and civic leaders, then set up a school-wide recycling program with their assistance.

Summary Activity

For each numbered blank, pick the answer from the list on the right that makes the most sense in the entire passage. Write your answer on a separate sheet of paper. No answer will be used more than once.

Communication is enhanced by technology such as pictures, printed messages, or machines. It helps us make decisions and solve __1__.

In any communication __2__, your message is the input, how you move your message is the process, and the reception of your message at the __3__ is the output. Comments from people about a communication system provide __4__ to the system.

Communication subsystems include a source, __5__, transmitter, receiver, decoder, storage, retrieval, and destination.

Communication systems can be grouped by the modes in which they carry your __6__. Biological communication occurs without technology. You communicate by speaking, facial expressions, and gestures. In graphic communication you __7__ messages visually through the use of drawings, paintings, and printed words and pictures.

Wave communication includes communication where messages __8__ through air, water, outer space, or solids as waves. Sound waves move through the __9__ as vibrations. Radio and TV signals may be carried by electromagnetic waves.

Long distance communication is called telecommunication. Telecommunication satellites travel around the earth in a geosynchronous __10__.

People and machines communicate in many ways. One example of machine-to-people communication is a(n) __11__. Your computer giving instructions to its printer is an example of __12__ communication. Can you think of other examples?

Communication technology has had many __13__. One example is learning how people of different cultures, religions, and nationalities live. Communication technology may help you be more __14__.

Answer List

- atmosphere
- encoder
- feedback
- impacts
- machine-to-machine
- message
- orbit
- problems
- receiver
- send and receive
- smoke detector
- system
- tolerant
- travel

Comprehension Check

1. What is the purpose of an encoder in a communication system?

2. Give several examples of graphic communication.

3. Suppose you write a letter on your computer, use the spell checker to correct it, and then print it. Which modes of communication did you use?

4. What is personal privacy?

5. What is the purpose of biometrics?

Critical Thinking

1. Infer. Why is personal privacy important for an individual to maintain?

2. Design. Draw a map showing the route from the school to your home. Identify the symbols that you used.

3. Classify. Make a list of the ways in which you have communicated during the past 24 hours. Identify the mode of communication for each example.

4. Appraise. Do some research about Helen Keller, who was both deaf and blind from the time she was a baby. Appraise the methods used to teach her to communicate. Do you think they were effective? Why or why not?

5. Compose. Write a paragraph evaluating the social impacts of communication technology on your life.

Visual Engineering

Step by Step. To communicate well, it is important to word your message carefully. Have you ever bought a product at the store and found the directions confusing? Writing step-by-step instructions can be harder than you might think. You might understand how something works, but will your instructions be clear? Try this. On a piece of paper draw a simple design using a rectangle, a circle, and a triangle. Save the drawing for later reference. Now, on another piece of paper write step-by-step directions on how to draw your design. Be as precise as possible with your instructions. Then, without showing the person the design, give your directions to a classmate and ask him or her to draw it. Compare the drawing to yours. Were your directions effective? Revise your directions as needed and try it on another classmate.

TECHNOLOGY CHALLENGE ACTIVITY

Developing an Ad for a Communication Invention of the Past

Equipment and Materials

- markers, pens, pencils
- posterboard
- scissors
- glue
- computer and printer
- PowerPoint software
- graphics software
- photocopy machine
- reference books with appropriate illustrations of inventions

Background

For any invention to become important, people must become aware of its existence. People must also be convinced they need the new invention in their lives.

If you wanted to inform everyone about a new invention, what mass communication system would you use? Be prepared to design an ad for the communication system of your choice.

Goal

Your goal for this activity is to create an advertisement for a communication device of the past, such as the first printing press, radio, or TV. The choice is yours.

Criteria and Constraints

▸▸ You should work in groups of two or three, but no more than four.

▸▸ You may choose to create a print ad, a Web page, or a PowerPoint presentation. A print ad must be a full page in size.

▸▸ Your advertisement must include the inventor of the device, when it was invented, what it does, and why it is important to the development of technology.

FIGURE A

FIGURE B

FIGURE C

Design Procedure

1. Determine the invention that your group is going to advertise.
2. Determine the communication method that you will use.
3. Brainstorm the theme of your ad. Study advertisements that you have seen. Be creative and have fun.
4. In your ad, state who invented the device, when the invention was invented, what it does, and why it is important to the development of technology.
5. Produce your ad using the appropriate equipment. This means pasting up your artwork and text for your graphic ad or using a computer to make a Web page or PowerPoint presentation. See Figures A, B, and C.
6. Share your advertisement with the class.
7. Make copies of the advertisement for each member of your team, so you all have a copy to take home.

Evaluating the Results

1. What makes a new technological development financially successful?
2. What are the input, process, and output phases of the communication system that carried your ad?
3. If you were to make a new ad, what would you do differently to improve your final product?

SUPERSTARS OF COMMUNICATION

Ts'ai Lun (50-121)

Imagine what your schoolbooks would weigh if they were made from chunks of bamboo. Bamboo books were used in China until Ts'ai Lun, a member of the Emperor's court, mixed tree bark, bamboo, cloth fibers, and other materials to produce the first paper. His methods were kept a secret for more than five centuries before other areas of the world discovered his invention.

Samuel Morse (1791-1872)

Samuel Morse was an artist, as well as an inventor. During a return voyage from Europe in 1832, where he had been studying art, he got the idea for the electric telegraph. Other telegraph systems had already been invented, but Morse based his device on the electromagnet, which made it more practical. Morse also devised a code, consisting of long and short signals, to be used with his device. Morse Code was printed out as a series of dots and dashes. Morse's work with electromagnetism became the basis for radio and television transmission and other electronic communication devices.

Dennis Gabor (1900-1979)

Dennis Gabor credited his success to serendipity—the art of finding something while looking for something else. It was while working to improve an electron microscope in 1947 that the communications engineer developed the idea for holography. However, holography was not successful until after the invention of the laser and the contributions of many other researchers. Regardless, Gabor was awarded the Nobel Prize for his invention in 1971.

Louise Kirkbride earned her degree in electrical engineering because she hoped to join NASA. When that didn't work out, she and her husband eventually started their own company. Kirkbride was interested in helping companies communicate better with their customers, and she designed and patented the first problem-resolution software, enabling businesses to give automatic feedback to customers through their Web sites. By building custom databases of responses to frequently asked questions, companies can answer up to 98 percent of repeat questions at a much lower cost than by e-mail or telephone.

Louise Kirkbride
(b. 1953)

Philo Farnsworth   (1906-1971)

Philo Farnsworth, the man who invented the television picture tube, came up with the idea at the age of fourteen while he was tilling a potato field. The Utah teen loved science, and by the time he was twenty-one, Farnsworth had built an electronic television system and found investors for his invention. In 1927, he finally got it to work. Unfortunately, another inventor, Vladimir Zworykin, who had emigrated to the U.S. from Russia, had interested the Radio Corporation of America in a similar device. Zworykin and RCA fought Farnsworth in court for the patent rights, and Farnsworth won.

Tim Berners-Lee   (b. 1955)

Millions of people use the World Wide Web, an organizational system that makes finding information easier. Designed by British software engineer Tim Berners-Lee in 1980, the Web keeps track of random associations using links. There is no central managing computer, just one computer linked to another and another in an open-ended way, with no limit on growth. For Berners-Lee, computers were second nature. He grew up with them. His parents both helped develop early computer systems, and while in college, he built his own computer out of spare parts and an old TV set.

Can you name a field of study that isn't influenced by the use of computers? If you were asked to name the greatest inventions of all time, would you choose the wheel and the computer?

Computers are now woven into every part of our society. In order to do well in that society, it is important that you become computer literate. You now use computers to work on mathematics, science, and technology problems. You use them to find information, write reports, and play video games. Computers already control automobile engines, DVD players, digital watches, and cell phones. If a device is *digital*, it means a computer controls its functions. In the future, you will use computers to make informed business, industry, governmental, or personal decisions. Don't you think that these are good reasons to study computers and computer technology?

SECTIONS

10A Computer Systems

10B Computers on the Cutting Edge

Astronauts train using this NASA flight simulator flight deck (cockpit). It's all computerized.

249

▶▶ **Identify** the main parts of a computer system.

▶▶ **Explain** why binary code is important to computer function.

▶▶ **Name** several computer input and output devices.

Terms to Learn

- binary code
- CPU
- computer virus
- integrated circuit
- operating system
- program
- RAM
- ROM

Standards

- ◼ Core Concepts of Technology
- ◼ Relationships & Connections
- ◼ Cultural, Social, Economic & Political Effects
- ◼ Engineering Design
- ◼ Design Process

integrated circuit
a tiny chip of silicon having many electrical circuits burned into it; also called *microchip*

CPU
central processing unit; the part of the computer that processes information; or, the "brain" of the computer

Computer Systems

◔ What are the main parts of a computer system and how do they work?

A computer is an electronic device that can calculate, store, and process data. The computer's actions are controlled by step-by-step processes coded in its software programs. Your home and school computers are general-purpose machines that can be used for word processing, accounting, image processing, games, and other activities just by installing the correct software.

Computers can be large or small. However, size has little to do with how powerful they are. A computer system is made of different components. See Figure 10-1. Each has a special job to do.

Central Processing Unit (CPU)

◔ What are the three main parts of the CPU?

Inside your computer are many devices called integrated circuits, or microchips. **Integrated circuits** are tiny pieces of silicon with many electrical circuits burned into them. The circuits act like switches. Sometimes they let electricity flow. Other times they shut it off.

The **CPU** (central processing unit) is the largest and most important integrated circuit. See Figure 10-2. It is the part of the computer system that performs all basic operations and follows procedures. You can compare a CPU to a highway system. Many roads must sometimes be taken to get from one place to another. Information travels the circuits (roads) of the different parts of the CPU to get processed.

The CPU's control unit guides the processes and flow of information. The arithmetic/logic unit performs mathematical calculations with the data that the control unit sends. The memory unit stores the information before and after processing. These three parts of the CPU work together to accomplish whatever task is assigned to the computer.

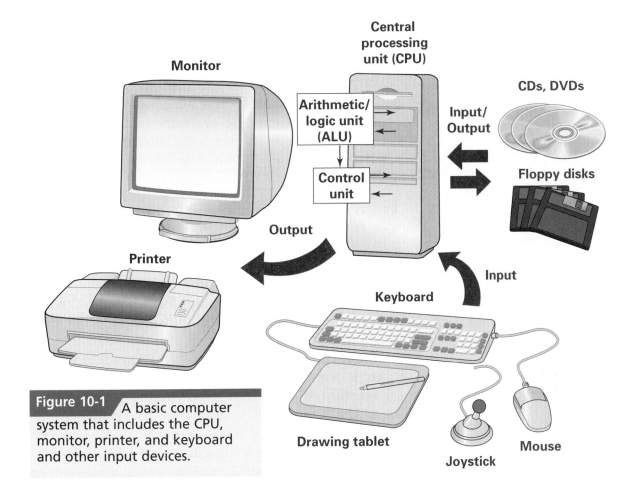

Figure 10-1 A basic computer system that includes the CPU, monitor, printer, and keyboard and other input devices.

The CPU has two types of memory. **ROM**, or read-only memory, contains the basic information that the computer needs to perform any operation. This information is permanent. You can't change it. When you shut off the power, this memory is not erased.

The other type of CPU memory is called **RAM**, or random access memory. All data fed into your computer is put into random access memory. The data is temporarily held in the RAM locations of the computer while you are using them. When the computer is shut off, this information is lost.

Computers are not the only machines that have integrated circuits in them. Integrated circuits help a machine's mechanical and electronic systems work together.

ROM
read-only memory; permanent memory that cannot be deleted or changed

RAM
random access memory; memory in the CPU that stores data temporarily

Figure 10-2 The CPU is located on the computer's motherboard. Other devices, such as memory chips and controllers, are also on the motherboard.

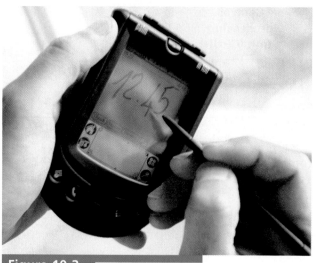

Figure 10-3 A Palm Pilot can do many of the same operations as a desktop computer.

Examples include DVD players, cell phones, digital watches, MP3 players, video game machines, household appliances, automobiles, industrial machines, and medical machines. These integrated circuits may be designed to do only one special task. However, many now allow the user to alter some functions. See Figure 10-3.

Computer Programs

▷ What is the function of a computer program?

A computer **program**, or software, is a set of instructions that the computer follows to do its work. The program controls the computer. It tells the CPU exactly how to handle all the data that is entered into the machine. The program turns your computer into a game machine, word processor, artist's drawing board, or teaching machine. The program even controls which devices the CPU will recognize, such as a printer. Computer programs are usually stored on CDs (compact discs), on DVDs (digital video discs), or on the computer's hard disk drive.

How does a computer know how to run a program or take information from the keyboard or other devices? Many microchips inside the computer have special **operating system** programs permanently burned into their circuits. Other programs are placed on the hard drive, start-up CDs, or DVDs. When the computer is turned on, these programs immediately tell the computer how to run certain systems. In a sense, the computer reads an entire instruction book on how to operate a computer each time it is turned on.

Binary Code

▷ How does a computer use binary code to process data?

The computer program and all the information that the computer will use must be changed into **binary code**, which is code that the computer can understand.

Two numbers are used in binary code—1 (one) and 0 (zero). Each 1 or 0 is a bit. See Figure 10-4. A bit is the smallest piece of information a computer can use. A computer sends or receives these 1s and 0s in the form of small

program
a set of instructions which the computer follows

operating system
the program a computer follows that tells it how to operate

binary code
an electronic code based on the binary number system that the computer can understand

Figure 10-4

The binary code uses two numbers, 1 and 0. Why do you think binary code uses only these two numbers?

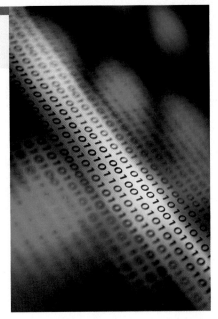

electrical pulses. A 1 means a pulse flows through the circuit; a 0 means no pulse. Stringing eight of these bits together forms a byte. Each byte is code for a letter, number, or punctuation mark represented as 1s and 0s.

The computer converts everything that you type into these binary bytes. When the computer reads the binary code 00000100, it is reading the number 4. Figure 10-5 shows how numbers can be represented in binary code.

Why does a computer have to use binary code? *Electricity on* and *electricity off* are the only two messages that a computer can sense. The computer does all of its work by having the power quickly turned on and off in parts of its circuits. Although using binary code seems slow to us, electricity travels very fast. Supercomputers can make over twenty billion calculations per *second*.

Computer Viruses

▶ How are computer viruses created?

A **computer virus** is a set of destructive instructions that someone has written and placed inside an innocent-looking computer program. Unsuspecting people then

computer virus
a set of destructive instructions that "infects" the computer system and can cause damage

TECH CONNECT
MATHEMATICS

Computer Algebra
George Boole, an English mathematician, developed a system for processing logical statements in a mathematical (symbolic) way. His system is called Boolean algebra and is the basis for the way computers work.

ACTIVITY Find out how words can be as important as numbers in mathematics. What three key words in Boolean algebra allow computers to work?

Figure 10-5 Putting together the binary code for the number 55. What is the code for the number 11?

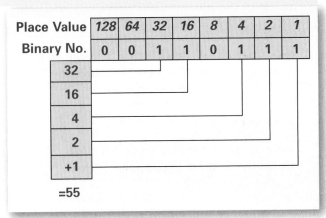

Place Value	128	64	32	16	8	4	2	1
Binary No.	0	0	1	1	0	1	1	1

32	
16	
4	
2	
+1	
=55	

Figure 10-6 A virus can be dangerous to a computer system. Extra care should be taken to avoid a virus.

place the program in their computer system. Many people get viruses by downloading files from the Internet, by opening files that have been sent as part of an e-mail, or by trading discs with other people. See Figure 10-6.

When one type of virus enters your computer, it hides in your computer's memory. When you save your data, you also save the virus. Slowly but surely, the virus causes problems. If you switch from one program to another, the virus might attach itself to the new program. In this way, it can infect all your programs before you know that it exists.

Viruses can take different forms. Some are merely annoying. Others do serious damage. Today millions of dollars are being spent to get rid of and protect computer systems from these virus programs. Anti-virus software can find and destroy viruses that exist on your computer. These anti-virus programs must be updated frequently to protect against new viruses.

Disk Drives

▶ What does a disk drive do?

The computer's disk drive allows data to be written to storage (memory) or read from storage. Most computers have a hard disk drive and a CD drive. Some also have a DVD drive. The hard disk is permanently installed in the computer. CDs and DVDs can be inserted or removed.

ETHICS in ACTION   Are Viruses Funny?

Some people who create computer viruses do not mean serious harm. They may be bored and see it as a challenge or a prank. They may want to get attention. However, even viruses that do not actually cause damage still cost individuals and companies money in lost work time and other factors. Like viruses that cause diseases in humans, computer viruses are no joke. They are a kind of vandalism—the deliberate destruction of or harm to someone else's property—and they are illegal.

ACTIVITY Do some research to find out how computer security specialists track down the creators of computer viruses and other suspects.

Willingness to Learn

Having a willingness to learn will help you throughout your life, whether you're in school, at work, or at home.

Employers need people who are willing to improve and update their skills. They often promote workers who show a willingness to learn. Fields in which you must obtain a license or certificate often require more education to renew them. Continued learning is especially important in computer technology. New programs and hardware appear almost daily. Are you interested in a career in which you must stay up to date? Here's how to find out:

Computer Technician

Computer technician wanted for large doctor's office. Our computer system currently handles medical, scheduling, and billing information. The successful candidate must have a strong background in computer technologies and a (willingness to learn) and grow as computer systems change.

✔ **Look up the career on the Web site for the *Occupational Outlook Handbook*. Check under "Education."**
✔ **Talk to someone who is currently working in the field. Ask the person how important updating your skills is to success.**

Hard disk drives use electromagnetism to write messages onto the disks. Signals are sent to the recording head. The recording head passes over the disk while a constantly changing signal is sent through the recording head. The metal oxide surface is magnetized into the coded message.

When the playback head passes along the disk, it picks up this magnetic coded message. The message is then converted back into an electronic signal.

CDs record information in tiny pits in the surface of the plastic. A laser on your CD drive reads the information. See Figure 10-7. Information can be recorded on CD-Rs once, and it can be read many times, but it cannot be changed. Information recorded on CD-RWs can be altered as often as desired.

DVDs are also encoded using a laser. The laser changes its focus so it can read pits on different layers of the DVD. DVD-Rs and DVD-RWs are also available. DVDs hold much more data than CDs. DVD-RAM drives supplement built-in computer RAM. A disc can be read and written to repeatedly.

Figure 10-7 A laser can read the tiny pits in a CD.

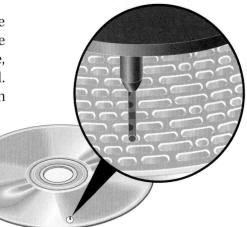

Figure 10-8 A joystick and a mouse are two kinds of input devices for a computer. What are some other input devices for a computer?

Input Devices

▶ What kinds of devices send data to computers?

Input devices are those that send information to the computer. The keyboard, mouse, joystick, digital camera, and video camera can all be considered input devices. Scanners can be used to scan printed material or photos into the computer's memory.

All inputs to the computer must be converted to binary code. When you press down on a keyboard letter, you cause contacts that are under the key to send the binary code for that letter. This coded message is sent to the monitor screen and into a memory location in the CPU. The screen shows you what you have just typed, and the CPU holds and uses the information according to the instructions from the program.

The surface of a computer screen is divided into horizontal and vertical coordinates. They are similar to the horizontal and vertical coordinates on a world map that we call latitude and longitude. The computer uses these coordinates to locate things on the screen. We can identify any location on the screen by giving particular row and column coordinates. When you move the lever of a joystick, move a mouse, or roll a track ball, the row and column location is changed, and these changes are fed into the computer. The computer responds to these changes by moving your cursor or other locator on the screen. Figure 10-8 shows a mouse and joystick. They use mechanical and electrical systems to convert your physical movement into electronic signals that the computer can understand.

Output Devices

▶ What devices are used for output?

Many kinds of output devices are used with computers. Almost every computer has a monitor and displays infor-

LASER PRINTER

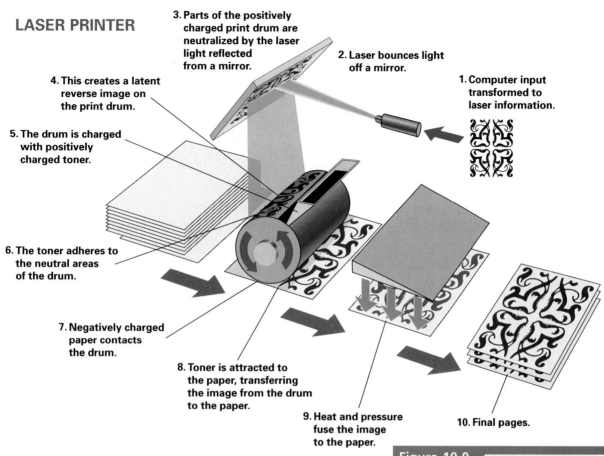

3. Parts of the positively charged print drum are neutralized by the laser light reflected from a mirror.

2. Laser bounces light off a mirror.

1. Computer input transformed to laser information.

4. This creates a latent reverse image on the print drum.

5. The drum is charged with positively charged toner.

6. The toner adheres to the neutral areas of the drum.

7. Negatively charged paper contacts the drum.

8. Toner is attracted to the paper, transferring the image from the drum to the paper.

9. Heat and pressure fuse the image to the paper.

10. Final pages.

Figure 10-9 A laser printer places dots so close together that it is impossible to see any one dot.

mation on its screen. Music synthesizers can create sound using computer data. One of the most common output devices, however, is a printer.

The image on your computer screen is called a soft copy because it is only temporary. Many different types of printers are attached to computers to make permanent copies of computer-generated material. These permanent copies are called hard copies.

Printers usually form letters as a series of dots. Take a pencil and just tap the point on a piece of paper. Try to keep hitting the same space over and over again. After a while, the dots will form a large dark spot. It will be very difficult to tell that this spot was actually made up of only dots. The CPU tells the printer the exact pattern and how many dots to print.

The dots produced by a laser printer are spaced so close together that it is impossible to see their individuality. See Figure 10-9. The signal that determines the letter is the trigger for a laser beam.

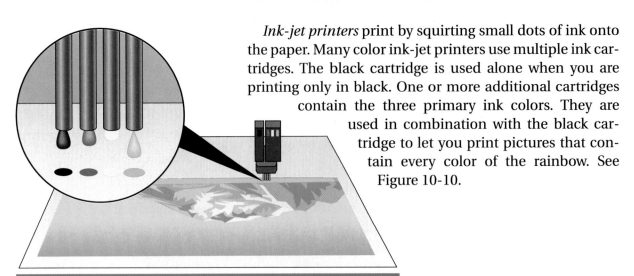

Ink-jet printers print by squirting small dots of ink onto the paper. Many color ink-jet printers use multiple ink cartridges. The black cartridge is used alone when you are printing only in black. One or more additional cartridges contain the three primary ink colors. They are used in combination with the black cartridge to let you print pictures that contain every color of the rainbow. See Figure 10-10.

Figure 10-10 Color ink-jet printers have three primary color ink jets plus a black jet.

10A SECTION REVIEW

Recall »

1. What does CPU stand for, and what does a CPU do?
2. What is the purpose of a computer program?
3. What two numbers are used in binary code?
4. What is the difference between ROM and RAM?

Think »

5. What are some positive and negative effects of computers on our society?
6. Why aren't all computer data and programs put in a ROM format?

Apply »

7. **Teamwork** Construct a human computer using the members of your class as individual bits. Then add two bytes (numbers) together. Brainstorm ways to make it work.
8. **Design** Working with two or three of your classmates, design the ideal computer. List all of the computer's features. Create a mock-up of what the computer would look like. Create an advertising poster. Prepare a five-minute presentation for the class.

Computers on the Cutting Edge

⏵**What is the future of computers?**

SECTION 10B

Objectives

▸▸ **Describe** artificial intelligence and identify ways in which it can be used.

▸▸ **Discuss** wi-fi and distributed computing.

Terms to Learn

- **AI**
- **distributed computing**
- **expert system**
- **wi-fi**

Standards

- Characteristics & Scope of Technology
- Core Concepts of Technology
- Relationships & Connections
- Cultural, Social, Economic & Political Effects
- Influence on History
- Engineering Design
- Design Process
- Use & Maintenance
- Information & Communication Technologies

How do you think the computer might evolve during your lifetime? Many people are fascinated by computers in the form of robots. Robots are used to do tasks people can't or don't want to do. Mobile robots carry food and medicines to people in hospitals to decrease the spread of any illnesses. Robots that mow lawns and vacuum carpeting are available to help around the house. If computers continue to develop at their current rate, will they surpass humans at most tasks? Could a future computer turn against us? Could the world shown in science fiction movies in which we battle our own machines ever become reality?

Computing will continue to grow with the development of artificial intelligence, immobots, speech recognition, wireless networks, and distributed computing. Let's look at the cutting edge.

Artificial Intelligence

⏵**What is artificial intelligence?**

People sometimes talk about how intelligent computers are. They are not really intelligent at all. A computer gets no satisfaction when it solves a problem. It is only a machine. It can only run programs and process data. You know now how computers work, so you know that computer error is often really human error. "Thinking" computers are, at this time, just science fiction.

AI (artificial intelligence) programs, however, give the impression that a computer can think. The programmer has provided the computer with a number of answers that will be triggered by certain requests.

Some artificial intelligence programs are called **expert systems**. Information from experts in a particular field has been stored in the computer's memory. A medical

AI
artificial intelligence; a computer program that can solve problems and make decisions ordinarily handled by humans

expert system
form of artificial intelligence in which information from experts is collected and stored in the computer's memory

Figure 10-11 A diagnostic scanner can determine what is wrong with a car's engine.

expert system might diagnose disease. If you provide it with a list of symptoms, it will match that list against all known diseases.

An automotive diagnostic program can match engine problems with causes and suggest repairs. The sensors in an engine constantly report engine conditions to the car's central computer. When a mechanic plugs a diagnostic computer into the electrical system, the car's computer reports what's wrong and what needs to be fixed. See Figure 10-11.

The most powerful AI system is a chess-playing system named *Deep Junior*. A number of computer processors are wired together for power and speed. *Deep Junior* is able to look at hundreds of possible chess moves every second. It draws on the stored knowledge of many expert chess players to figure out its next move. In 2003, *Deep Junior* played chess champion Garry Kasparov. *Deep Junior* and Kasparov both won an equal number of games. See Figure 10-12.

Figure 10-12 Grand Master Garry Kasparov starts a six-game tournament against the world champion computer chess program *Deep Junior*.

Immobots

▷ What is an immobot?

Another new development in AI is immobots, or "immobile robots." Developed by researchers at the Massachusetts Institute of Technology, most immobots don't move around themselves, but they control a machine that probably does.

Presently, AI software follows a long list of complex rules to solve problems. Unfortunately, it's hard to think of all the possible difficulties that could occur with complex machinery and write a rule for them. Immobot software is different. It includes a model of the machine's system. When a problem occurs, the immobot studies the system, finds the source of the problem, and determines a way around it. The goal is for the immobot to be able to respond to unexpected situations on its own and learn from its experience.

Speech Recognition

▷ Why is speech recognition difficult?

AI is also being used for speech recognition. Imagine being able to talk to your computer instead of typing on a keyboard.

Currently, speech recognition software does not work well because background noise confuses it. However, researchers are now combining vision inputs along with sound, and that seems to help. Not only does the computer hear what you say, it has a camera and software that can read your lips. By combining technologies, computer recognition of speech is improved by over fifty percent. See Figure 10-13.

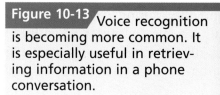

Figure 10-13 Voice recognition is becoming more common. It is especially useful in retrieving information in a phone conversation.

Language Challenged

U.S. military troops in Afghanistan and Iraq were equipped with a handheld computer that uses speech recognition software. When a word in English is spoken into the Phraselator, it shows the word on its small screen, then translates it and broadcasts it in the chosen language.

ACTIVITY Do you think a common language spoken everywhere in the world would be a good thing or not? Write a paragraph giving reasons for your opinion.

wi-fi
wireless connection to your network components and the Internet

distributed computing
network of computers that takes advantage of downtime

Figure 10-14 This watch called the Wrist Net connects to the Internet using wi-fi. It obtains the news, weather, sports, stock quotes, and instant messaging.

Wi-Fi

▷ Why would you want to add wi-fi to your computer?

Wi-fi, or "wireless fidelity," is a wireless connection to the Internet. Based on wireless radio and cell phone transmission, wi-fi allows you to access your network and the Internet from anywhere within range of a base station. See Figure 10-14. With the right equipment and software, computers with wi-fi can be online outside, in stores, and in public areas. Wi-fi also allows computers to be networked (connected together) without wires.

Does a school, bookstore, library, airport, or hotel in your area have wi-fi? Wi-fi is also becoming available on airplanes. People in all walks of life are going wireless because it is an easy way to hook up several computers to one fast Internet connection. For companies it is much cheaper than wiring computers together into a local area network (LAN). A wi-fi connection is also called a wireless local area network (WLAN). Computers that are networked on a WLAN or LAN need to be protected so that outsiders cannot access them.

Distributed Computing

▷ How can your computer help search for extraterrestrial intelligence?

At any moment in time worldwide, almost a billion desktop computers stand idle, waiting to be used. Millions of people are now donating this computer downtime for science, mathematics, and technological research. They are using a type of computer networking called **distributed computing**.

Do you want your computer to look for life on other planets, perform cancer research, search for prime numbers, help develop new drugs, or even help to improve weather prediction? You can. See Figure 10-15. Your computer must be hooked up to the Internet. Then all you need to do is choose a topic and contact the proper agency to download a special screen saver. While your computer sleeps and this screensaver is running, your computer analyzes a small amount of the data under research. The next time you're online, the analyzed data is sent back to the research agency and new data is picked up. Millions of computers like yours bring to the problem more power than the most powerful supercomputers. To find a research problem you can help with, type *internet based distributed computing project*s into your favorite search engine.

Figure 10-15 This large radar dish is scanning outer space. Data from it could be networked through distributed computing to many desktop computers.

SECTION REVIEW 10B

Recall ▸▸

1. Describe how immobots are different from ordinary AI systems.
2. What is an expert system?
3. What is wi-fi?

Think ▸▸

4. How is improved speech recognition similar to the way humans understand speech?
5. How does artificial intelligence differ from human intelligence?

Apply ▸▸

6. **Brainstorm** Set up a wi-fi network in your technology classroom. Brainstorm ways to make it work.
7. **Research** Working in groups, research different distributed-computing projects. Select the topics that interest the majority of students and set up distributed-computing screen saver programs in the classroom.

Summary Activity

For each numbered blank, pick the answer from the list on the right that makes the most sense in the entire passage. Write your answer on a separate sheet of paper. No answer will be used more than once.

A computer is an electronic device that calculates, stores, and processes __1__. Today we find computers in every area of our society. Integrated __2__, or microchips, are tiny pieces of silicon with electrical circuits burned into them. The most important is the __3__, but they also exist in your telephone, television, and automobile.

A computer system consists of hardware and software. The __4__ of an average computer system includes the central processing unit (CPU), keyboard, monitor, disk drive, joystick, mouse, and printer. The instructions that tell the computer what to do are called a computer __5__.

__6__ devices include the keyboard, mouse, and joystick. __7__ devices include monitors and printers. All computers have two types of __8__ called ROM and RAM. ROM contains the information that the computer needs to run programs and __9__ its own system. RAM is temporary, and it is the place where the computer stores data received from input devices. Data in RAM must be saved to your disc or hard drive or it will be __10__ when the computer is shut down.

Researchers are now writing programs to help a computer think. These programs are called __11__. A(n) __12__ gathers information from people in a particular field and stores it in the computer's memory. Immobots can study a(n) __13__ of a machine's system in order to solve problems. Wi-fi is a(n) __14__ connection to the Internet. Computers can be __15__ without wires. When people donate their computers downtime to research, it is called __16__.

Answer List

- artificial intelligence
- CPU
- circuits
- distributed computing
- expert system
- hardware
- information
- input
- lost
- memory
- model
- networked
- operate
- output
- program
- wireless

Comprehension Check

1. What is a computer virus and how does it get into your computer?
2. What does the term digital refer to?
3. How is an integrated circuit made?
4. What is distributed computing?
5. What is artificial intelligence?

Critical Thinking

1. **Judge.** Some people don't like computers because they think machines will take over the world. Write a paragraph giving your opinion.
2. **Criticize.** Old, discarded computers have become an environmental problem. Research and evaluate what happens to them in your community.
3. **Propose.** How do you think computers could be used at your school to promote efficiency?
4. **Infer.** Would you want to play a game with *Deep Junior* as your opponent? Why or why not?
5. **Hypothesize.** Some scientists are working to develop bio-computers in which at least some parts would be made from living things. Research bio-computing and explain how you think it will change what computers can do.

Visual Engineering

Simulating Experience. NASA, the military, and commercial airlines use flight simulators to train their pilots. Today's simulators are very realistic. Simulators put a flight crew through a dangerous situation to give them experience before they actually have to fly the plane. Several companies produce simulators that can be run on your home computer. The controls are like those you would find in a small airplane. These simulators are so realistic that in some cases the time spent on them can be counted toward a pilot's license. Do a little research to find out how simulators actually work. How can they be so realistic? What basic principles of science and physics are applied to make a simulator work? Share your findings with the rest of the class.

TECHNOLOGY CHALLENGE ACTIVITY

Programming a Computer to Control a Machine

Equipment and Materials

- LEGO Mindstorms™ Invention System
- computer system

FIGURE A

Background

The United States landed the first people on the moon using less computer power than is found in today's automobiles. Today, computers control the fuel system, engine, and other parts of your car as well as many other technology systems. Have you ever programmed a computer to control a motor-powered machine?

Goal

For this activity you will build a motorized robot machine using a LEGO Mindstorms™ Invention System. The motors and sensors that are part of your machine will be controlled by a computer program that you will write. See Figures A, B, C, and D.

Criteria and Constraints

▶▶ You will use the problem-solving process to create your robot machine.

▶▶ Your computer program must be thoroughly tested before downloading.

▶▶ You will download your program into your robot machine using the RCX transmitter.

FIGURE B

FIGURE C

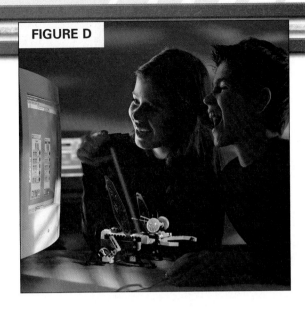

FIGURE D

Design Procedure

1. Read the user's guide that comes with the LEGO Mindstorms Invention System. You might want to complete the control experiment before trying to build and program a machine of your own.

2. Each user's guide provides a step-by-step procedure for a different computer-controlled project. Keep in mind that you should use the problem-solving process as you work. Define your problem, generate ideas, select a solution, test the solution, make the item, evaluate it, and present the results.

3. Pick one machine and follow the directions for assembly.

4. Test motors and sensors following the user's guide.

5. Use the LEGO manuals to learn the programming language for the RCX controller. You must use different commands for controlling the motors and sensors on your machine. The programming language lets you "talk" to the computer using phrases that the machine (through the software) can understand.

6. Plan out the sequence of commands that will tell the computer how to control your machine.

7. Test each command one at a time before downloading to your machine's microprocessor. Troubleshoot solutions to any malfunctions.

8. Use the infrared transmitter to transfer your program to your machine.

9. Demonstrate the operation of your machine to the class.

Evaluating the Results

1. In computer control systems you often find that the machine under control is equipped with optic sensors, touch sensors, and motors. How do these subsystems play a part in the control of the machine?

2. If you were to design and program another machine, what would you do differently?

CHAPTER 11

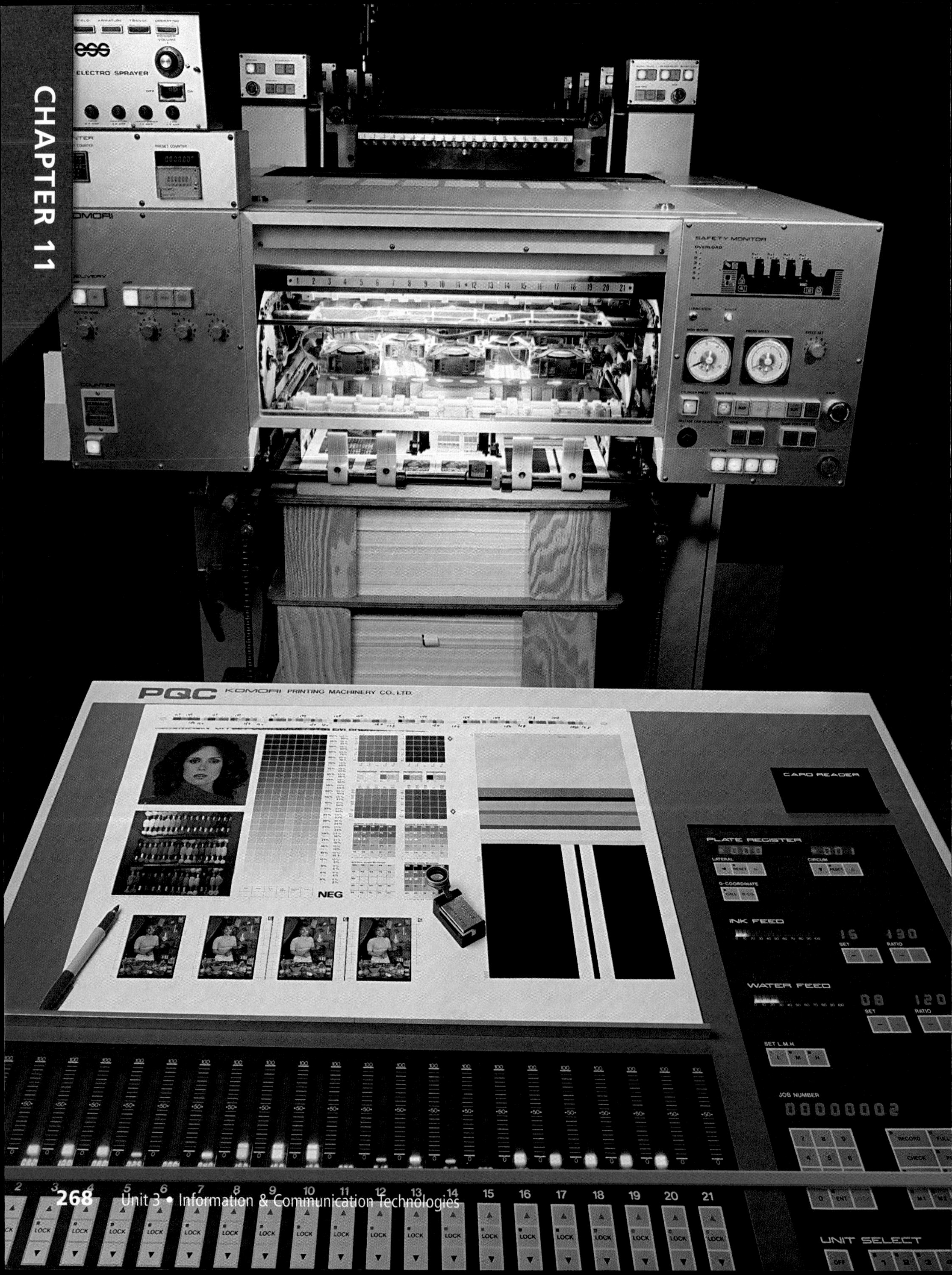
ELECTRO SPRAYER
PRESET COUNTER
KOMORI
DELIVERY
COUNTER
SAFETY MONITOR
OVERLOAD
PQC KOMORI PRINTING MACHINERY CO., LTD.
NEG
CARD READER
PLATE REGISTER
LATERAL
CIRCUM
INK FEED
SET
RATIO
WATER FEED
SET L.M.H
JOB NUMBER
RECORD
CHECK
UNIT SELECT

GRAPHIC COMMUNICATION

Without printing there would be no books, magazines, or newspapers to read. A supermarket would contain thousands of items that had no labels. There would be no paper money and no circuit boards for computers and TVs.

With information no longer available in printed form, would we know less? Would the workings of businesses and governments change? The answer is probably yes.

Graphic communication includes all the inputs and processes that transform artwork, photographs, and text into printed messages, or outputs. In the past, this transfer was always printed in the form of ink on surfaces such as paper, cardboard, metal, fabric, and glass. However, the computer has changed graphic communication. It now includes new forms of publishing, such as CDs, DVDs, e-books, and Internet pages.

This unit controls ink coverage for a large four-color printing press.

Printing Processes

⊙ What are the five basic printing processes?

▶▶ **Describe** common printing processes.

▶▶ **Explain** how dynamic digital printing differs from more traditional printing methods.

Terms to Learn

- **dynamic digital printing**
- **flexography**
- **gravure printing**
- **ink-jet printing**
- **letterpress printing**
- **lithography**
- **serigraphy**
- **xerography**

Standards

- Relationships & Connections
- Cultural, Social, Economic & Political Effects
- Environmental Effects
- Design Process
- Use & Maintenance
- Information & Communication Technologies

Most printing processes involve transferring a message from one medium, such as a printing plate, to another, such as paper. A soda can and the pages of a book could not be printed on the same machine because they have different shapes and are made of different materials. However, the basic printing process may be the same. All printing can be grouped into five basic processes:

- Letterpress printing, also called relief printing
- Gravure printing, also called intaglio printing
- Lithography, also called planographic printing
- Serigraphy, also called screen process printing
- Dynamic digital printing

Letterpress Printing

⊙ What part of a letterpress printing plate does the printing?

When your grandparents were your age, about forty percent of all printing was done on letterpress printing presses. These presses used metal type that was very similar to the type created by Johannes Gutenberg in the fifteenth century. Letterpress printing with metal type now produces only seven percent of the material that is printed each year.

In **letterpress printing**, the printing surface is raised above the rest of the plate. See Figure 11-1. Another name for letterpress printing is relief printing. A relief surface is raised above its surroundings just like the letters on a rubber stamp. Letterpress is used to print on a variety of materials using metal, rubber, and plastic plates.

When the printing plates are made of rubber, the process is called **flexography**. See Figure 11-2. This form of letterpress printing is still popular and produces eighteen percent of materials printed each year. It is often used to print packaging.

letterpress printing
a process that uses raised letters, symbols, or designs that are inked and then pressed against paper

flexography
a relief printing process that uses a raised, rubber printing plate

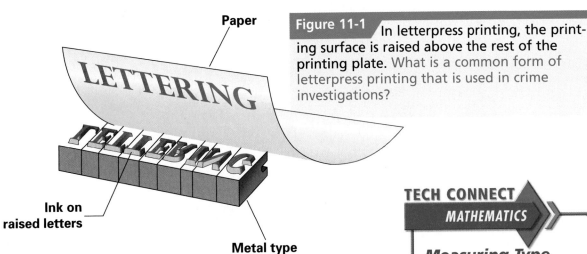

Paper

Figure 11-1 In letterpress printing, the printing surface is raised above the rest of the printing plate. What is a common form of letterpress printing that is used in crime investigations?

Ink on raised letters

Metal type

Gravure Printing

▷ What part of a gravure printing plate does the printing?

Gravure printing produces twenty percent of all printed products. In **gravure printing**, the printing plate has depressions in its surface. See Figure 11-3 (page 272). Hand-engraving or etching creates thousands of pits in the copper or steel plates. The pits vary in depth and thickness and serve as miniature inkwells. The deeper the inkwell, the more ink it will hold and the darker it will print. Gravure is also called intaglio printing. It is used for printing magazines, paper money, and postage stamps.

gravure printing
(grah-VYOOR)
a printing process in which letters and designs are etched or scratched into a metal plate; ink fills these grooves and is then transferred to paper

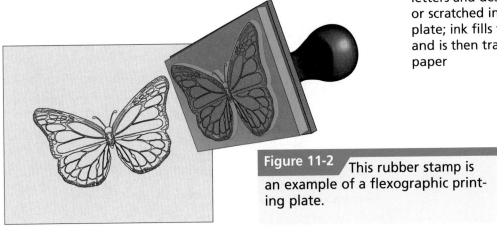

Figure 11-2 This rubber stamp is an example of a flexographic printing plate.

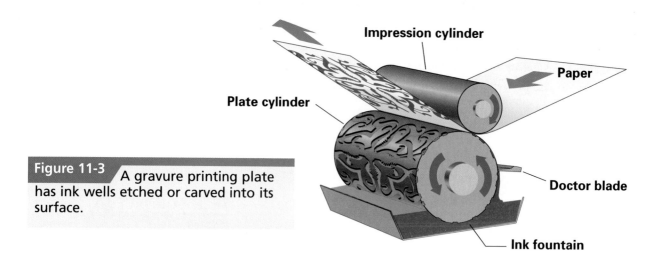

Figure 11-3 A gravure printing plate has ink wells etched or carved into its surface.

Labels: Impression cylinder, Paper, Plate cylinder, Doctor blade, Ink fountain

Lithography

▶ What is the difference between stone and offset lithography?

lithography
a printing process that is based on the principle that oil and water don't mix

In **lithography**, sometimes called planographic printing, the surface of the printing plate is perfectly flat. In stone lithography, the entire process is done by hand with a limestone slab. An image is created on the slab with an oily medium. Water is applied to the slab, then ink. Ink sticks only to the oily image. It doesn't stick to the wet stone. Paper is pressed against the slab and the ink transfers to the paper.

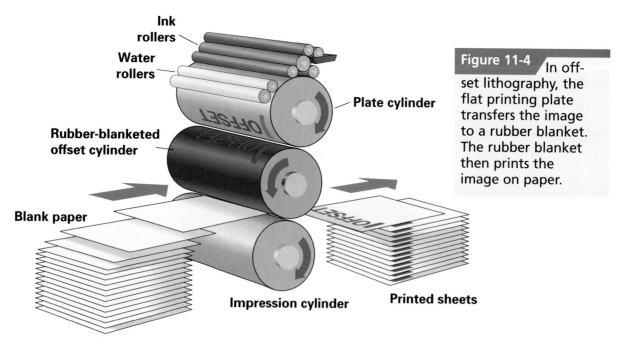

Labels: Ink rollers, Water rollers, Rubber-blanketed offset cylinder, Blank paper, Plate cylinder, Impression cylinder, Printed sheets

Figure 11-4 In offset lithography, the flat printing plate transfers the image to a rubber blanket. The rubber blanket then prints the image on paper.

In offset lithography, the image is usually placed on a thin metal plate through a photographic process. The plate is mounted on an offset press, which prints the image to a rubber blanket. Then the blanket transfers the image to the paper. See Figure 11-4.

Offset printing is used to print business forms, magazines, newspapers, and books. Today, about forty-seven percent of all commercial printing in the United States is done by this process. This textbook was printed on an offset lithographic press. A web offset press can print up to 3,000 feet of paper every minute. This is then cut into about 35,000 book-size pages.

Serigraphy

▷ How is a serigraphy printing plate used?

In **serigraphy**, or screen process printing, the printing plate is an open screen of silk, nylon, or metal mesh. See Figure 11-5. The openings in the screen are sealed in the non-printing areas. This is done with a paper stencil, a special film stencil that is cut by hand, or a photographically prepared film stencil. Ink is then forced through the open areas of the screen by the movement of a tool called a squeegee.

serigraphy
(sih-RIG-rah-fee)
a printing process that uses a printing plate made of an open screen of silk, nylon, or metal mesh

Figure 11-5 In serigraphy or "screen printing," the ink is forced through openings in a screen on to the surface to be printed.

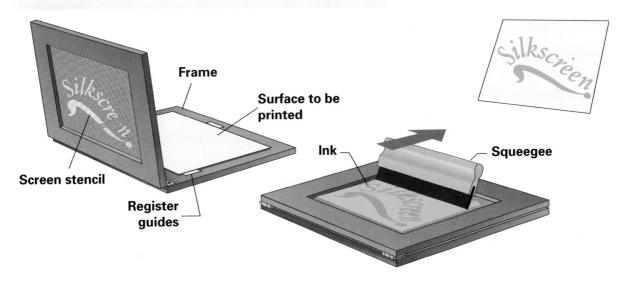

Frame

Surface to be printed

Ink

Squeegee

Screen stencil

Register guides

Figure 11-6 In this production shop, workers are screen printing a logo onto many T-shirts.

dynamic digital printing
the use of printing machines that print directly from a computer file instead of a printing plate

Serigraphy can be used to print on paper, glass, metal, wood, and other materials. Objects of every size and shape have been printed by this process, including shirts, fabrics, toys, banners, and bottles. See Figure 11-6. About three percent of printed products are printed using this method.

Dynamic Digital Printing

▷ How does dynamic digital printing differ from the other printing processes?

The newest type of commercial printing is called **dynamic digital printing**. These new machines print directly from a computer file without using a traditional printing plate. The newest machines are cheaper to operate for short printing runs than conventional printing presses. You perform dynamic digital printing every time you send a document from your computer to your printer. However, your small printer isn't fast enough to meet the needs of a commercial printer.

All digital printing machines need to refresh the image before each page is printed. As a result, every page in a continuous printing run can be different without stopping to change printing plates. This allows an entire book to be printed one copy at a time.

This process is currently the most economical for short printing runs of up to 2,000 copies of books, magazines, and brochures. See Figure 11-7. It is also used for longer printer runs that require personalized mailing labels. The labels are printed along with the rest of the page.

Dynamic digital printing processes include xerography, electron beam imaging, electrophotography, magnetography, and ink-jet printing.

Figure 11-7 This dynamic digital printing press will print, collate, and bind a book all in one operation.

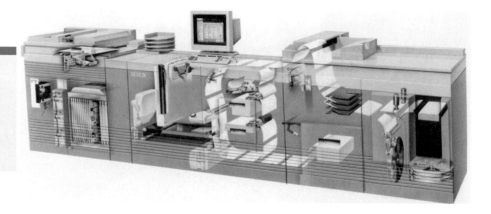

Xerography

▶ **How does xerography work?**

You use **xerography**, or electrostatic printing, whenever you use a laser printer or make a photocopy. It uses a special metal plate in the shape of a cylindrical drum, static electric charges, and a powder that is called a toner.

The drum (printing plate) conducts electricity when exposed to light but acts as an insulator when the light is off. An image is projected onto this statically charged drum. The lit area of the drum conducts electricity and loses its static charge. The image area remains dark and therefore still charged. The image is made visible when the negatively charged toner sticks to the drum. A static charge causes the toner to transfer to the paper. Then the paper travels through a heater, causing the powder to melt onto the page and become permanent. The image on the drum is refreshed for the next print. In laser printers, information from a computer controls a laser that creates the image area on the drum.

Color electrostatic printing is also available. A color printer operates exactly like a black and white printer except that it separates the original into four images. These images each receive their own color toner in perfect alignment on the same piece of paper. When the different colors of toner are fused to the paper, they work together to produce all the colors of the original.

Electrostatic printing lends itself to quick printing of newsletters, advertisements, and business forms. It is becoming the printing process that is preferred by the small printing shop. See Figure 11-8. However, it is used for less than five percent of all printed products.

xerography
(zee-RAH-grah-fee)
a printing process that transfers negatively charged toner to positively charged paper

Figure 11-8 Small printers, or "quick copy" shops, use electrostatic printing more than any other printing process.

Electron Beam Imaging

▷ **How does electron beam imaging differ from laser printing?**

Electron beam imaging uses an electron beam instead of a laser to create the image on the drum. Streams of electrons carry toner to the drum, instantly forming the image. The toner is then transferred and fused to the paper. The paper is then heated, which makes the image permanent. The electrical charge is removed from the drum and the process begins again for the next image.

This equipment can print the front and back of the paper at the same time. Some machines, with almost no human help, can produce a 400-page book every ten seconds. They can switch from one book to another without stopping. At the moment, they are the fastest dynamic digital printing presses. They can print 78,000 (8½ by 11) pages per hour.

Electrophotography

▷ **What is electrophotography?**

Electrophotography machines use 600 tiny light-emitting devices to expose a light-sensitive, rotating electronic film. The flashing devices and moving film produce an image by drawing a static electric charge from the non-printing area. Toner then sticks to the static charge remaining in the image area. The rest of the process is very similar to laser printing.

Magnetography

▷ **What is magnetography?**

For magnetography, the drum is covered with chemical salts that can be magnetized. In a thousandth of a second, a magnetized image is created on the drum. The image then picks up magnetized toner and transfers it to the paper. A flash fusing system melts the toner into the paper.

Ink-Jet Printing

▷ **How is ink applied to a surface in ink-jet printing?**

In **ink-jet printing**, there are no printing plates or drums. The printer actually has very tiny spray guns that shoot their ink to the printing surface. A static-electric charge is also used to help direct the movement of the ink. Since the guns never touch the material being printed, this method can be used to print on very uneven or fragile surfaces. See Figure 11-9.

ink-jet printing
printing process that uses spray guns to spray ink on the printing surface

Figure 11-9 Ink-jet printers can produce attractive photos.

With this process, a number of colors can be printed at the same time. In multicolor ink-jet printing, a separate ink gun is used to apply each color ink. Since the guns don't touch the paper, there is little chance of smearing.

Impacts of Graphic Communication

▷ **Have printing processes caused negative impacts?**

Not all of the changes brought about by graphic communication have been positive. Economic factors and environmental concerns have often been in competition with one another. For example, at one time waste products from bleaches and other chemicals used to make paper were dumped into rivers and streams, causing pollution. Doing so was less costly than finding a better way of disposing of the wastes. Printing inks, too, can be harmful to the environment and must be disposed of properly.

Technology can often be used to develop ways to protect the environment while also providing us with the things we need and want. For example, new technologies have had to be developed to break down wastes from graphic communication industries and protect waterways.

SECTION REVIEW 11A

Recall »

1. How does letterpress printing work?
2. How does gravure printing work?
3. How does lithography work?
4. How does dynamic digital printing differ from traditional methods?

Think »

5. What social and cultural impacts do you think the first printing presses had? Explain.

6. When do you think quality would be very important in a printing job? When would it be less important? Give examples.

Apply »

7. **Design** Create designs for T-shirts on a computer. Using special transfer sheets, print your designs as iron-on transfers with an ink-jet or laser printer.
8. **Research** Look up the impacts of graphic communication on society or the environment. Write a report on your findings.

Producing a Graphic Message

▷ **How are graphic messages designed?**

copy
a graphic message ready for reproduction

print
a printed copy of a graphic message

layout
arrangement of elements on a page to be printed

Before it can be printed, every graphic message must be designed. The design is influenced by such things as the intended audience and the medium used. (The medium is the means used for the message, such as a book or a sign.) The purpose and nature of the message are also important. A message designed to warn about road hazards would be handled differently from a message in the form of a valentine.

After the message is designed, steps are taken to prepare it for printing. These are called prepress operations. At this stage, the message is called **copy**. After printing, the copies of the message are called **prints**. This section will discuss how pages are designed and illustrations are prepared for printing.

Message Design

▷ **What steps must a graphic designer take to develop an idea?**

The design principles that you learned in Chapter 7, "Design and Problem Solving," can be applied to graphic messages. The first step is to identify the message. You must then decide how to use type, artwork, and photographs to convey the message. The arrangement of these elements on a page is called a **layout**.

Designers often create a number of thumbnail sketches of possible layouts. Then they select the best layout or brainstorm improvements. Finally, they create a comprehensive layout, which includes all the details.

The last step in graphic page design is the creation of camera-ready copy. This is an exact copy of what the item will look like after it is printed. The steps in this design process are shown in Figure 11-10.

FIGURE **11-10**

Message Design

Sketches

Rough layout

SENIOR DAYS

Helping people
makes a
difference to
others as
well as
yourself.

Donate
time and
call
555-2121
today.

VOLUNTEERS
N E E D E D

Camera-ready copy

Comprehensive layout

Making Things Fit

Pictures often have to be reduced or enlarged to fit the space in a layout. When you change one dimension, the other dimension changes proportionately. If you change the first dimension of a 3 × 5 photo to 6 inches, the second dimension becomes 10 inches.

ACTIVITY Calculate the missing number in these photo sizes: 4 × 5 = 6 × ?; 5 × 7 = 7.5 × ?; 11 × 14 = 1.57 × ?.

line art
drawings and other art elements made of solid lines and shapes

halftone
an image reproduced using a series of dots

Today, ninety-five percent of all the prepress operations are done on computers. This includes preparation of the text, drawings, photographs, and layout; creation of the camera-ready copy; and conversion of the page to match the method of printing.

Illustrations

▷ What is the difference between photographs and line art?

When original drawings and charts are made of solid lines and shapes they are called **line art**. See Figure 11-11. Today, most line art is drawn directly on a computer using paint or drawing programs. If the line art wasn't created on a computer, a scanner can be used to create a computer file.

If the original illustration is a photograph, then the tones of the picture must be reproduced. When you look at a black-and-white photograph, for example, you see black areas, white areas, and shaded gray areas. See Figure 11-12. For printing, these tones are converted into a series of dots. The result is called a **halftone**. A halftone is an image made completely of dots. The spacing and size of the dots determine if you see black, white, or gray.

The simplest way to create a halftone is to scan the photograph into a computer file. You can also convert pictures into halftones by re-photographing the picture through a halftone screen.

Digital cameras shoot all photographs as halftones. This means that every picture you take with a digital camera

ETHICS in *ACTION* — Honoring Copyright

Have you ever seen a funny story or joke on an Internet Web site and wanted to copy it? The Internet makes it easy to copy and transmit such material. However, when material is copyrighted, it means it should not be reproduced without permission. You must assume that everything you see on the Internet is copyrighted. Using copyrighted material without permission is stealing. Lawmakers are now working on laws to regulate copying material found on the Internet.

ACTIVITY Research the laws being considered to protect material found on the Internet.

has a dot structure. The dots are so close together that the human eye sees them as solids, shades, lines, or points.

Reproducing Colors

▷ How are colors produced?

Color adds life to anything that is printed. How do you suppose printers are able to print all the colors we see in the world around us? If color printing needed hundreds of ink colors and printing plates, not much would be printed in color.

All colors in nature can be reproduced through the printing of three primary ink colors plus black. These colors are magenta (similar to red), cyan (blue), and yellow. See Figure 11-13 (page 282). With black, these colors make up what is called **four-color process printing**.

When you look at a printed object, you see the image because light is reflected from the inks. Remember, white light contains the wavelengths of all the colors. The inks on printed objects absorb (subtract) some of the wavelengths of white light and reflect the color wavelengths that you see. This is why the inks used in four-color process printing are called subtractive colors.

To create color images, a process called color separation is used. In the past, four halftones were created in the four process colors. Then the halftones were printed on top of

Figure 11-11 Line art has no gray tones. The entire drawing is made up of solids or lines that are all of equal density.

four-color process printing using magenta, cyan, yellow, and black to produce all the other colors used on printed materials

Figure 11-12 These motorcycles were photographed through a halftone screen. The screen breaks the picture into tiny dots.

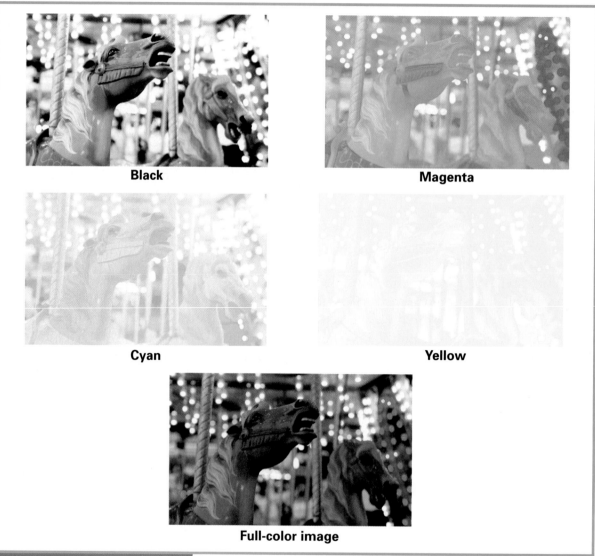

Black

Magenta

Cyan

Yellow

Full-color image

Figure 11-13 All the colors of the rainbow can be reproduced on a printing press using magenta, cyan, yellow, and black inks.

one another to produce the illusion of all colors. Today, ninety-five percent of all color separations are created using computer programs and scanners.

Preparing the Printing Plate

▷ What are flats?

The last prepress operation is photographing pages of copy to produce negatives. The negatives are then arranged into large sheets called flats. See Figure 11-14. The flats are large because the paper that printers use comes in huge rolls. The flat is then used to create the image on a printing plate.

Self-Discipline

When you have a job, you need self-discipline to make sure the job gets done right.

Graphic Designer

We are looking for graphic designers who are creative problem solvers and skilled in using design software. Successful candidates must be self-disciplined and able to work with minimum supervision. Responsibilities include meeting with clients to determine their needs and developing exciting proposals.

Working with self-discipline and minimum supervision means that no one needs to watch you constantly to be sure you do a proper job. You show the necessary self-control to handle a job well. Unfortunately, many people feel it is easier to supervise others than to supervise themselves. They may not have much self-discipline. They may be irresponsible. Here are some tips that may help with self-discipline:

✔ **Concentrate on the task at hand. It can be very easy to get distracted.**
✔ **Be responsible. Not only are you responsible for yourself, but others will be depending on you as well.**
✔ **Pace yourself. Don't rush to get the job done by sacrificing quality. Likewise, don't put things off that need to be done by a set time.**

You may already use self-discipline in school, whether you're taking notes or doing homework. Using self-discipline now will better prepare you for a future career.

Figure 11-14 This large flat of negatives is used to burn the images onto a plate that is then placed on the printing press.

Flats continue to be used in some printing shops. However, most printers today use a computer file to create the image on a plate with laser technology.

Desktop Publishing

▷ What is desktop publishing?

Have you ever used a computer to work on a school newspaper or magazine that was duplicated and then passed out to your classmates? This kind of activity is now called **desktop publishing**. It was used in schools before making its way to the business world.

Computers and small, high-quality printers have made it possible for many companies to create their own printing shops. Employees create sales reports, booklets, stationery, telephone pads, and interoffice memos without ever leaving their desks.

When you use desktop publishing software, you can set up your page exactly the way it will look when it is printed. See Figure 11-15. Your headings, text, and illustrations are all combined right on the computer screen. You do not have to cut and paste any articles and pictures by hand.

Desktop publishing software is now also being used to convert electronic images into the formats for publishing CDs, DVDs, e-books, and Internet pages.

desktop publishing
the use of desktop computers and small printers for publishing

Figure 11-15 Using a computer and special desktop publishing software, you can design an entire publication. The electronic files are then sent to the printer.

The E-Book Advantage

E-books have been available for several years. An e-book is downloaded into your computer or a hand-held reading device. Instead of reading a paper copy of the book, you read it from a screen. However, consumers have been slow to accept e-books, and some stores no longer sell them. In the future, e-books may be popular for certain uses. For example, instead of carrying many pounds of schoolbooks around, students may prefer to have all their books inside a device no larger or heavier than a single paperback. What do you think? Would you rather switch to e-textbooks?

SECTION REVIEW 11B

Recall »

1. Film photographs and digital photographs have to be handled differently for printing. Why?
2. Name the ink colors that are used by printers in four-color process printing.
3. What is line art?
4. What is a layout?

Think »

5. How is a halftone screen similar to a window screen?
6. How do you think the computer affected the printing industry? Suggest some positive and some negative impacts.

Apply »

7. **Mixing Colors** Obtain red, yellow, and blue watercolor paints. They represent the primary ink colors. Use them in different combinations to create a color wheel having at least twelve colors. Label each color with the combination you used to produce it.
8. **Research** Do some research and then make a poster or display showing the evolution of the basic printing processes. How did people's needs and wants change at the same time?

Summary Activity

For each numbered blank, pick the answer from the list on the right that makes the most sense in the entire passage. Write your answer on a separate sheet of paper. No answer will be used more than once.

People often refer to graphic communication as printing. Graphic communication includes printing with ink on paper, as well as publishing CDs, DVDs, e-books, and **1** . It is the eighth largest **2** in the United States.

All printing processes require the preparation of **3** and illustrations before you can start to print. Many of the same procedures are used to prepare printing **4** , even though they might be used in different processes.

The type of printing process used is determined by the part of the printing plate that does the **5** . **6** prints from the raised part of the plate. Gravure **7** from inkwells etched into the plate. Lithography prints from a plate that is perfectly flat. Screen process printing prints right through little holes that exist in the plate. The newest form of printing is called dynamic **8** printing and it prints directly from a computer file. The process is **9** for short printing runs.

10 uses an electron beam to create the image. These machines can print 78,000 pages per hour. Electrostatic printing prints from a plate that is charged with **11** electricity. Ink-jet printing sprays the **12** directly on the material that you are printing. A(n) **13** directs the spray guns and tells them when to spray.

14 publishing is used to print newspapers, stationery, booklets, and other items without having to go to a printer. Small office computers and **15** have made this type of printing possible.

Answer List

- computer
- desktop
- digital
- economical
- electron beam imaging
- electrostatic printers
- industry
- ink
- letterpress
- plates
- printing
- prints
- static
- text
- Web pages

Comprehension Check

1. Name the five basic printing processes.
2. Which type of printing requires no printing plate?
3. Which type of pollution was once caused by paper-making technology?
4. List the steps in designing a message.
5. What is a halftone?

Critical Thinking

1. **Analyze.** List at least five ways your school would change if printing technology suddenly did not exist.
2. **Compare.** Joe wants to print a brochure using gravure printing. Sue wants to produce a brochure using ink-jet printing. Who do you think will be done first and why?
3. **Decide.** Suppose you have been asked to design an ad for an auto repair shop and an ad for a flower shop. List at least three factors regarding the nature of each message that you would consider.
4. **Infer.** Why do you think skill is important when designing a graphic message?
5. **Critique.** Select a magazine advertisement and evaluate its message. Do you think it is effective? Why or why not?

Visual Engineering

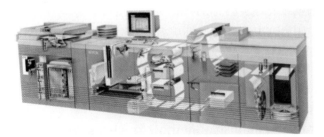

Brainstorming about Printing. This printing machine is capable of printing, collating, and binding, all in one operation. It does not use a printing plate. Each page is printed from an electronic file sent from the computer. You can customize each book, magazine, or brochure that is printed. With two or three of your classmates, brainstorm ways in which this technology could be used. For example, could each student have a textbook customized for his or her learning skills? Does it mean that someone could publish a magazine tailored to each subscriber's interests? Think about what this technology might mean for the future. In what form will published materials be presented? Share your ideas with the rest of the class.

TECHNOLOGY CHALLENGE ACTIVITY

Creating Graphic Communication Products

Equipment and Materials

- button machine
- button parts
- button print cutter
- computer and printer
- linoleum blocks
- linoleum block cutters
- proof press
- printer's ink
- heat transfer ink for letterpress printing
- padding cement
- paper cutter
- spiral binding machine
- rubber stamp machine
- rubber stamp supplies
- piece of plastic

- paper
- spirals
- type
- engraver

Reminder
In this activity, you will be using tools and machines. Be sure to always follow appropriate safety procedures and rules. Remember, safety is an attitude that you must develop and maintain at all times.

Background You can use graphic communication processes to develop many different products to keep or sell. Buttons, rubber stamps, and notebook covers are only a few of the possibilities. Buttons can be designed with funny sayings or drawings on them. Rubber stamps can be used to decorate letters, cards, and reports. Notebooks with covers in school colors are always popular.

Goal For this activity you and your classmates will create products using graphic communication processes. If you choose, you may then decide to form a company to produce large quantities of the products and sell them.

Criteria and Constraints

▶▶ You may choose to make all three products—buttons, rubber stamps, and notebooks—or only one of them.

▶▶ As a class, you should brainstorm ways to organize the tasks involved for maximum efficiency. See Figure A.

▶▶ You should establish a method for checking product quality.

FIGURE A

FIGURE B

FIGURE C

FIGURE D

Design Procedure

Buttons

1. Draw designs for buttons or create them using a computer and a graphics program. Your school emblem would be a good choice.
2. Take orders for photographic buttons. People can provide you with photographs, and you can turn the photos into buttons.
3. Give people the opportunity to design their own buttons. Print a special form that gives them a circle that is the exact size that the button design must fit.
4. Cut the design out using the button print cutter.
5. Place the design and button materials into the machine and press the button.
6. Remove the completed button. See Figure B.

Notebooks

1. Prepare a linoleum block design for use as the cover for a small assignment pad or spiral notebook.
2. Transfer the design to the block, and carve it out using linoleum block cutters and a bench hook.
3. Print the design on cover stock at the proof press. Your linoleum block can also be printed as a heat transfer by using special heat transfer inks.
4. Cut paper to size at the guillotine paper cutter.
5. Pad or spiral bind the covers and paper together. See Figure C.

Rubber Stamps

1. Set type or engrave the copy into a piece of plastic. The plastic mold must be engraved very deep into the plastic.
2. Lock the type into the special tray or frame that goes with your rubber stamp machine.
3. Prepare a matrix (mold) from the type, following your teacher's instructions. Your engraved plastic is your matrix.
4. Place gum rubber on top of your mold.
5. Place these materials on the machine's tray and follow all safety and processing procedures in the machine's instructions.
6. After it has cooled, separate the mold from the vulcanized rubber and mount the stamp. See Figure D.

Evaluating the Results

1. What are safety considerations that you must take into account when operating printing equipment?
2. What printing project did you enjoy the most? Why?
3. If you were to repeat this project, what would you do differently?

How would you catch a moment in time so that it can be shared with others? Would you rush to canvas with oil paints? Would you chisel a picture into a stone tablet? Methods such as these were all that were available to our ancestors throughout most of history.

Joseph Niepce changed all that when he created the first photograph in 1826. Many photographic inventions and innovations have evolved since then. People tested and refined them using slow and methodical processes. Over time, photographic systems were improved. Today, photography is an important tool of communication technology.

When they were your age, your parents probably took photos with film. Do you and your friends take photographs using digital cameras? Do you have a digital camera attached to your computer, cell phone, or wristwatch? Photography is changing from a chemical technology into a digital technology.

A professional photographer at work. How would you like a job like this one?

Objectives

▶▶ **Identify** the main parts of a camera.

▶▶ **Explain** some differences between film and digital cameras.

Terms to Learn

- **aperture**
- **CCD**
- **focus**
- **lens**
- **shutter**

Standards

- Characteristics & Scope of Technology
- Core Concepts of Technology
- Relationships & Connections
- Cultural, Social, Economic & Political Effects
- Environmental Effects
- Influence on History
- Design Process
- Assessment

lens
a piece of glass used to focus and magnify light

shutter
a covering that opens to let light into the camera

Cameras

▷ What are the main parts of a camera?

Photography is the use of light reflected from a scene to record an image of that scene. All cameras are basically the same. They consist of a light-tight box, a lens, and a shutter. The **lens** focuses the light rays. When the **shutter** is open, the light is allowed through an opening into the box.

Film cameras record the image on light-sensitive film. Digital cameras record the image electronically. Otherwise, digital and film cameras share most of the same basic components. See Figure 12-1.

Lens

▷ What is the purpose of a camera lens?

In addition to focusing light, the lens can magnify the size of the object. Special telephoto lenses make it possible for objects that are very far away to appear close in the photograph. Wide-angle lenses take in more of the surrounding area. Figure 12-2 shows the effect of using different camera lenses.

Figure 12-1 A film camera and a digital camera have a lens, aperture, and shutter. One camera records the image electronically and the other uses film. Can you tell which camera is digital?

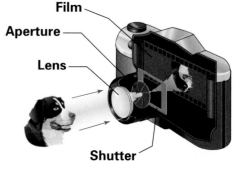

Film

Aperture

Lens

Shutter

Figure 12-2 These photos were shot from the same location using a wide angle, normal, and tele-photo lens.

Shutter

▶ What does the camera shutter do?

The shutter blinks open for a very short period of time, allowing only a certain amount of light to enter. On some cameras, its speed can be adjusted. The shutter speed helps determine if moving objects will appear sharp or blurred. See Figure 12-3. It can also help compensate for low-light conditions.

TECH CONNECT
SCIENCE

Looking at Lenses

Lenses are curved on one or both sides. The curvature determines the lens's effect on light waves. Concave lenses spread the light waves. Convex lenses bring them together.

ACTIVITY Obtain a series of photo lenses. Use them to observe the same object in order to determine their effect.

Figure 12-3 Shutter speed helps determine what is in focus. For the photo on the left, the shutter speed was slow. For the photo on the right, the shutter speed was fast. How did the photographer manage to keep the car in focus in the first picture?

aperture
(APP-ih-chur)
the opening that controls the
amount of light that enters
the camera

focus
term used to describe the
sharpness of an image

TECH CONNECT
· MATHEMATICS

Measuring Light

Ten feet away, a light
source is one fourth as
bright as it would be at
five feet. At twenty feet,
it's only one sixteenth as
bright. This is the *inverse
square law* of physics, and
it is used to determine
how bright a flash or extra
lighting has to be for a
photograph.

ACTIVITY Using a light
meter and a consistent
light source, devise an
experiment and take
measurements to test the
inverse square law.

Aperture

▷ What effect does the aperture of a camera have on the way you take pictures?

The adjustable opening that controls how much light will enter the camera when the shutter opens is called the **aperture**. The size of the aperture is controlled by a device called the diaphragm (DI-uh-fram). See Figure 12-4. When the aperture is opened very wide, a lot of light can enter the camera.

The aperture on many cameras automatically adjusts to lighting conditions. Some inexpensive cameras have an aperture that can't be adjusted. The aperture also affects how much of your picture will be in **focus**. The smaller the aperture, the more of the image is in focus. See Figure 12-5. By opening the aperture wider, you can take pictures with less light without the need of a flash attachment. However, a wide-open aperture reduces the area of the picture that will be in focus in your picture. That is, a smaller area will be sharp.

Figure 12-4 The smaller the aperture, the larger your field of focus. The relationship between the f-stop and the size of the opening is shown here.

APERTURE SETTINGS

Figure 12-5 The photo on the left was taken using a large aperture. Only one car is in sharp focus. The photo on the right was taken using a small aperture.

The size of the aperture is indicated by units called f-stops. The larger the f-stop number is, the smaller the aperture opening. The smaller the number is, the larger the opening.

Viewfinder

▶ What is the purpose of a viewfinder?

The viewfinder on a camera allows you to view your picture before you actually take it. If you don't carefully check your picture in the viewfinder, your photographs might not include the heads of your subjects!

In some cameras, the viewfinder looks directly through the camera's lens. Most digital cameras have a small screen that displays the picture.

Film or Camera Memory

▶ What is used to record the image?

In a film camera, film records the image. Photographic film is produced with different levels of sensitivity to light.

TECH CONNECT

SCIENCE

Extending Vision
Cameras can show us places and things we may never see any other way. For example, cameras have gone far beneath the ocean to levels too deep for human divers.

ACTIVITY Find out about other things photography has shown us that we could not have seen otherwise. For example, think about very small things, things very far away, and things hidden from us.

Figure 12-6 With a digital camera, you can see your photo as soon as you shoot it.

CCD

charge-coupled device; a special microchip inside a digital camera that converts light into an electrical signal

This light sensitivity is referred to as film speed. The International Standards Organization (ISO) uses a number system to rate the speed of film. The higher the number, the "faster" the film. If you know you will be taking photos where the light levels are low, you should select faster (higher number) films.

A digital camera does not use film. Light from the image falls on a microchip device that converts the light into an electrical signal. See Figure 12-6. The device may be a **CCD** (charged-coupled device) or a CMOS (complementary metal-oxide semiconductor). The signal is stored in the camera's memory. Unlike film, the camera's memory can be erased and reused many times.

In a film camera, film winders move the film forward to ready the camera for the next picture. Most film cameras now have motorized film winders. Digital cameras automatically advance to the next available place in their memory storage system.

12A / SECTION REVIEW

Recall ▸▸

1. Name at least three camera components.
2. What is the purpose of a camera aperture and shutter?
3. What is used to create the electrical signal in a digital camera?

Think ▸▸

4. What impacts do you think photography has had on the following: law enforcement, manufacturing, the environment?

5. Assess current trends in photography, such as combining cameras and telephones. Forecast what these trends might lead to in the future.

Apply ▸▸

6. Construction The first camera was the *camera obscura*. Do some research and make a model of it that works.

Recording the Image
How are images captured?

Film, digital, and movie cameras record images using different recording media. Let's look at how each technology performs its function.

Images on Film
How does film use colors to record an image?

Photographic film consists of a sheet of thin plastic coated with chemicals that are very sensitive to light. See Figure 12-7. Color film has three layers of chemicals, each one sensitive to one of the three primary colors of light—red, green, or blue. When the film is exposed, these chemicals undergo changes that record the image.

However, the image is invisible until after the film is developed. This invisible image on the film is called a **latent image**.

Different kinds of film are made with different properties. These properties can produce different effects. For example, some fast films require very little light to produce a satisfactory picture.

APS is a hybrid digital/chemical photographic technology. The image is stored as a standard latent image along with information about the scene for use during processing. Some camera models can also record captions, the number of prints you want for each individual shot, and the time and date when the picture was taken.

Objectives

▸▸ **Describe** how images are captured on film.
▸▸ **Describe** how images are captured with a digital camera.

Terms to Learn

• latent image
• photosite

Standards

■ Characteristics & Scope of Technology
■ Core Concepts of Technology
■ Information & Communication Technologies

latent image
the invisible image produced on exposed film

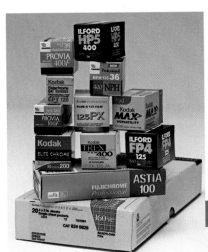

Figure 12-7 Why are there so many different types of film?

Digital Images

▶ **How does a digital camera record a picture?**

In digital photography, pictures are recorded on the CCD. The CCD contains millions of tiny light-sensitive cells called **photosites**, which convert the light into an electrical charge. The CMOS has image sensors that do the same thing. The brighter the light that strikes a single photosite or image sensor, the greater the electrical charge. These charges become the signal stored in the camera's memory.

To record colors using an ordinary digital camera having a CCD, filters are placed in front of the photosites. The filters separate the light into red, green, or blue—the three primary colors of light. Each photosite records only one color, and each color produces a different electrical signal. These varying signals create the range of tones and colors that appear in the final photograph. The image from each photosite appears as a tiny dot called a pixel. All pictures shot with a digital camera are collections of these dots. See Figure 12-8. Cameras equipped with a CMOS also use filters to record colors, and the images also appear as pixels.

photosite
tiny light-sensitive cell that converts light into an electrical charge

Figure 12-8 In a digital camera, the light goes through filters and light sensors, which create the tones and colors that appear on the final photograph.

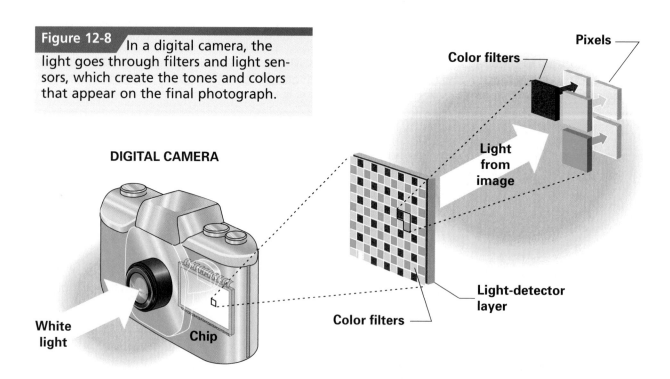

DIGITAL CAMERA

White light

Chip

Color filters

Pixels

Color filters

Light from image

Light-detector layer

Figure 12-9 These flash cards are used for storing pictures taken with digital cameras.

The camera images are stored in the camera's flash memory, memory card, a computer floppy disk, or on a recordable CD. See Figure 12-9.

Motion Pictures

▷ How are motion pictures recorded?

The invention of the film camera led to the invention of the movie camera. The movie camera produces movies using the same components that are found in a still camera. It's designed to photograph twenty-four still pictures each second of someone or something in motion. When played back at the same speed, the images create an illusion of motion. Our eyes cannot see the change from one image to the next. It happens too fast. See Figure 12-10.

LOOK TO THE FUTURE

Snapshots of the Future

You've probably seen holograms on credit cards or in science fiction movies. Holograms are three-dimensional images created using mirrors and lasers. Until recently, the images did not look very real. French holographer Yves Gentet has changed that. He has developed a film coating that produces holographic images so true to life you have to touch them to know the difference. One day in the future, you may have hanging on your walls three-dimensional images of the people and places you want to remember.

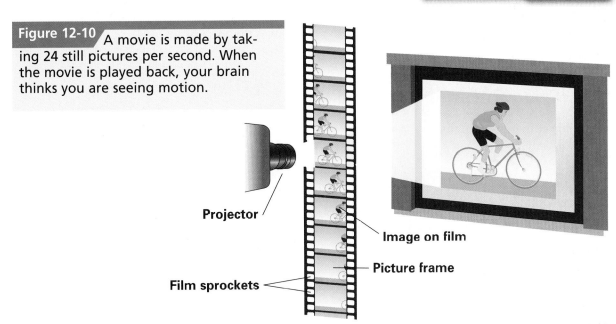

Figure 12-10 A movie is made by taking 24 still pictures per second. When the movie is played back, your brain thinks you are seeing motion.

Projector

Image on film

Picture frame

Film sprockets

Creativity

Creative people come up with new ideas and solve problems.

If you can come up with a new idea or a new way of doing things, then you are creative. Although some people are more creative than others, we are all creative in some ways. Develop your creativity by trying to see things in a new way. Try the following exercises:

✔ **Some people ask themselves the question, "What if...?" For example, what would you do if Martians took over your community and cut off all communications? How would you call for help?**

✔ **Try to think of new ways to use familiar items. For example, how many unusual uses can you think of for a brick?**

After you practice looking at things in a different way, you will find that being creative is much easier.

12B SECTION REVIEW

Recall ▸▸

1. What effect does exposure to light have on film?
2. How does a digital camera record a picture?
3. How many images does a standard movie camera record if you shoot for 10 seconds?

Think ▸▸

4. How is a CCD similar to the microchip in a computer?

5. The term *photography* means "drawing with light." Why do you think it was given that name?

Apply ▸▸

6. **Research** The fastest camera can shoot 20 million frames per second. Research how photographers use stop-action photography.
7. **Creativity** Make a display showing at least five unusual uses of photography.

Processing the Image

▶ Do all photographs require processing?

What happens between the time you snap the picture and the time you finally see the prints? Photographic film is usually taken to a store where it is developed using a wet chemical process. Digital pictures appear instantly on the display screen of the camera. However, they, too, need to be further processed if you want to print them on paper.

Film Processing

▶ How is film developed?

Commercial film processing centers use automatic equipment. You can see these machines in action at most stores that do on-site processing. See Figure 12-11.

The first step is chemical processing of the film into **negatives**. Special chemical **developers** are used to reveal latent images. After developing, areas that were light colored in the original scene will appear dark on the negatives, and dark areas will appear clear. Other solutions are then used to make the image permanent and to end the film's sensitivity to light. This process usually needs to take place in total darkness using special equipment.

SECTION 12C

Objectives

▶▶ **Explain** how film negatives are processed.
▶▶ **Explain** how digital images are processed.

Terms to Learn

- developer
- negative

Standards

- Core Concepts of Technology
- Cultural, Social, Economic & Political Effects
- Design Process
- Information & Communication Technologies

negative
image produced on exposed film after processing; light areas appear dark and dark areas appear clear

developer
chemical used during processing to reveal the latent image

Figure 12-11 In a commercial film processing center, the film is developed and the prints are made automatically.

After the negatives have dried they are ready to be printed. This is done by projecting light through the negatives onto photographic paper. See Figure 12-12. The distance between the negative and the paper determines print size, and enlargements can be very big. Photographic papers used for color prints also have layers sensitive to the three primary colors of light. Light passing through the negative causes a chemical reaction in the paper and the proper color is produced.

Digital Processing

▶ What equipment is found in a digital darkroom?

Chemicals are not needed to process digital pictures. That makes digital photography much friendlier to the environment. Computer software is used to transform digital photographs into prints. You just need to download the pictures from the digital camera into your computer. Because the information in the camera is already encoded as an electrical signal, the computer can interpret it. No further changes are necessary.

The download process usually involves placing the camera or a removable memory card in a cradle attached to a port on the computer. The software copies the digital files into computer memory. You can then use other software to alter the pictures in some way, such as to improve color or focus. When you are happy with them, you can send them to your printer. For quality printing, a color laser printer is better than an inkjet. Most software packages also have features to make e-mailing pictures easier.

Figure 12-12 An enlarger is used to make the photo the desired size. Does a darkroom have to be dark all of the time?

ETHICS in ACTION

Is Seeing Really Believing?

As you know, you can use your computer and special software to alter a digital photograph. In 2003, a newspaper photographer lost his job for doing just that. The changes were minor. In a photograph of soldiers fighting in Iraq, he altered one soldier to make the picture look more balanced and pleasing to the eye. However, the photo was no longer true to the facts.

ACTIVITY Hold a panel discussion on the ethics involved. What are the possible impacts of altering photos that appear in the news? For example, can altered photos be used as evidence in a court of law?

It is also possible to simply give a memory card to a commercial company for processing while you wait. New home printers have also been designed that print pictures without a computer. Just plug in the memory card, view the pictures on a screen, and pick the pictures you want to print. See Figure 12-13.

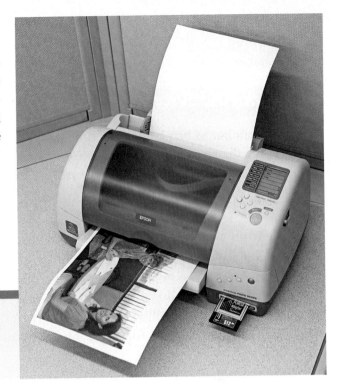

Figure 12-13 This printer does not have to be connected to a computer to print photos. The printer needs only the memory card.

SECTION REVIEW 12C

Recall ▶▶

1. What is developer used for?
2. How does the image on a negative differ from the original scene?
3. Why is digital processing easier on the environment than film processing?

Think ▶▶

4. What effect do you think digital cameras and processing have had on film processing businesses?

5. Have you ever made changes in digital photos using special software? If so, what kind of changes did you make and why?

Apply ▶▶

6. **Photography** Using either a film or a digital camera, create a photo essay having to do with technology and the future. Keep in mind your audience, medium, purpose, and the nature of your message as you work.

Summary Activity

For each numbered blank, pick the answer from the list on the right that makes the most sense in the entire passage. Write your answer on a separate sheet of paper. No answer will be used more than once.

The first photograph was created by __1__. Since then, photography has become an important tool of __2__ technology. It is changing from a chemical to a __3__ technology.

A camera has several main parts. The __4__ focuses the light on the film. It may also magnify the size of the object. When open, the __5__ allows light to enter the camera. The size of the __6__, or opening, is controlled by a diaphragm. A device called a(n) __7__ allows you to view the picture before you actually take it.

Photographic film consists of a sheet of thin plastic coated with __8__ that are very sensitive to light. When the film is exposed, changes occur that record the image.

In a digital camera, tiny devices called __9__ convert the light into an electrical charge. The brighter the light, the greater the electrical charge.

A(n) __10__ camera uses the same components as a still camera. It photographs twenty-four pictures per second. When played back at the same speed, the eye is fooled into thinking it sees motion.

Chemical __11__ is used to reveal the latent image in exposed film, creating a negative. In a negative, areas that appeared light in the original scene now appear dark, and dark areas appear __12__.

Information from a digital camera can be fed directly into a(n) __13__. The information does not have to be changed, because it is already encoded as electrical charges.

Answer List

- aperture
- chemicals
- clear
- communication
- computer
- developer
- digital
- Joseph Niepce
- lens
- movie
- photosites
- shutter
- viewfinder

Comprehension Check

1. What is the purpose of a camera lens?
2. What is the purpose of a viewfinder?
3. What is a latent image?
4. What is the purpose of a photosite?
5. What device is used to process digital photos?

Critical Thinking

1. **Distinguish.** If you were shown a camera and asked if it was a film camera or a digital camera, what is the first clue you would look for?
2. **Analyze.** The f-stop on camera A is set at 5.6. The f-stop on camera B is set at 2.8. Which setting will produce a picture with more of the image in focus? When do you think you would use each of the settings?
3. **Relate.** In what ways do you think photography is similar to graphic arts?
4. **Appraise.** Find a photo in this book that you think was taken using a special lens or altered with computer effects. Why do you think so?
5. **Compare.** A film camera costs $150. The film you want to use costs $5 a roll for 36 pictures. Development costs $12. A digital camera costs $220, including software. Which camera would you choose to buy and why?

Visual Engineering

Uses for Photography. Most people use photography for vacation pictures and for photos of family and friends. Some people also like to take pictures as a form of artistic expression.

Photography can also be a great tool in the service of science or technology. By adjusting the shutter speed, aperture, and lighting, interesting studies of wildlife, plants, and even machine movements can be made. Using motor drives and strobe lights and making multiple exposures are just a few of the techniques you can try.

Do some research to find ways to take different kinds of pictures. Use your creativity. Share the results with your classmates.

TECHNOLOGY CHALLENGE ACTIVITY

Processing Photographs

Equipment and Materials

Film Processing
- a completed roll of black and white film
- developing tank
- photo enlarger
- developing trays
- photographic paper
- print dryer or hanging clips and wire
- darkroom
- developer
- stop bath
- fixer

Digital Processing
- digital photos
- computer and photo-editing software
- old photo needing repair
- scanner
- printer

! SAFETY

Picturing Safety
Wear gloves, wear OSHA-approved safety glasses, and use tongs when working with photographic chemicals. Don't bring them in contact with food. Follow all safety rules for their use.

Background The darkened room where light-sensitive materials can be developed into photographs is called a darkroom. The film is developed in a tank without exposing the film to light. Special safelights are on during the part of the process that creates photographic prints from the negatives. The light from safelights does not affect the prints. Chemicals are used to develop the film and print the pictures. Enlargers are used to make big prints out of little negatives.

Digital photos are processed using such computer software as *Photoshop®*, *Picture It®*, and *Hyperstudio®*. Software can also change the size of the photograph.

Goal For this activity, you will develop and print a roll of black-and-white film. You will also process digital photographs or scanned pictures using special software on a computer and then learn how to repair an old damaged photo.

Criteria and Constraints

▸▸ You must select three of the best film photos for submission and indicate any changes you would make if you could alter the negative.

▸▸ You must select three of your best digital photos for submission and indicate how you digitally changed the picture to improve its content (removed red eye, etc.).

▸▸ You must submit before and after prints that show how you removed scratches, cracks, and other defects from the damaged photograph.

▸▸ You must dispose of all chemicals properly when you have finished the activity.

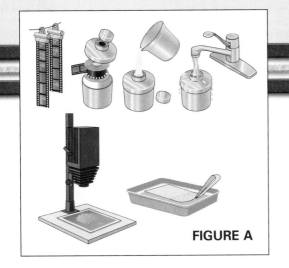

FIGURE A

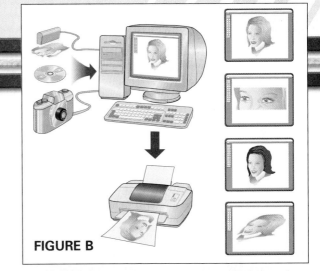

FIGURE B

Design Procedure

Film Processing—Figure A

❶ Set up the trays and other equipment, following your teacher's instructions.

❷ Load the film into a developing tank in total darkness.

❸ Develop your negatives according to the manufacturer's instructions.

❹ Place the film in the stop bath for the proper time.

❺ Place the film in the fixer for the proper time.

❻ Wash the film and dry it.

❼ Follow your teacher's instructions for making prints.

Digital Processing—Figure B

❶ Scan the photo needing repair and create a computer file for it. Following the instructions for the software, make repairs on the photo. Print copies of the original photo and the retouched photo. See Figure C.

❷ Download new photos into the computer. Select three for processing. Use the software tools to process the new photos. Remove red eye, crop unwanted areas, and fix backgrounds. Print copies of the original photos and the retouched versions.

❸ Select a series of photos that you can "morph" together. Place the photos in the desired order and use the morphing tools to make your changes. Show the results.

FIGURE C

Evaluating the Results

1. What part of these processes did you find most interesting?

2. Describe any problems that you experienced and the way you solved them.

3. If you were asked to repeat this activity, what would you do differently?

Multimedia technologies involve the use of more than one medium in order to communicate. (*Multi* means many. *Media* is plural for medium.) The word *media* has two meanings. It can refer to the forms into which messages are put, such as pictures, or the technologies used to create the messages, such as photography or video. The use of multimedia is not new. An artist who adds printed words to a drawing has used multimedia. However, the development of computers has made combining media easier and more interesting.

Video games and computer games are good examples of the use of multimedia. They include still and moving pictures, music and other sounds, and printed text. Today, the use of multimedia technologies enhances movies and television programs, makes advertising more eye catching, and makes learning more interesting.

In earlier chapters, you learned about graphic communication and photography. In this chapter, you will learn about audio and video technologies and how they are combined with other technologies to create multimedia productions, including Internet Web pages.

This television-broadcasting technician is working in the control room of a TV station.

Objectives

▸▸ **Explain** how radio and TV signals are transmitted.

▸▸ **Describe** how tapes and discs are recorded.

▸▸ **Discuss** the use of computers for animation.

Terms to Learn

- animation
- audio
- digital compression
- pixel
- video

Standards

- Core Concepts of Technology
- Relationships & Connections
- Cultural, Social, Economic & Political Effects
- Information & Communication Technologies

audio
the recording and reproduction of sound

video
the electronic recording and reproduction of moving images

Audio and Video Technologies

⊙ **What is the difference between audio and video technologies?**

The term **audio** refers to sound. Your radio, MP3 player, and CD player are all audio devices. **Video** refers to electronic images, or pictures. Your TV, VCR, and DVD player are video devices that have built-in audio devices. Both audio and video technologies are used for multimedia communication.

Radio

▷ **How does a radio work?**

To transmit a voice signal by radio, a microphone at the radio station converts speech into a variable electrical signal. This type of microphone works just like the one in your telephone.

The variable signal is then attached to a radio carrier wave. You can compare the joining of these two signals to a class trip on a bus. The voice signals are like passengers on the bus. The bus is the carrier wave. Just as the bus carries you to your destination, the carrier wave carries the signal to its destination (the radio).

When you tune your radio, you are adjusting which radio carrier wave will make it to the circuits of your radio. The chosen frequency creates a weak but exact duplicate of the signal that left the radio station. This signal is amplified (strengthened). The signal is then sent to the speaker of your radio. See Figure 13-1.

Figure 13-1 Radio signals are sent from this transmission tower.

Television

▶ How does a television work?

In your television, the picture transmission and sound transmission are handled separately. The picture and sound come to your home through completely different and separate systems that function together.

The audio system includes the television station's microphones and transmitter and your home television set's built-in radio. The video system includes the television station's cameras and transmitter and your home television set's tuner, video decoder, and picture tube. If your television uses an antenna, the signal reaches it by means of a carrier wave. If you have cable, it reaches your television through wire or fiber optic cables. Your television picture tube displays the electronic signal.

The TV image consists of tiny dots of light called **pixels**. (Remember, pixels also make up the image from a digital camera.) You might be surprised to learn that TV images are really a series of still pictures. These images are replaced thirty times each second. This happens so fast that your eyes and brain are fooled into thinking that you see a continuously moving image.

In digital HDTV (high-definition television), the camera's analog image is converted into a compressed digital signal. This signal is then transmitted to your TV receiver. The computer in your HDTV converts the signal into moving images.

The main difference between the two systems is the use of a digital signal. Digital images are sharper and more detailed than ordinary TV images. See Figure 13-2.

> **pixel**
> one of many tiny dots of light used to create a video image; also called a *picture element*

Figure 13-2 The digital images on HDTV are much clearer and more detailed than on ordinary television.

Audio and Video Recording

▶ How are audio and video recorded?

All audio and video recording systems have three things in common. 1) They need to convert sound or images into a recordable signal. 2) These signals must be stored on some type of material. 3) They must have a way of converting the stored information back into sound and/or images in a playback unit. Both audio and video can be converted into analog or digital signals. Different materials are used for storage.

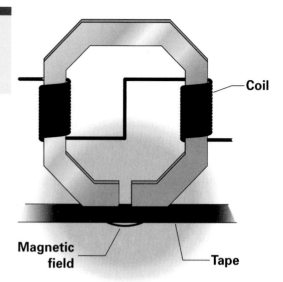

Figure 13-3 The recording tape is magnetized as it passes through the magnetic field. This determines how an image or sound will be recorded.

Coil

Magnetic field

Tape

Tape Recorders

▷ How are sound and images reproduced on tape?

An audio tape recorder, video tape recorder, and non-digital video camera all use the same basic system to record and play back their recordings. When you press the record button, a variable electrical signal is sent to the coils of the recording head. See Figure 13-3. This signal creates a magnetic field in the coils. As the tape moves past the head, it becomes slightly, moderately, or fully magnetized. The sound or image is recorded onto the tape.

The only significant difference between audio and video recording is that video pictures contain more information that needs to be recorded. Videotape recording heads record at an angle on a tape that is much wider than audiotape because it must hold more data.

When you listen to the tape or watch the recorded video, the tape passes a playback head, and coils of wire "read" the magnetic field and turn it back into a weak, variable electrical signal. This signal is then strengthened and played back through the speakers or video screen of the playback unit.

Digital Disc Recorders

▷ What media are used for disc recording?

Digital recording systems use similar technology to store digital information on CDs or DVDs. A hard drive disk has magnetic tracks that spiral from the outer edge down to the center of the disk. The system first converts sound and images into a digital electrical signal. This signal is then compressed (squeezed) so it will take up less space. The "write" head of the driver converts the compressed digital signal into a magnetic pattern and records this pattern on the disk. As you learned in Chapter 10, the signal is recorded on CDs and DVDs using lasers instead of magnetism. During playback, another laser "reads" the signal, which is then sent to the speakers or video screen.

Animation

▷ How are individual drawings used to create animation?

Animation creates the illusion of movement. A series of slightly different drawings or models is filmed. When played back, the images move past your eyes at twenty-four frames (separate images) per second. See Figure 13-4. This is too fast for your eyes to see the switch from one to the next. Instead, you see what appears to be movement.

Have you ever received an e-mail that contained an animated picture? If the picture moved at no less than twenty-four frames per second, you saw smooth movement. If not, the motion was choppy.

▶ **animation**
creating a series of slightly varying drawings or models so that they appear to move and change when the sequence is shown

TECH CONNECT
MATHEMATICS

Too Fast to Count
Animation moves past your eyes at twenty-four frames per second. Each frame is a separate drawing.

ACTIVITY If you were to draw all the individual frames needed for a 9½-minute animated feature, how many drawings would you have to draw?

Figure 13-4 For animation, still drawings pass rapidly in front of your eyes and give the sensation of motion. What happens if the drawings are shown slower than twenty-four frames per second?

For computer-drawn animation, the beginning and ending images in a sequence are drawn and then saved to the animation program's memory. The computer then fills in the images in between.

When you download an animation from the Internet, it takes a long time because all the slightly changed images must be transferred to your computer. Animation files are very large.

Animation that appears three-dimensional is created using complex mathematical formulas. The process starts with a drawing that combines various shapes, lines, and arcs to create solid surfaces. This drawing is then rendered, which gives it depth and produces reflections. See Figure 13-5. Many movies today use animated characters as well as real actors. Feature-length films with only animated characters can contain over one hundred thousand separate digital images. Without the aid of computers, those films would take years to draw.

Video and Computer Games

▶ How is animation used in video games?

In video and computer games, the action is always changing. It is you against the computer or another player. The graphics of the game change with each action sequence. The animation consists of many possibilities that are put together, as you play, to create a "story."

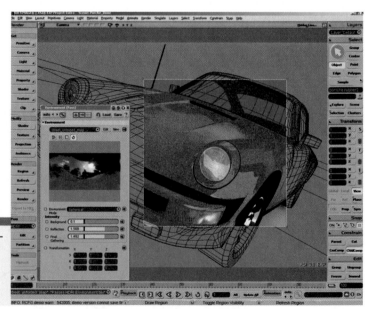

Figure 13-5 By rendering animation, objects like this car appear more realistic with added depth and reflections.

Figure 13-6 Joysticks make input for video game action much faster and easier.

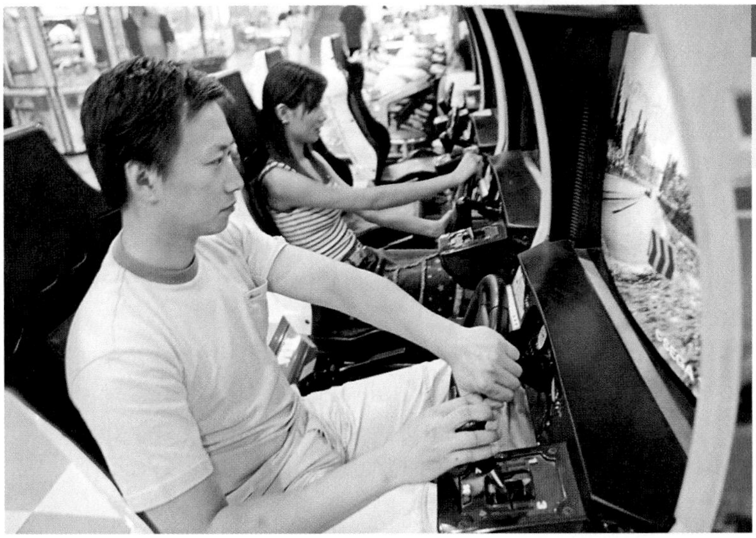

A joystick is often used to move objects around the screen. See Figure 13-6. The joystick sends a varying electrical signal to the computer. The signal transmits your location.

Presentations

What do multimedia presentations often include?

Teachers and speakers at meetings sometimes create multimedia presentations for their audience. The presentations often include a series of slides combined with text, graphics, animation, and sound. The use of multimedia makes the information more interesting and helps the audience remember it better.

Presentation software, such as PowerPoint and HyperStudio, enables you to use a computer to organize a presentation. Text, graphics, sound, and other media are easy to format and arrange. Some of this software can also be used to create tutorials, which are teaching programs. The student moves at his or her own pace through the material, one slide at a time.

Digital Compression

What is digital compression?

All the thousands of images in a feature-length animation contain too much data to store conveniently. **Digital compression** reduces the physical size of a digital file by removing bits of information that a computer can recreate

digital compression reducing the size of a digital file by removing bits of data that can be recreated later

when the file is processed. Digital compression is also used for ordinary audio and video files. The most common digital compression for video is called MPEG. The audio compression for music is commonly known as MP3.

The computer is told to keep the part of the video picture that remains the same. Only new data is compressed. Audio compression drops out sound values that are beyond human hearing or that are often drowned out by other sounds.

When MP3 was first released, it could shrink a music file to ten percent of its original size. The latest version can shrink a file to ten percent of that amount. Large files can be stored in less space and transmitted more easily.

13A SECTION REVIEW

Recall ▸▸

1. What form of transmission is used for radio signals?
2. Why is a wider tape required for video recording than audio recording?
3. How is HDTV different from ordinary TV?
4. How many frames per second are needed to see smooth movement in an animation?

Think ▸▸

5. Do you think computers have reduced the need for skilled artists in making animated films? Why or why not?
6. If computer memory and computer storage are now fairly cheap, why do you think digital compression is still important?

Apply ▸▸

7. **Design** Create your own flip-pad animation. Obtain a small pad or notebook that contains at least fifty pages to draw on. On the first page, use a compass to draw a 1-inch circle. On the next page, draw a $1\frac{3}{8}$-inch circle. Repeat this process, slowly increasing the circle diameter by eighths of an inch, to the limits of your compass. Then reverse the process to decrease the circle size. Then, once again, increase the size of the circle until the pad runs out of paper. Use magic markers to color the circles. Flip the pages and watch the circles change.

Multimedia Applications

▶ In what ways can multimedia be used to make videos and Web pages?

As you know, multimedia productions combine several different media, such as graphics, photography, audio, and video. In this section you will learn how they can be combined into a video production and a Web page.

Producing a Video

▷ How is a video produced?

Although this section is about producing a video, feature films, TV programs, and commercials are usually made using the same basic processes. Except for camera work, the same processes are also used to produce radio programs.

The **producer** is usually in charge of the entire project and oversees the planning. He or she hires the workers and manages the money being spent. Principal workers include the writer, the director, camera operators, performers, and the editor. In a small production, one person may do several jobs. Based on when they are done, these jobs can be grouped into three main categories: pre-production, production, and post-production.

Pre-Production

▷ What tasks are done during pre-production?

Pre-production includes all those jobs that must be done before the cameras start rolling. If you think about the many movies you have seen, you will probably recognize that big-name actors and special effects can't save a boring story. The **script** is usually the foundation on which the entire production is built. It lists the different characters and any lines of dialogue they will speak. See Figure 13-7 (page 318). Scripts also indicate the action that takes place during each scene and what the scene should look like.

producer
the person responsible for an entire production

script
the written version of a production that contains a list of characters, their dialogue, and descriptions of action and sets

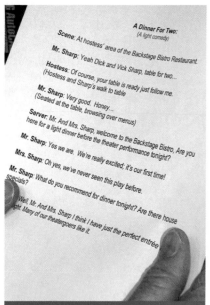

Figure 13-7 The script has all the information actors need for their parts. It includes dialogue and any action.

director
the person in charge of instructing the performers and guiding the camera work

Some scripts are done in storyboard form. A storyboard contains sketches of what the scenes will look like from the camera's point of view. They show the camera operators how to shoot each scene.

The **director** is in charge of actual taping of the video. He or she instructs the performers and guides the camera work. The director decides how each scene will be carried out and how it will look. During pre-production, the director is in charge of rehearsals, which give the performers a chance to practice and to learn their lines.

Many other tasks are also accomplished during pre-production. Stage sets and backgrounds may have to be created. Costumes and props, such as furniture, may be needed. Graphics, such as signs, may have to be drawn. Cameras, lights, and microphones have to be set up. See Figure 13-8. When everything is ready, production can begin.

Production

▷ **How is production different from pre-production?**

Not all the scenes in a video are taped in the same order in which you see them. Some scenes are taped in a studio. Some videos may include scenes done on location—the real setting of the story. For a video about Chicago, for example, the scenes done on location would actually be

Figure 13-8 This street scene is being taped on location. What do you think would be the advantage of shooting on location?

Self-Esteem

When you are not afraid to express your ideas, you are demonstrating that you have good self-esteem.

Having good self-esteem means that you respect yourself and your own abilities. It means that although you may have much to learn, you are confident and willing to try new things. Good self-esteem is often part of a positive attitude. Here are some ways to help develop your own self-esteem:

✔ **Accept the fact that you will make mistakes and remember that everyone makes them. Learn from them.**
✔ **Focus on your abilities and successes.**
✔ **Set goals that are reachable and then work to achieve them.**
✔ **Set aside time to do things for others. It will make you feel good about yourself.**

Scriptwriter

Advertising agency is looking for a scriptwriter who loves to write, is motivated, and has good self-esteem. You will write scripts and narrations and help develop commercials for our clients. A good understanding of technical aspects of TV and radio production would be helpful.

---

shot in that city. All scenes shot in a particular location would be done together. All those shot in a studio would also be done together.

The director may ask for a scene to be done over until it's the way he or she wants it to be. Placement of cameras and lights may be changed. Last-minute changes may also be made in the script.

ETHICS in ACTION — Asking Permission

For legal reasons, before a company can use your photograph for one of its products or advertisements, you must give your written permission. This is also becoming important for individuals who are taking pictures in public places. People have a right to privacy. Before you include someone in a video, ask his or her permission. Do not include something in a video that might embarrass the person.

ACTIVITY Create a form you could ask people to sign giving you permission to use their photographs for a school project.

Figure 13-9 The evening news broadcast is typically taped in a television studio similar to the one shown here.

During some TV shows, such as news broadcasts, the director sits in a control room along with audio and video engineers who supervise signals coming from cameras and microphones. The director can see what each camera sees by watching that camera's monitor, or view screen. What do you think the cameraman sees in his monitor in Figure 13-9?

Post-Production

▶ What work takes place during post-production?

editing
cutting and arranging material in order to decide its final sequence and content

When taping is finished, post-production begins. This is the final phase. The most important post-production task is editing. **Editing** is the cutting and arranging of the taped material to decide its final order and content. See Figure 13-10. Careful editing can often make the difference between a successful production and a confusing one.

The editor copies scenes or parts of scenes in the proper order onto a master tape. Special graphics, music, animation, and other elements may be added at the same time.

The title and credits are also included. When the editor is finished and the director is satisfied, the project is complete.

Would you like to create your own video? It is the creativity of the filmmakers, not the cost of the equipment, that determines how good a video is. One or two video cameras, two VCRs, and a computer are all that is needed. Using the computer's output jacks, you can transfer your computer work to the VCR and add that material to your tape.

Also, special computer hardware and software can turn a computer into a desktop video studio. These systems allow you to transfer video and audio from any source directly into your computer. All of your editing can also be done directly on your computer. When the project is completed, you can burn it directly to a CD or DVD or output it to a VCR to record it on tape.

Figure 13-10 Often, scenes in a video are not recorded in the same sequence that they will be shown. The editing room is where the video is arranged into its final form.

Developing an Internet Web Page

▷ **What is the Internet?**

The Internet is a vast network of many computers. The Internet is constantly changing. Everyday, some sites are removed from the network; others are added. When you explore the Internet by clicking on a link or typing in an address, you move from computer to computer until you reach the site you are looking for.

You will find many examples of multimedia technology on the Internet. Audio, video, text, graphics, photography, and animation all come together in an interactive format. (The term *interactive* means that you can participate in what goes on at many sites.)

The World Wide Web

▷ **What is the World Wide Web?**

The World Wide Web is only one part of the Internet. It is a huge set of files linked in such a way as to make information easy to locate. That's why so many people and organizations use it. The code language used for the Web is called **HTML**, which stands for hypertext markup language. Each

HTML
hypertext markup language; the code in which information on the World Wide Web is written

LOOK TO THE FUTURE

Non-Stop Internet

Researchers predict that by 2020, the Internet will be so common we won't notice it anymore. You'll even be able to turn on your dishwasher and other appliances over the Internet. In fact, more appliances, buildings, and vehicles will be online than people. Most Internet access will be wireless, and you'll be able to go online from many different places, as long as you're within range of a wireless signal. Enormous amounts of information will be available to everyone. Will there be a downside? Much privacy will be lost, and some information may be misused. Can you think of other impacts?

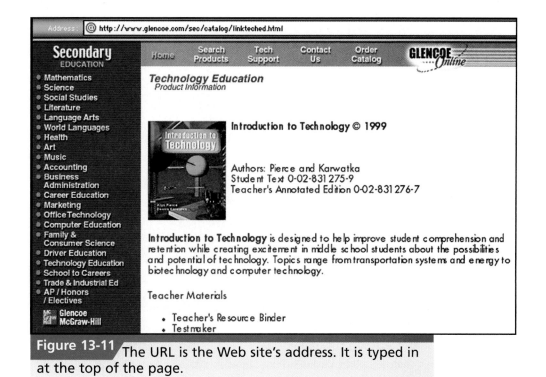

Figure 13-11 The URL is the Web site's address. It is typed in at the top of the page.

site on the Web may have many "pages" like pages in a magazine. Each Web page has its own address, called a **URL**, or uniform resource locator. See Figure 13-11.

An Internet service provider (ISP), such as America Online, gives you access to the Internet. A Web **browser**, such as Internet Explorer or Netscape, is a software program that allows you to view pages on the Web. Other programs, such as Google or All the Web, are called **search engines**. They help you find information quickly.

Web Page Production

▷ How are Web pages created?

Web sites of large companies or other organizations are usually created by professional Web page designers. These designers are skilled in combining different media to produce an exciting result. If a particular site involves complex features, they may even write a special software program for it. However, you don't have to be a professional designer to create your own Web page. See Figure 13-12 (page 324). You can use a software program that will

URL
uniform resource locator; an address on the World Wide Web

browser
software program that provides access to the World Wide Web

search engine
software program that helps users quickly find information on the Internet

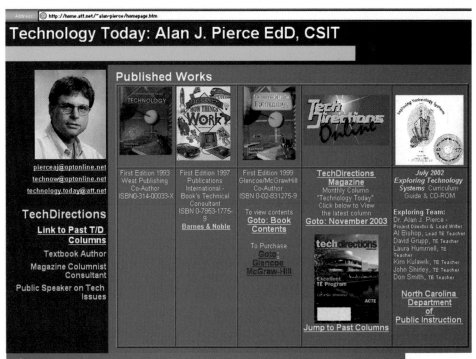

Figure 13-12 With the right software, Web pages are fairly easy to create.

automatically convert your work into HTML. Most building programs also have a preview button that allows you to see what your page will look like when it is changed to HTML.

Cute Site Builder, which is very user friendly, and *Netscape Communicator*, which is very powerful, are examples of Web site building programs. The software that you use will determine how you add text, graphics, audio, animation, video clips, and links to your page. Just follow their directions. The elements that you use can be of your own design or copied from sites that make them freely available. Caution is needed before you copy material that was designed by someone else. Copyright laws protect the material on the Web. You must assume that anything you find there is copyrighted. However, many sites do make their graphics, templates, and animations available for anyone's use. It is important that you respect copyright laws.

You can practice creating a Web site on a computer that isn't connected to the Internet. However, to place your page on the World Wide Web, you need an Internet connection. Software allowing you to insert the page is probably built into the program that you use. The program will determine how you upload your page to the Web.

After you upload your page, it will receive its own URL. If you enter this URL into a browser on any computer that has Internet access, your own page will appear.

Many schools now provide space for students to upload their Web pages. Most Internet providers have Web page space set aside for their customers.

You can design a page to share information with relatives, friends, or even strangers who share your interests. You can use text, audio, graphics, animation, links to other sites, e-mail links, and even video clips to create a page other people will enjoy.

SECTION REVIEW 13B

Recall ▸▸

1. What provides the dialogue in written form for a video production?
2. Who is the person in charge of an entire video production?
3. What is editing?
4. What does HTML stand for and what is it used for?
5. Why does a Web site need a URL?

Think ▸▸

6. Why do you think it would be important to know and understand the intended audience for a multimedia presentation?

7. If you were the director of a video, would you want a storyboard created? Why or why not?

Apply ▸▸

8. **Write** Develop a script for a one-minute radio commercial that uses two performers, music, and at least two sound effects. Indicate on the script where the music and sound effects will be inserted.
9. **Produce** Develop a video and enter it in the Technology Student Association's video challenge. Ask your instructor to check the *TSA Curricular Resources Guide* for rules and procedures.

Summary Activity

For each numbered blank, pick the answer from the list on the right that makes the most sense in the entire passage. Write your answer on a separate sheet of paper. No answer will be used more than once.

Multimedia technologies involve the use of more than one medium in order to __1__. The development of __2__ has made combining media easier and more interesting.

The term *audio* refers to __3__. Your radio, CDs, and CD player are all audio devices. __4__ refers to images shown with devices such as a TV, VCR, or DVD player.

To transmit a voice signal by radio, a microphone at the radio station converts speech into a variable __5__. If your television uses an antenna, the signal reaches it by means of a(n) __6__. The TV image consists of tiny dots of light called __7__.

__8__ creates the illusion of movement. A series of slightly different drawings or models is filmed. __9__ reduces the physical size of an animation file.

Videos, feature films, TV programs, and commercials are usually made using the same basic processes. __10__ includes all those jobs that must be done before the cameras start rolling. During __11__, not all the scenes in a video are taped in the same order in which you see them. Some scenes are taped in a studio, some on location. The most important post-production task is __12__, the cutting and arranging of the taped material in order to decide its final order and content.

You will find many examples of multimedia technology on the Internet. Audio, video, text, graphics, photography, and animation all come together in an interactive format. The __13__ is one part of the Internet. The code language used for the Web is called __14__, which stands for hypertext markup language.

Professional Web page designers are skilled in combining different __15__ to produce an exciting result. You can use a software program that will automatically convert your work into HTML. After you upload your page, it will receive its own __16__, or address.

Answer List

- animation
- carrier wave
- communicate
- computers
- digital compression
- editing
- electrical signal
- HTML
- media
- pixels
- pre-production
- production
- sound
- URL
- video
- World Wide Web

Comprehension Check

1. How are TV signals transmitted?

2. How does drawing animation with a computer differ from doing it by hand?

3. Name the three stages of producing a video.

4. What is a browser?

5. What is a search engine?

Critical Thinking

1. Propose. In what ways could multimedia be used to create more interest in one of your school's sports teams?

2. Appraise. What impacts do you think the use of multimedia has had on the effectiveness of TV commercials? Do you think the impacts are positive or negative?

3. Connect. In which ways are language arts and mathematics used to make an animated film?

4. Classify. Visit three teacher-approved Web sites. Identify the media used.

5. Decide. What impacts has the Internet had on your life? Write a short paragraph describing those impacts.

Visual Engineering

Could You Be a Screenwriter? Before a video or movie is produced, there must first be a story or message to tell. That is the job of a screenwriter. Think of a story or message that interests you and that you think would make a good movie or video. For example, you might develop an idea for a home improvement TV show in which you show people how to paint bookshelves. Write a three-page script for a part of the movie or video and be sure to include what the performers should be doing.

TECHNOLOGY CHALLENGE ACTIVITY

Creating a Web Page Using Multimedia

Equipment and Materials

- computer
- Web page design software

Optional
- drawing materials or software
- video recording and editing equipment
- photography equipment
- props and/or costumes

Background

Sites on the Internet can have many purposes. Commercial sites wish to sell visitors their products. Government sites provide information about government policies and services. School sites often aim to educate. The most effective sites are usually designed with their audience in mind.

Goal

For this activity, you and the members of your team will create a Web page using at least three different media. The purpose of your page will be to inform. The theme of your page will be exploring technology.

Criteria and Constraints

- ▸▸ You must use at least three of the following media to create your Web page: printed text, photographs, drawings or graphics, animation, audio (speech, music, or special effects), and video.
- ▸▸ The purpose of your page will be to inform.
- ▸▸ Although you may choose your own topic(s) and approach, the theme of your page will be exploring technology.
- ▸▸ Your page will contain links to other sites, including those developed by other teams in your class.

1. Brainstorm with your team to select one or more topics for your Web page and to decide on the approach you will take. Try to choose a subject that is of manageable size. For example, your page might be titled "Exploring Technology: Cars for the 21st Century." Then you might show photos of three new car designs and links to manufacturers' sites. You could inform your visitors about what makes the new cars special.

2. Do some research on several topics before selecting those you will use. Consider what information and resources will be available to you.

3. Decide which media will present your topics best and who your audience will be.

4. Divide the work so that everyone on the team is responsible for some part of the project. Tasks will include designing the appearance of the page, finding links to use, creating any audio or video sequences, taking any photographs, or preparing any artwork.

5. Follow the instructions in your Web page design software to prepare your page.

6. Before you install your Web page on the Internet, obtain your teacher's approval. Then show it to your classmates.

Evaluating the Results

1. What was the easiest part of creating the Web page? What was the most difficult? Why?

2. If you could add something more to your page, what would it be and why?

3. What other technology subjects might make interesting Web pages?

Genetics
Breaking the Code

Genetics is the science of heredity—how traits are passed on from generation to generation.

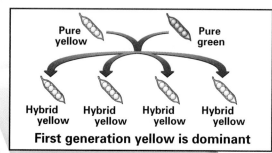

1

Pure yellow

Pure green

Hybrid yellow | Hybrid yellow | Hybrid yellow | Hybrid yellow

First generation yellow is dominant

Hybrid yellow

Hybrid yellow

Second generation both colors occur

Genetics began with Gregor Mendel, an Austrian monk, who began experimenting with heredity **in 1857**. Mendel was interested in what happened when plants with one set of characteristics were crossed with plants having another set of characteristics. He theorized that cells from the parent plants contained certain elements that influenced heredity. Today, we call these elements genes.

In 1907, Thomas Hunt Morgan used fruit flies for his gene experiments because they are fairly simple creatures and they reproduce quickly. He published research confirming the fact that each gene is responsible for a particular characteristic.
Investigate: *How many chromosomes do humans have in their cells and how many come from each parent?*

2

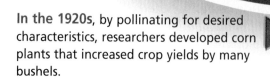

In the 1920s, by pollinating for desired characteristics, researchers developed corn plants that increased crop yields by many bushels.

3

Then, in 1953, genetic researchers made headlines when they worked out the structure of DNA (deoxyribonucleic acid). DNA is a long chain of four molecules within each gene that is twisted into a coil resembling a double helix. The DNA molecules are the actual carriers of genetic information.

▼4 How the molecules are arranged and linked to one another creates the code that determines the type of living organism.

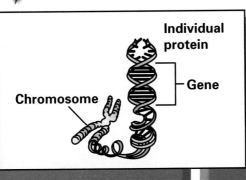

Individual protein

Gene

Chromosome

▶5 The first free-living organism's DNA to be successfully analyzed in 1995 was a bacterium, *Haemophilus influenzae*. Other results soon followed. In the United States, the Human Genome Project is an effort to analyze the DNA of human beings. Much information still remains to be discovered about DNA. Investigate: *Search on the Internet for the most recent work being done on the Human Genome Project.*

▶6 The existence of genetic codes among the great majority of organisms suggests that they were formed more than three billion years ago. Some researchers believe that all humans can be traced back to an original parent they have nicknamed "Eve." Scientists hope the secrets hidden in the genetic code will help them cure diseases.

Medical technology is a biotechnology. Biotechnologies use knowledge about living things to meet human needs. Like other technologies, medical technologies can be studied as systems. They have inputs, processes, outputs, and feedback. For example, a company that makes aspirin requires resources, such as chemicals (materials), people, and machines. It uses processes to mix the chemicals and fill the bottles. Its outputs include plain aspirin and aspirin combined with other ingredients, such as caffeine. Feedback from customers may indicate that a particular combination is not popular. The company may decide to change it.

Medical technologies have been around a long time. Ancient Egyptian doctors drilled into people's skulls in order to relieve headaches. As medical knowledge improved, so did advances and innovations in medical technologies.

General health care technologies will be discussed in this chapter along with medicine. They can be classified into three main groups: disease prevention, diagnosis, and treatment.

◀ In the future, tiny submarines like this one will be placed inside our bodies to make repairs and fight disease.

Terms to Learn

- ergonomics
- immunization
- irradiation
- pasteurization
- pathogen
- sanitation
- vaccine

Standards

- Characteristics & Scope of Technology
- Relationships & Connections
- Cultural, Social, Economic & Political Effects
- Environmental Effects
- Role of Society
- Attributes of Design
- Design Process
- Medical Technologies
- Agricultural & Related Biotechnologies

pathogen
organism that causes disease

pasteurization
heat treatment used to destroy pathogens in food

Disease Prevention
▷ How can disease prevention improve health?

If people don't get sick in the first place, they don't need medicines and other treatment. That is the goal of disease prevention. Prevention involves many technologies. Some seek to remove or kill **pathogens**, organisms that can cause disease. Pathogens include bacteria, viruses, and fungi. Other technologies help strengthen the body so it can better resist illness.

Pasteurization and Irradiation

▷ How are pathogens destroyed in foods?

Pasteurization and irradiation kill pathogens in foods. **Pasteurization**, a heating process, is used to kill bacteria that can turn milk sour and make you sick. In 1864, Louis Pasteur, a French chemist, developed the process, which is named after him. Pasteurization is still used today.

Figure 14-1 These astronauts and Russian cosmonauts are eating a meal on board the International Space Station.

Since 1963, a variety of foods have been approved for **irradiation**. During this process, the food is briefly exposed to a source of radiation, such as X rays or an electron beam. This cooks the food slightly, kills parasites, insects, and bacteria, and controls molds. Astronauts on the International Space Station eat irradiated foods. See Figure 14-1. However, the general public has been slow in accepting this technology. Some people are concerned it might make the food radioactive or change its nutritional value.

> **irradiation**
> treating products with radiation to destroy pathogens

Sterilization

▶ When did cleanliness become important to health care?

Until the late 1800s, doctors performed surgery in their street clothes without washing their instruments or their hands. Hospitals were seldom clean. Pathogens thrived, and many patients died from infections. Joseph Lister, an English surgeon, discovered that infections could be prevented with disinfectants—chemicals that killed pathogens. The disinfectants that he developed saved many lives.

Today, sterile (very clean) methods are required for many medical procedures. Surgical instruments are put in an autoclave, which uses steam and pressure to kill pathogens. Doctors scrub their hands before surgery and wear special garments. See Figure 14–2. People also understand the importance of everyday cleanliness.

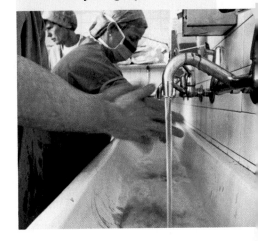
Figure 14-2 Doctors must be as germ free as possible before any surgery.

Water Treatment

▶ How is water purified?

Clean drinking water is important to health. Local governments constantly check the water supply for pathogens, dirt, and other contaminants. A community's water supply is purified before it is allowed to pass through the distribution system.

Depending on the quality of the untreated water and local regulations, the water may be given several treatments. First, chemicals are added to the water that stick to bacteria, mud, and other impurities. Clumps of impurities sink to the bottom. The water is then pumped through a combination of sand, gravel, and other minerals and the impurities are filtered out. If the water is still not clean enough, disinfectants may be added.

Some communities add fluoride to the water to help prevent tooth decay. Fluoride is a chemical compound that hardens tooth enamel and improves dental health.

Sanitation

▷ What is sanitation?

sanitation
removal of waste products or contaminants that could cause disease

When a place is sanitary, it is clean. **Sanitation** usually involves the removal of waste products that could cause disease or contaminate the environment. Our sewerage systems dispose of wastes from homes, factories, businesses, and public streets.

Sewage is mostly water and goes through several treatment stages. The water is removed, purified, and released back into the environment. See Figure 14-3. The solids are treated to make them safe to transport to landfills or to be burned.

Hazardous (dangerous) wastes from medical facilities require special treatment to protect people from harmful organisms and disease. If possible, the waste is treated to make it harmless. Otherwise it is burned or buried. Proper disposal of medical products contributes to medical safety.

Figure 14-3 This aerial view shows a sewage treatment plant.

Figure 14-4 Is this how you felt the last time you were immunized?

Immunization

▶ How do vaccines prevent disease?

Immunization helps the body fight disease by causing the immune system to produce cells that attack certain pathogens. By making people immune, it also prevents diseases from spreading.

Before modern medicine, people recognized that if you survived certain illnesses you often didn't catch the same disease again. Edward Jenner, an English doctor, noticed that people who caught cowpox, a mild disease, seemed to be safe from smallpox, an often fatal illness. In 1796, using killed or weakened cowpox organisms, he created a vaccine for smallpox. **Vaccines** do not cause disease, but they stimulate the body's immune system to recognize and attack the pathogen. See Figure 14-4.

Pharmaceutical (far-mah-SOO-tik-uhl) companies use special technologies to grow the organisms used to make vaccines. Environments are created that support speedy growth in order to produce enough vaccine for all who may need it. The organisms are then treated and purified.

The latest vaccines that are being developed use only part of a pathogen cell rather than the entire organism. This is much safer and protects the immunized person more effectively.

immunization
process for making the body resistant to disease, usually by vaccination

vaccine
preparation containing dead or weakened pathogens used to stimulate the immune system

Food Additives

▷ **How does good nutrition help prevent disease?**

Not all diseases are caused by pathogens. Some, like heart disease, may be related to unhealthy habits, such as smoking, eating diets high in certain types of fat, or lack of exercise. Proper nutrition helps keep the body strong and able to defeat or survive pathogens and other causes of disease.

Many food manufacturers add vitamins and other nutrients to their products to ensure that people receive adequate amounts. Major nutrients contained in processed foods are listed on food labels. See Figure 14-5. Manufacturers also add preservatives to foods. Preservatives are chemicals that prevent spoilage.

Figure 14-5 Many food products are now labeled with nutrition facts. Why do you think this information is important?

Ergonomics

▷ **Why is comfort important to health?**

Ergonomics, or human factors engineering, is the design of equipment and environments to promote human safety, health, and well-being. The people who work in this field design products around people's limitations and comfort needs. See Figure 14-6. Have you ever

ergonomics
the design of equipment and environments to promote human safety, health, and well-being

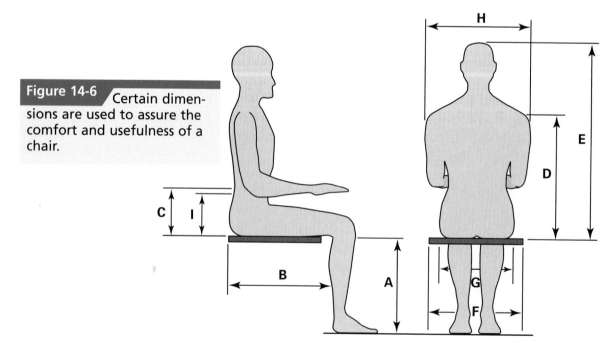

Figure 14-6 Certain dimensions are used to assure the comfort and usefulness of a chair.

seen certain types of kitchen utensils with handles specially designed for easy gripping? They are examples of ergonomic products. The special handles allow the user to have better control over the utensil. Because they are comfortable to use, they are safer and can prevent injury. This is especially important in the workplace when people may do tasks for long hours, such as typing at a keyboard. Repetitive stress on certain muscles and joints can result in painful injuries.

Ergonomic engineers also develop such products as car instrument panels, comfortable seating for offices, safety devices, protective clothing, and people-friendly machines and appliances. They also help create the life-sustaining environments needed by astronauts, deep-sea divers, and airplane travelers.

SECTION REVIEW 14A

Recall ▸▸

1. What is a pathogen?
2. What is the purpose of pasteurization and irradiation? What is the difference between them?
3. How does immunization work?
4. How does ergonomics support human health?

Think ▸▸

5. Have you ever been camping? How do campers ensure a supply of clean water?

6. Think of at least two products you have used that were designed with safety and health in mind. Explain why you think this was part of their design.

Apply ▸▸

7. **Design** Select an item that you use frequently, such as a utensil or chair. Sketch ways in which it could be redesigned to improve it ergonomically.

Objectives

▸▸ **Discuss** genetic testing.

▸▸ **Describe** several imaging technologies.

▸▸ **Identify** tests used to read electrical impulses created by the body.

Terms to Learn

- CT scan
- endoscope
- genetic testing
- MRI
- ultrasound

Standards

- Characteristics & Scope of Technology
- Core Concepts of Technology
- Relationships & Connections
- Cultural, Social, Economic & Political Effects
- Design Process
- Medical Technologies

Diagnosis of Disease

▶ **How do doctors diagnose an illness?**

Doctors diagnose (identify) an illness by examining the patient and conducting tests. The doctor's experience is especially important. However, technology has also been very helpful.

Has your doctor ever listened to your heart and lungs with a stethoscope? See Figure 14-7. A diaphragm in this tool picks up the sound vibrations from your chest and transmits them to the earpieces. Since the invention of the stethoscope in 1816, many tools and machines have been developed to aid doctors in diagnosing their patients. This section will discuss a few of them.

Laboratory Tests

▶ **How does the microscope help diagnose illness?**

Microscopes make it possible to see pathogens and the cells that make up our bodies. Optical microscopes use lenses to magnify objects up to 2,000 times. Electron microscopes use a beam of electrons to magnify the objects.

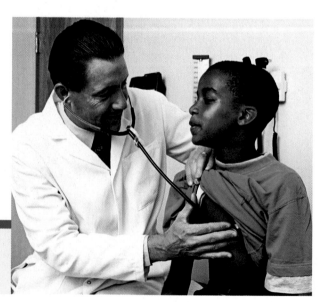

Figure 14-7 Doctors use a stethoscope to listen to your heart and lungs.

Using microscopes, doctors and technicians can examine body fluids for pathogens and other signs of disease, such as a reduced supply of red blood cells. Along with other equipment, the microscope makes laboratory testing especially important to diagnoses.

In recent years, **genetic testing** has become more interesting to doctors. In genetic testing, the DNA (deoxyribonucleic acid) within a person's genes is examined. DNA is the carrier of heredity. See Figure 14-8. Through testing it is possible to discover if the person is at risk for a particular disease. If so, he or she might be able to make certain lifestyle changes that would reduce the risk.

Imaging Body Structures

▷ **How do imaging machines work?**

Imaging machines show doctors the inside of your body. The first imaging device was the X-ray machine. Fast-moving electrons give off X-ray radiation, which

genetic testing
evaluation of a person's genes to discover if the person is at risk for a particular disease

---

ETHICS in ACTION   Disease Discrimination

DNA testing may soon make it possible to determine a person's risk for disease. What if employers obtain such information and decide not to hire a person because he or she might get the disease? What if insurance companies refuse to insure the person because treatment for that disease will cost them too much money? So far, rules have not been established for the use of genetic information. Who has a right to it and who doesn't? Can you think of other abuses that might occur?

ACTIVITY Hold a panel discussion on the ethics of genetic testing.

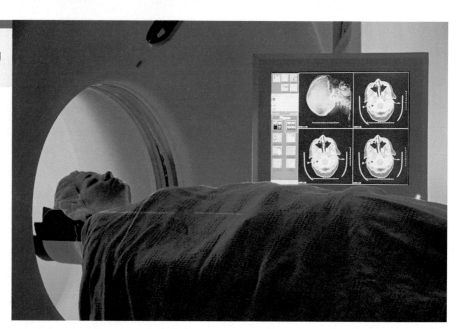

Figure 14-9 This patient is undergoing a CT scan.

CAREERS

Emergency Medical Technician (EMT)

Licensed and state-certified EMT wanted for position with ambulance service. You will be working closely with police and fire departments and using the most advanced medical equipment. Position requires someone who works well under pressure and who has good decision-making skills.

Decision Making

Careful decision making is important to making a successful life for yourself.

You have been making decisions since you were very young. Some decisions, such as what to buy for lunch, are simple. Others, such as what career to choose, are more complicated. Some decisions can have a lasting effect on your life and the lives of other people. If possible, make important decisions only after you've considered them carefully. Here are some guidelines to making good decisions:

✔ **Clearly identify what needs to be decided upon.**
✔ **Gather information about different alternatives.**
✔ **Ask for advice from people whose knowledge and understanding you respect.**
✔ **Consider as many consequences of each alternative as you can.**

passes through some parts of the body but not others, and exposes the X-ray film. This type of X ray is especially good for viewing bones.

To view places deep inside the body, a **CT scan**, also known as a "CAT" scan (computerized axial tomography), is often used. See Figure 14-9. The patient is X-rayed in order to take thousands of measurements of body structures. The information is then processed by a computer and an image is formed on the screen.

Ultrasound imaging uses sound waves too high for humans to hear. Structures inside the body reflect the sound waves back as an echo. A computer interprets the echoes to create images. Ultrasound is popular for determining whether or not a baby is developing properly inside the mother. See Figure 14-10. The newest form of ultrasound produces three-dimensional images.

In an **MRI** (magnetic resonance imaging), the patient is placed inside a magnetic field. See Figure 14-11 (page 344). As the field encounters internal body structures, a signal is generated and sent to a computer. The computer interprets the signal to create an image.

Sometimes to be sure of the diagnosis, doctors must examine the problem more directly. In these cases, an **endoscope** may be used. This small, flexible instrument

CT scan
computerized axial tomography; an X-ray image enhanced by a computer and shown on its screen

ultrasound
the use of sound waves to create an image of internal body structures on a computer screen

MRI
magnetic resonance imaging; the use of a magnetic field to create an image of body structures

endoscope
a small, flexible instrument, having a camera attached, that is inserted into a patient through an incision or by some other means

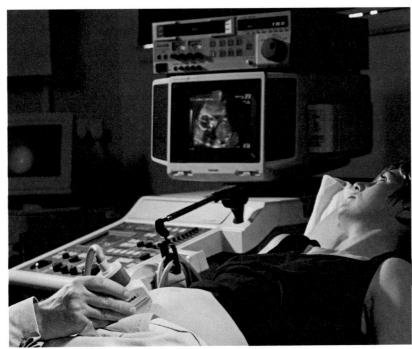

Figure 14-10 Unborn babies are scanned using ultrasound to check for any problems that might be present.

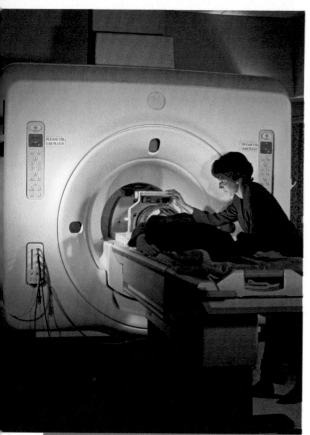

Figure 14-11 An MRI is another way for doctors to view the inside of our bodies.

has a camera attached. It may be inserted into the patient through a small incision or by some other means, such as through the mouth. The doctor views the image on a special screen. Endoscopes are commonly used to look inside a patient's stomach for ulcers and other problems. See Figure 14-12.

Other Screening Methods

▷ What other screening methods are commonly used?

Some machines can read electrical impulses created by the body. For an electrocardiogram (EKG), wires are attached to the patient's body, and a machine records the heart's electrical impulses. See Figure 14-13. This information is printed out on a paper chart. Doctors who read the chart can tell if the heart is working normally.

A similar machine, called an electroencephalograph (EEG), records the activity of the brain. By reading the machine's printout, doctors can tell if the brain is functioning properly.

Figure 14-12 This surgeon uses an endoscope to explore the inside of the patient's stomach for possible ulcers. What the camera sees is shown on the monitor.

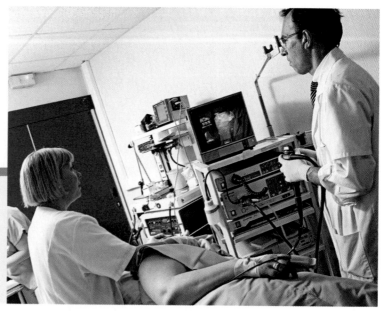

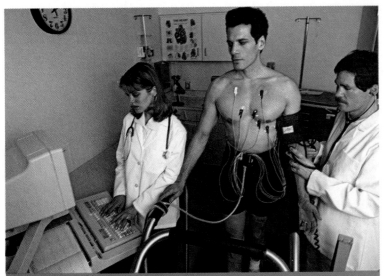

Figure 14-13 An electrocardiogram is a diagnostic tool for possible heart problems. Here, the patient's heart is being put under stress and its action observed.

SECTION REVIEW 14B

Recall ▸▸

1. Why is genetic testing done?
2. What is the difference between a CT scan and an MRI?
3. What is an endoscope?
4. What machines are used to read electrical impulses produced by the body?

Think ▸▸

5. Would you want to be genetically tested for a particular disease? Why or why not?

6. Which instrument do you think was more important to the development of modern medicine, the stethoscope or the microscope? Give reasons for your answer.

Apply ▸▸

7. Design The first stethoscope was just a perforated wooden cylinder. Design and make your own stethoscope and test it.

antibiotic
substance that kills bacteria

Treatment of Disease

▶ What technologies are available for treating disease?

Some diseases can be cured with the help of medicines and other treatments. For non-curable diseases, doctors try to relieve symptoms and make patients more comfortable.

Medicines

▶ How were effective medicines first developed?

As you know, bacteria, viruses, and fungi cause many diseases. In 1928, Alexander Fleming, an English physician, discovered that a substance he called penicillin killed certain bacteria. Penicillin was the first **antibiotic**, and it was effective against many life-threatening bacterial infections, such as pneumonia. In 1944, researchers developed another antibiotic, streptomycin, which was effective against other kinds of bacteria. Since then, many additional antibiotics have been created. However, because antibiotics have been overused, some strains of bacteria have grown resistant to them.

Although an antibiotic has been developed that can kill fungi, no antibiotic works against viruses. Some common viral infections include colds, flu, and AIDS. Researchers have developed some antiviral drugs that suppress the infection, and they are working to develop vaccines.

The causes of other diseases, such as some forms of cancer, have yet to be determined. Many other important medicines have been developed to treat such illnesses. Some, such as aspirin, relieve pain. Others promote health, such as those that regulate blood pressure, reduce the spread of cancer cells, or control allergies.

Surgical Procedures

▶ What new surgical procedures are being used?

Over 40 million surgeries are performed in the United States each year. Some doctors have begun using simulations to plan difficult surgeries. Using a computer and an MRI scan, the doctor enters suggested surgical changes, and the computer analyzes the effects they will have on the patient. The doctor can then decide if the plan is safe and potentially effective.

Some surgeries can now be done using very small incisions. An endoscope with a cutting tool attached is inserted into the incision and guided to the right location. The endoscope's camera allows the doctor to guide the cutting tool. Scars are smaller, and patients require less healing time. Endoscopes are also being used to operate on babies while they are still in the mother's womb. Birth defects and other problems can be repaired before the child is born.

In recent years, **laser surgery** has become very important. Lasers are frequently used to stop ulcers from bleeding and to correct vision defects. See Figure 14-14. Unlike the scalpels (knives) used in ordinary surgery, lasers don't cause bleeding. The surrounding tissue is vaporized. Because the laser is computer-controlled, cuts can be very precise.

Sound waves are also being used for surgery. For example, high frequency sound waves are often used to break up kidney stones.

Would you want to be operated on by a robot? Robotic arms were first developed for use in manufacturing. Perhaps the most interesting new technology is the use of robotic arms for surgery. An arm developed at Ohio State

laser surgery
surgery done with a laser beam instead of a scalpel

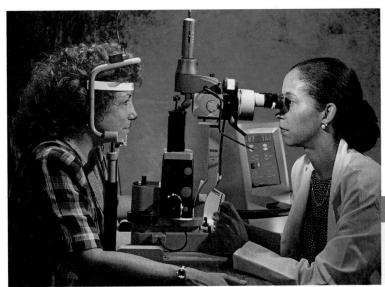

Figure 14-14 After laser surgery to correct vision defects, some patients no longer have to wear glasses.

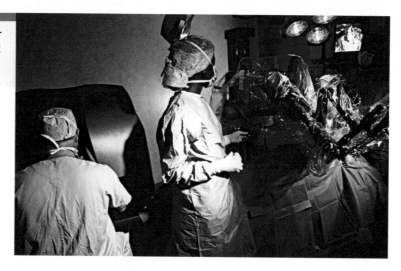

Figure 14-15 Because it is controlled by a computer, a robot surgeon can be very precise.

TECH CONNECT
SCIENCE

Robot Efficiency

A robot's flexibility is measured in "degrees of freedom." These have to do with how the robot's hand or arm can move—up or down, in or out, and so forth. The greater the number of degrees the robot possesses, the more flexible it is. Most robots have no more than six.

ACTIVITY Research the joints in a robot's hand and wrist and compare them to the joints in a human's hand and wrist. How are they alike? How are they different?

implant
a small device inserted into the body to treat a medical problem

University performs heart bypass surgery. See Figure 14-15. The doctor controls it remotely using a computer console. The incision is small, and recovery is faster than with ordinary surgery. With ordinary heart surgery, the heart must be stopped. One type of robotic surgeon can work on a heart while it continues to beat, which is much safer for the patient. Other robotic surgeons are used to prepare bones for joining to artificial limbs.

Electronic Implants

▷ What are implants?

Implants are small devices inserted into the body. They have been used successfully to treat many medical problems. Many implants are electronic

People who are blind or who have lost their hearing may recover partially or completely with the use of electronic implants. Vision implants may be inserted into the eye or directly into the brain. The implant sends electronic signals to the visual cortex of the brain, which creates the image. Some implants require a TV camera mounted in a pair of glasses. Vision implants are still experimental. Devices to restore hearing have already been implanted successfully in about 100,000 people. See Figure 14-16. The restored level of hearing is often excellent.

Pacemakers are implants used to stimulate a heart with electrical impulses. Researchers are working to develop pacemakers that deliver drugs to the heart, radio for help in an emergency, and provide the doctor with a computerized report.

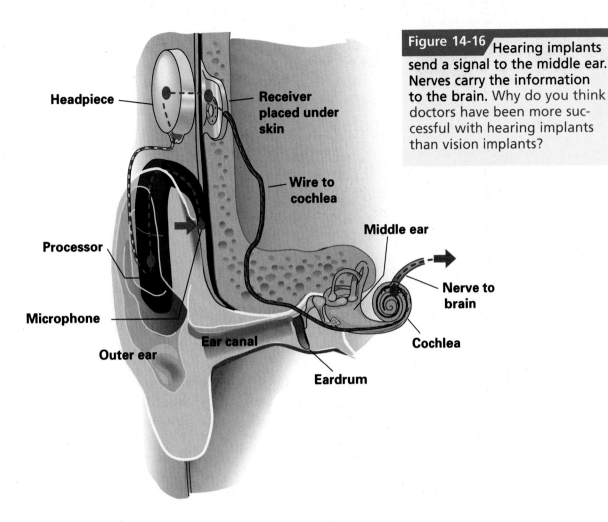

Headpiece

Receiver placed under skin

Wire to cochlea

Processor

Microphone

Outer ear

Ear canal

Middle ear

Nerve to brain

Cochlea

Eardrum

Figure 14-16 Hearing implants send a signal to the middle ear. Nerves carry the information to the brain. Why do you think doctors have been more successful with hearing implants than vision implants?

Genetic Engineering

▷ **What is genetic engineering?**

As you know, your genes can influence your risk for a particular illness. **Genetic engineering** involves altering or combining the genetic material in DNA in order to treat a disease or modify body characteristics. For example, if a person has a defective or missing gene, a normal gene is inserted into a harmless virus that delivers the new gene to the affected area. This is called gene therapy. So far gene therapy has been experimental. However, researchers hope it will one day prove useful in treating such illnesses as cystic fibrosis and certain types of cancer.

Other genetic engineering research is being done to develop drugs for treating disease. Genes influencing a particular disease are placed into other organisms. This

genetic engineering altering or combining genetic material in order to treat a disease or modify body characteristics

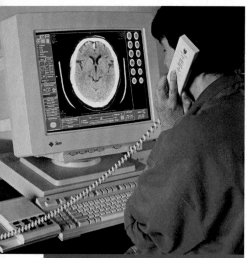

Figure 14-17 This doctor is viewing a scan from a patient in another city. He then gives his diagnosis over the phone.

telemedicine
medicine at a distance

bionics
creating replacements for human body parts

TECH CONNECT
MATHEMATICS

Metric Measures

Most scans and other medical procedures are done based on the metric system of measurement.

ACTIVITY Suppose you needed an artificial arm. Measure the length of your forearm and hand, from elbow to fingertip, using millimeters and centimeters. Then measure the lengths from joint to joint in your index finger.

causes the organism to produce substances that can treat the disease. These medicines include vaccines and treatments for burns and some cancers.

Telemedicine

▷ **How can doctors treat patients who are far away?**

Telemedicine is medicine at a distance. It brings medical advice or treatment to patients through the use of computers and telephone or cable connections. A medical practitioner attaches electronic diagnostic equipment to the patient. The readings are then sent to the doctor who is miles away—sometimes in a distant country. See Figure 14-17. The doctor then delivers instructions or consults with other doctors who themselves may be in distant places. With telemedicine, doctors who are experts in a particular field can be available to more people. People in remote areas, where no doctors are located, can still obtain treatment.

The use of robots is also considered a form of telemedicine. The doctor does not necessarily touch the patient, although he or she may be in the same room.

Bionics

▷ **What is bionics?**

The human body is a perfectly crafted machine that contains thousands of living parts. Medical engineers are developing systems that can replace many parts if they are missing, injured, or diseased. This area is called **bionics**.

Artificial organs are important because too few donor organs are available. However, creating replacement parts for the human body involves a number of problems. All construction materials must be biologically neutral, so that the person's immune system won't attack what it sees as a foreign body.

The new part must be able to work in the environment of the human body for many years. It should be able to duplicate the function of the part it replaces well enough that it doesn't cause a breakdown in other body systems. If a replacement heart valve causes blood clots, it can't be used. If a replacement joint causes bones to break, then it can't be used.

The *Jarvik-7* artificial heart passed all kinds of animal tests before it was placed into the body of a dying man. The heart kept Barney Clark alive for 112 days. The

LOOK TO THE FUTURE

Tissue Engineering

Many people today receive transplant organs. A kidney or heart removed from one person is placed into another. Problems arise when the patient's body tries to reject the new organ. Tissue engineering is an attempt to grow a new organ in the laboratory from the sick person's own cells. Rejection would not be a problem. Skin transplants for burn victims are already grown this way. Although tissue engineering is still experimental, researchers hope that in the future lab-grown replacement parts will be commonly used.

AbioCor artificial heart is one of the latest products that has been approved for experimental use. Tom Christerson had an AbioCor replacement heart installed in his chest in 2001. The unit kept him alive for seventeen months. Research will continue until the most effective design and materials are found.

SECTION REVIEW 14C

Recall ▸▸

1. Against what kind of organisms are antibiotics effective?
2. What are implants?
3. How does laser surgery differ from surgery with a scalpel?
4. What is telemedicine?

Think ▸▸

5. If genetic engineering is perfected, people might be able to have "designer" babies. They could choose eye color, intelligence levels, and other characteristics. Do you think this is a good idea? Explain your answer.
6. If you needed surgery, would you want a robot to do it? Why or why not?

Apply ▸▸

7. **Research** Visit a Web site that shows how things work and do some research on an artificial heart or other bionic device. Make a poster or a model of the device showing its operation.

Summary Activity

For each numbered blank, pick the answer from the list on the right that makes the most sense in the entire passage. Write your answer on a separate sheet of paper. No answer will be used more than once.

Medical technology is a biotechnology. Biotechnologies make use of information about __1__.

Prevention involves many technologies. Some seek to remove or kill __2__, the organisms that can cause disease. Local governments remove pathogens, dirt, and other __3__ from the water supply. __4__ usually involves the removal of waste products that could cause disease or contaminate the environment.

__5__ helps the body fight disease by causing the immune system to produce cells that attack certain pathogens. __6__ is the design of equipment and environments to promote human safety, health, and well-being.

Using microscopes, doctors and technicians can examine body fluids for pathogens and other signs of disease. In __7__, a person's genes are examined. Genes are the carriers of __8__.

Imaging machines show doctors the inside of the body. The first imaging machine was the __9__. Imaging technologies include CT scans, __10__, ultrasound, and endoscopes.

Penicillin was the first __11__, and it was effective against many life-threatening __12__ infections. Many other important medicines have been developed to treat illnesses.

In recent years, laser surgery has become very important to health. Unlike the scalpels used in ordinary surgery, lasers don't cause __13__.

__14__ are small electronic devices inserted into the body. People who are blind or who have lost their hearing may recover partially or completely with their use.

__15__ alters or combines genetic material in order to treat a disease or modify body characteristics. Other genetic research is being done to develop drugs for treating disease.

__16__ is medicine at a distance. It brings medical treatment to patients through the use of computers and telephone or Internet connections.

Answer List

- antibiotic
- bacterial
- bleeding
- contaminants
- ergonomics
- genetic engineering
- genetic testing
- heredity
- immunization
- implants
- living things
- MRIs
- pathogens
- sanitation
- telemedicine
- X ray

Comprehension Check

1. What is genetic engineering?
2. What is the purpose of bionics?
3. How do computer simulations help surgeons?
4. Who discovered penicillin?
5. What does EEG stand for, and what does it do?

Critical Thinking

1. **Classify.** Categorize the following as prevention, diagnosis, or treatment: pasteurization of milk, genetic testing for heart disease, genetic engineering for liver disease, MRI for auto accident injuries, antibiotics for pneumonia, ergonomics in designing chairs, bionics, immunization against flu.

2. **Connect.** For what purposes do you think a doctor might need language arts skills?

3. **Evaluate.** Suppose a family used genetic engineering to create a child with superior skills as a skier. Should this person be allowed to compete in the Olympics against "normal" players? Explain your answer.

4. **Infer.** Why do you think it is better to prevent a disease than to cure it?

5. **Propose.** As you know, certain habits such as smoking can lead to serious disease. If you knew a young person who had begun to smoke, what would you say to that person to persuade him or her to quit?

Visual Engineering

A Design for You. No two people are the same. We all have our unique dimensions. Designers must consider these dimensions when designing a product. Most products are designed for the "average" person.

This illustration (also on page 338) shows where measurements are taken (A through I) for a chair design. Take these same measurements from three or four of your classmates and calculate the average for each. Compare your averages with those of other students in your class. How similar are they? Calculate the averages for the class. If possible, compare your class averages with those from other classes.

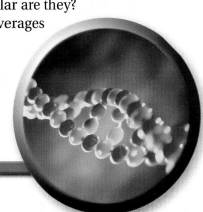

TECHNOLOGY CHALLENGE ACTIVITY

Egg-citing Safety Crash Testing

Equipment and Materials

- lumber for ramp and vehicle
- technical drawing equipment or computer and CAD software
- 2 axles for each vehicle
- 4 wheels for each vehicle
- 2 eye hooks for each vehicle
- 1 egg, uncooked
- zip-shut sandwich bag
- string
- protractor to determine ramp angle
- glue
- foam rubber
- paper
- rubber bands
- balloons

! SAFETY

Reminder

In this activity, you will be using tools and machines. Be sure to always follow appropriate safety procedures and rules. Remember, safety is an attitude that you must develop and maintain at all times.

Background

Your family car is designed to protect you in case of a crash. This system includes the front and rear bumper, steel-reinforced door panels, lap and shoulder seatbelts, airbags, padded instrument panels, headrests, and a collapsible steering column.

The people who originally developed these systems had to take into account the physical features of the human body to determine what was needed. Then the system was tested using dummies in actual car crashes before it was approved for commercial use.

Goal

For this activity, you and your team of three to five students will design a safety restraint system that can protect a raw egg from breaking during a vehicle collision.

Criteria and Constraints

▶▶ All vehicles must be standardized so that only the safety system is being tested.

▶▶ You can include anything in your design that will directly or indirectly protect the egg. However, you cannot interfere with the vehicle's speed.

▶▶ The egg or its shell *cannot* be strengthened in any way, including hard-boiling it. All eggs must be placed in plastic bags prior to testing.

▶▶ The vehicle containing the egg should be sent down a ramp at a 75° angle. It will then crash into a wall. Side collisions can also be included by arranging ramps at right angles so that cars crash into one another. If the egg survives the crash, the design is successful.

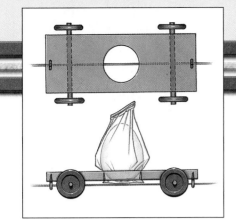

FIGURE A

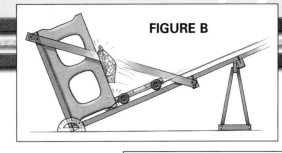

FIGURE B

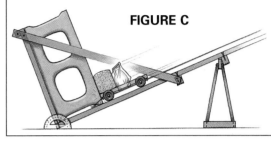

FIGURE C

Design Procedure

1 Brainstorm designs for your vehicle. Select the most promising design for prototype construction.

2 Prepare a set of plans to build the prototype.

3 Design the ramp. Build the ramp long enough so that you can increase vehicle speed by starting the cars higher up on the ramp. Build a protractor into the ramp design so you can increase the ramp angle by a known number of degrees. A higher angle will increase the speed of the vehicle at the moment of impact.

4 Place a string guide through eye hooks on the vehicle to guarantee that the vehicle runs the full track.

5 Test your vehicle prototype to determine if it meets the criteria and constraints.

6 Improve the ramp and the prototype vehicle if necessary.

7 As a class, select the vehicle that works the best. Mass-produce this vehicle so that every team has a regulation vehicle for the activity.

8 Run quality-control testing on the vehicles to make certain that they all function equally well when they run down the ramp.

9 Along with your teammates, design a safety restraint system.

10 Before testing the restraint system, place an egg in a zip-shut sandwich bag for easy cleanup. Run a test of your design and improve your vehicle if necessary.

11 As a class, hold a competition to see which system allows the egg to survive the crash. If possible, videotape the event.

Evaluating the Results

1. Did your egg survive the crash? What part of your design seemed to be most effective?

2. Did any team's egg survive more than one crash? What method was used to protect it?

3. Did the protections you used correspond to those used in an automobile? Explain.

SUPERSTARS OF BIOTECHNOLOGY

INVESTIGATION ////

In 1546, Girolamo Fracastoro proposed a theory about the cause of disease, but his theory was not proven true until 300 years later. Find out what he believed and who finally proved he was right.

Harriet Russell Strong (1844-1929)

Harriet Strong was an engineer who won several patents for dams and water storage systems. The first person to introduce winter irrigation in California, she was also an entrepreneur who raised walnuts, citrus fruits, pomegranates, and pampas grass. Eventually, her orchard grew to 25 miles in length and was the largest in the world. Other engineers recognized the value in her methods and soon adopted them. Her irrigation system design was responsible for the rapid growth of the food-producing regions in Southern California.

Jonas Salk

(1914-1995)

During the 1940s and '50s, a viral disease called polio, or infantile paralysis, struck young children. Some were even unable to breathe. Many researchers struggled to develop a vaccine quickly. While some made important contributions to the effort, no one managed to find results faster than Jonas Salk. Salk and his team announced success in 1955, and immunization of thousands of school children began. Today, polio is virtually unknown.

Norman Borlaug

(b. 1914)

Norman Borlaug, a U.S. biologist, won the Nobel Prize for peace in 1970 for helping begin the "green revolution" in agriculture. From 1944 to 1960, he tried to develop a hybrid strain of wheat that would yield larger crops than existing varieties did. He succeeded. His "miracle" wheat allowed Mexico to triple its grain production within a few years. His wheat and other successes helped to reduce world hunger.

George Washington Carver (1861-1943)

George Washington Carver was born a slave on a Missouri farm. Freed at the end of the Civil War, he began an education that led to a master's degree in horticulture. He spent his career researching ways to improve the soil. Along the way he developed over three hundred by-products from plants such as peanuts and sweet potatoes. They included cereals, oils, dyes, soaps, carpeting, synthetic marble, and food substitutes.

Virginia Apgar (1909-1974)

Virginia Apgar was one of the first female American doctors to specialize in surgery. However, her greatest achievement was the Newborn Scoring System, a method for evaluating the health of newborn infants. Up until 1949, when she first conceived of the system, babies were assumed to be healthy unless they showed obvious symptoms. Many internal problems were missed. The six criteria she used—breathing, reflexes, irritability, muscle tone, heart rate, and color—helped doctors quickly evaluate a baby's health and saved many lives.

Wilhelm Roentgen (1845-1923)

German physicist Wilhelm Roentgen was experimenting with light when he unexpectedly discovered a new form of electromagnetic energy that he called X rays, which could penetrate a person's hand. The outline of the bones could be seen on a screen coated with fluorescent chemicals that was placed behind the hand. By replacing the screen with photographic film, Roentgen made lasting pictures of the images and realized that these X rays could be used for medical purposes. He received the first Nobel Prize for physics in 1901.

CHAPTER 15 — AGRICULTURAL BIOTECHNOLOGIES

Agriculture is the science of cultivating the land to produce crops and raise livestock. Agriculture is a biotechnology because it is related to living things. It applies the principles of biology to create commercial products and processes, including textile fibers and fuels. However, the chief output of agriculture is food.

In 2003, the world population reached 6.3 billion people. At the current rate of growth, it is expected to reach 8 billion by 2020 and 10 billion by 2050. Can current agricultural and food production methods keep up with this growth and feed so many people?

In this chapter, you will learn how technology has affected traditional farming and the role genetics is beginning to play. You will also learn about other uses of agricultural technologies.

The global positioning system (GPS) can be accessed from the cab of a farmer's tractor.

Terms to Learn

- dehydrate
- fertilizer
- hybrid
- irrigation
- monoculture farming

Standards

- Characteristics & Scope of Technology
- Core Concepts of Technology
- Relationships & Connections
- Cultural, Social, Economic & Political Effects
- Role of Society
- Influence on History
- Troubleshooting & Problem Solving
- Assessment
- Agricultural & Related Biotechnologies

Farming Technologies

▷ What are some examples of farming technologies?

Like other technologies, farming technologies can be thought of as systems. They use the same seven resources: people, materials, tools and machines, energy, capital, information, and time. Suppose a farm produces grain used to make breakfast cereals. People are needed to plant the seeds and harvest the grain. Materials include such things as seed, fertilizer to prepare the soil, and water for the plants. Tools and machines include everything from giant plows to small hammers used to repair fences. Human energy guides the machines, and fuel energy powers them. See Figure 15-1. Capital is used to pay the workers and buy the materials and equipment. Information includes such things as instructions for running the machines and daily weather reports. Time, of course, is needed to get all the tasks done.

The Evolution of Farming

▷ How has farming changed since ancient times?

Very early humans obtained food by hunting and gathering. They traveled with the seasons and followed animal migrations. Eventually they began to domesticate animals and plant crops. They worked the soil with sticks and

Figure 15-1 In some parts of the world, this kind of farming is still done.

Figure 15-2 This farmer is ready to show how wheat was once cut and gathered. What tools is he carrying?

Figure 15-3 The GPS system, combined with a computer on board the tractor, allows the farmer to map the soil types and determine the flow-rate needed for the fertilizer.

primitive hoes. Then, around 3000 B.C.E., people developed the plow, which enabled them to work larger areas of land more efficiently. An animal usually pulled the plow, and the farmer walked behind to guide it. Other implements, including shovels, saws, axes, and grinding tools, were also developed. See Figure 15-2.

It took the Industrial Revolution to produce a real revolution in agriculture. After 1750, people began to leave the countryside to work in factories in the cities. With fewer workers available, farmers began to rely on machinery to help them prepare the soil, plant seed, and harvest crops.

The tractor was introduced to farming at the beginning of the 20th century. Now, at the start of the 21st century, it is undergoing some important changes. It is becoming part of the Information Age. Some tractor models are being equipped with robotic control systems. Guidance information comes from the Global Positioning System (GPS). See Figure 15-3. The GPS is a series of satellites orbiting the earth that tells the farmer where the tractor is located. Computers also help monitor crop quality and make planting the next crop more efficient.

hybrid
an organism bred from two different species, breeds, or varieties

irrigation
bringing a supply of water to crops

Figure 15-5 An irrigation system rescues this land from the desert.

Breeding Plants and Animals

▷ How are plants and animals bred traditionally?

The first farmers probably saved seeds from wild plants and planted them. Over time, seeds from crops with desirable characteristics, such as hardiness or tastiness, were prized. Eventually, certain plants were crossed with other plants to try to improve crop quality. Pollen from one was transferred to the flowers of another. When it worked, the seeds produced plants with the desired characteristics.

The first livestock were also found in the wild and raised for human needs. Herders learned to select the best mates for breeding better size, health, growth rate, and temperament. When animals of two different species or varieties are bred, the resulting animal is called a **hybrid**. See Figure 15-4.

These breeding processes, while controlled by humans, are natural. The animals themselves have not been altered. More advanced breeding technologies will be discussed in Section 15B.

Plant and Animal Maintenance

▷ What processes are used to maintain plants and animals?

After seeds have been planted and animals purchased or bred, they must be cared for, or maintained. They need water and food, and conditions have to be right for their growth. They must also be kept free of disease.

Irrigation

▷ How do farmers deliver water to crops?

Without water, rich farm soil quickly dries out, killing crops. Famines have been caused by too little rainfall. Too much water causes flooding, which can also destroy crops. **Irrigation** technology is used to supply the right amount of water to a farmer's field through an arrangement of pipes and sprinklers.

Many farmers grade their land to carry excess water away from their crops into canals that drain into ponds. During dry periods, water is pumped from these canals, ponds, or other sources to the crops. Some farmers use tractor watering systems or sprinkler systems. See Figure 15-5.

Fertilizing

▷ What is the purpose of fertilizer?

When a plant grows, it takes nutrients out of the soil. If you plant the same crop on the same plot of land repeatedly, crop yields will fall. Before the development of modern agriculture, farmers learned to rotate their crops. One year they might plant corn in a field and soybeans in that same field the next year. Some fields were left unplanted.

Then, in 1912, Fritz Haber, a German chemist, developed a process for making nitrogen fertilizer. **Fertilizer** is a chemical compound that restores nutrients to the soil. Fertilizer had an explosive effect on farming. See Figure 15-6. Mixing fertilizer that contains nitrogen, phosphorus, and potassium into the soil makes it possible to grow plentiful crops year after year on the same fields without rotating them. Today, fertilizer mixtures are made to match the needs of particular plants.

Fertilizer has caused some unexpected negative outcomes. It has produced big crops of weeds. Runoff into lakes and oceans has affected wildlife and increased the growth of algae. Success with fertilizers has also increased monoculture farming and made it very cost effective. **Monoculture farming** is growing one crop or one species over large areas. If pests or disease attack, an entire region's crop can be destroyed, and the income of many farmers is lost.

Figure 15-6 Today, special machines like this one are dedicated to providing the fertilizers needed for a farmer's fields.

fertilizer
a chemical compound used to restore nutrients to the soil

monoculture farming
raising only one crop or one species of plant

Use of Antibiotics

▷ What are the advantages and disadvantages of using antibiotics?

To keep costs low, many farm animals are raised in crowded conditions. Diseases can spread very quickly from one animal to the rest of the flock or herd. Many farmers have been giving their animals antibiotics to protect them from disease. Antibiotics also promote growth.

This technology, too, has negative outcomes. The antibiotics killed most pathogens, but unfortunately, antibiotic-resistant pathogens survived and multiplied. Now, most antibiotics do not work against these resistant pathogens. If the animals are exposed to these resistant pathogens, antibiotics cannot help. Because food from the animals contains antibiotics and is eaten by humans, human diseases, too, have become more antibiotic resistant. The antibiotics are present in animal wastes and have found their way into the soil and waterways, causing problems there as well.

Writing Skills

Developing your writing skills will help you communicate better at home, at school, and in the workplace.

Most jobs require workers with good writing skills. Why? Much communication, including online communication, takes place through writing. If workers cannot express themselves clearly, they may be misunderstood. Work may be delayed or done incorrectly. Here are some tips for developing better writing skills:

✔ **Organize your thoughts in a logical order. Revise your writing until your meaning is clear.**
✔ **Check and double-check spelling and grammar.**
✔ **Be aware of the tone of your writing. Don't be disrespectful.**
✔ **Proofread your writing before you send it out.**

Explore job opportunities that require good writing skills. Many such jobs can be found using the *Occupational Outlook Handbook*.

Pharmacologist

The U.S. Food and Drug Administration needs pharmacologists to help evaluate new drugs and drug production. Responsibilities include evaluation of data and research into new diseases, drug interactions, and side effects. Must have degree in chemistry and pharmacology. Good writing skills required for reports and records.

From Farm to Consumer

▷ What processes are used during harvesting?

At harvest time, large machines go into the fields and pick cotton, corn, beans, and other crops. See Figure 15-7. Waste, such as stalks, is separated out and the rest is sent to a market or a factory for processing. Livestock, such as hogs and cattle, are herded together and shipped to processing plants where they are butchered for their meat.

Before refrigeration, people cooked and canned foods at home. Animals were killed just before they would be eaten, and leftover meat was smoked or heavily salted to prevent spoilage. Then refrigeration became common, and food producers had more options. Today, food is often processed in very cold environments to keep it fresh. Many food products are shipped in refrigerated trucks and stored in refrigerated cases at the supermarket. Freezing is also used for some food products.

Some foods, such as rice and other grains, must be kept very dry. Other foods are **dehydrated**. This means the water they normally contain is removed. Raisins, which are really dried grapes, are an example.

dehydrate
remove moisture

Figure 15-7 A combine can harvest a field in just a few hours. Compare that with how long it would have taken 75 years ago with horses.

Recall ▸▸

1. How have tractors become part of the Information Age?
2. What is a hybrid?
3. Why are antibiotics given to animals who are not ill?
4. Name at least three methods for preserving foods.

Think ▸▸

5. What do you think the solution is to the overuse of antibiotics in meat-producing animals? How can farmers continue to keep their animals disease free and promote their growth?

Apply ▸▸

6. **Experiment** Design an experiment to test for the effects of fertilizers or light conditions on plant growth. Use photographs and data analysis to show your findings.

Objectives

▶▶ **Discuss** the use of genetic engineering in agriculture.

▶▶ **Describe** the cloning process.

▶▶ **Identify** the role of the USDA in approving new technologies.

Terms to Learn

- cloning
- DNA
- gene
- transgenic organism

Standards

- Characteristics & Scope of Technology
- Core Concepts of Technology
- Relationships & Connections
- Cultural, Social, Economic & Political Effects
- Role of Society
- Assessment
- Medical Technologies
- Agricultural & Related Biotechnologies

gene
the factor in cells that carries heredity

DNA
deoxyribonucleic acid; the molecules in a gene that carry genetic information

transgenic organism
an organism into which genes from another organism have been transplanted

New Breeding Technologies

▷ What contribution did Gregor Mendel make to agriculture?

In 1865, the Austrian monk Gregor Mendel discovered the basic principles of genetics by careful breeding of pea plants. Mendel theorized that characteristics were passed from one generation to the next by means of factors in the cells.

Today, these hereditary factors are called **genes**. Within the genes is all the information needed to reproduce an organism in the form of **DNA** (deoxyribonucleic acid).

In the past, improving a species of plant or animal was a slow process. Today, it has been sped up with genetic engineering and cloning. However, many questions about both processes have also been raised.

Genetic Engineering

▷ How is genetic engineering used in agriculture?

In the 1990s, researchers learned how to direct the evolution of plants and animals through genetic engineering. Genetic engineering, as you know, involves altering or combining the genetic material in DNA in order to produce desired characteristics or remove undesirable ones. In some processes, genes from one organism are transplanted into another to create a **transgenic organism**.

The U.S. Department of Agriculture (USDA) must approve any genetically modified species. The offspring of the altered plants or animals are raised in a laboratory to see if the new characteristics show up and can be passed on. Other tests are done to learn if the altered plant or animal could become a danger to the environment. Still other tests determine if food produced from these plants or animals would be safe for humans.

Improving plants genetically has produced crops that need less water, can grow in salty soil, are immune to certain diseases, can kill pests, can grow and ripen faster, and stay fresh longer after harvesting. See Figure 15-8. Improving livestock genetically has produced animals that grow faster and are larger and leaner. Their meat tastes better and stays fresh longer.

As of 2003, thirty-three percent of corn crops and fifty percent of soybean crops in the U.S. were grown from genetically engineered seeds. Genetically engineered material was found in sixty percent of all processed foods and 100 percent of all manufactured candy. Although researchers and the USDA claim foods made from genetically engineered products are safe for humans to eat, many people have doubts.

Also of great concern is that pollen or seeds from altered plants could escape into the wild and affect other plants. For example, the seeds of some useful plants have been given genes that make them safe from poisons used to kill weeds. Could these genes find their way into the weeds themselves? If so, those weeds might then be unstoppable. To prevent this, some seed companies have developed a "terminator" gene. This gene prevents a plant from reproducing. Its seeds will not grow. Any genetic changes would die with it. However, this might be even more dangerous. What if the terminator gene escapes into the wild? All affected plants would die out in one generation.

Figure 15-8 The tomato was one of the first foods to be genetically altered.

Cloning

▶ What is cloning?

Cloning is a process that produces an identical copy of a plant or animal. Some clones, such as identical twins, are natural. A single cell has divided in the mother's womb and produced two identical individuals.

Cloning of some species of plants has been common for a very long time. Cuttings from an individual plant are placed in soil. The cuttings take root and a new plant grows. Today, however, a single cell from the "parent" plant can be used to create a copy. See Figure 15-9 (page 368).

In recent years, animals, too, have been cloned. Animal cloning includes two different technologies. The older method produces identical twins by mimicking the natural accident that causes twin births. Two animal parents produce a fertilized egg. After fertilization, the egg is artificially divided into two separate eggs. The eggs then develop into two separate but identical animals.

TECH CONNECT
MATHEMATICS

Organic Foods

Organic farm products are those raised using only natural materials and methods. No artificial fertilizers, pesticides, or genetics are permitted. Organic foods are becoming more popular with consumers than ever before.

ACTIVITY Find out the sales figures for organic foods for the past five or ten years. Graph the results and estimate the percentage in growth.

cloning
producing an identical copy of an individual plant or animal

Figure 15-9 In plant tissue culture, one piece of a plant is used to grow a new plant. If the piece is a single cell, then the new plant will be a clone.

Cell is removed from plant and placed in growing medium

Figure 15-10 In 1996, Dolly the sheep was cloned. She and her parent had the same DNA. Dolly died in 2003.

In 1996, a sheep named Dolly was cloned in a new way. This method, which has produced many other mammals, including a horse, is used to create clones of adult animals. The single parent and baby have identical DNA. See Figure 15-10. A cell is taken from the animal to be cloned and placed in a weak nutrient culture. The lack of all the needed nutrients causes the cell to stop dividing and switch off its active genes. Then an unfertilized egg cell is taken from a female animal and its nucleus is removed. Although its genes are gone, the egg cell still has all the other things needed to produce a baby. The nucleus from the first cell is now removed and fused into the egg cell with a spark of electricity. Another spark "wakes up" the sleeping genes in the nucleus, and the cell begins to divide. After a few days, the embryo is placed into the womb of another female animal and allowed to develop normally.

Humans already use animals for food. The animals live in crowded, unnatural conditions. Now we can alter their genes to make them even more useful. One species of pig, for example, has been altered so its organs might one day be used for human transplants. Is it ethical to use animals in this way? Animal rights activists usually say no. Researchers who are looking for cures for human diseases say yes.

ACTIVITY Take a survey of at least ten people about the rights of animals used for food and other human needs. Graph your results.

Cloning animals this way is difficult. Most do not survive, and those that do often have problems. Many feel that this type of cloning is ethically wrong and that it should be outlawed.

Could this procedure be used to clone animals back into existence after they have become extinct? A Texas A&M University project, named Noah's Ark, is freezing tissue of endangered animals in the hope that someday clones of them can be created.

SECTION REVIEW 15B

Recall ▸▸

1. What is a transgenic animal or plant?
2. What is a clone?
3. What is a terminator gene?

Think ▸▸

4. The U.S. Department of Agriculture must approve new technologies that affect the foods we eat. Do you think this is a good idea or not? Explain.

5. Now that you know that all manufactured candy contains genetically modified organisms, will you refuse to eat it? Why or why not?

Apply ▸▸

6. **Research** The McIntosh apple is a cloned species. Do some research on the McIntosh. Where was the parent tree located? When was it first cloned? How many copies have been made? Write a report on your findings.

Other Agricultural Technologies

▶ What other agricultural biotechnologies are common?

Food production is only one of the processes for which we use living organisms. They are also used to manufacture materials, produce medicines, process wastes, produce fuels, and develop artificial ecosystems.

Biosynthesis

▶ What is biosynthesis?

Biosynthesis is the making of chemicals using biological processes. Technologists are working to develop living organisms that will produce chemicals that were once produced in factories. These organisms are usually genetically altered.

For example, one microbe can now manufacture polyester used to make clothing. Another can produce silk. A British company has developed biopolymers, which are plastics produced from living things.

Pharming

▶ How is biosynthesis used to create medicines?

Some crops, animals, and microbes have been genetically modified to produce medicines. This process is called **pharming**. (The term is a combination of the words *pharmaceutical* and *farming*.) Plants, animals, and microbes become bio-factories that convert food and water into medicine.

Certain transgenic animals, such as cows, are able to produce compounds in their milk that are active against diabetes, arthritis, hemophilia, emphysema, and gastrointestinal infections. A tobacco plant can produce vaccines and human growth hormones. Other transgenic organisms produce human blood components and anti-tooth decay compounds. Although pharming is still a new technology, researchers hope that one day it will be common.

Figure 15-11 A floating boom helps a team of workers stop further pollution of a beach. Why does the boom work?

Bioremediation

▷ What is bioremediation?

Bioremediation is the use of bacteria and other organisms to clean up contaminated land and water. Ordinary microbes work in landfills to break down garbage. Engineered microbes are being used to clean up the oil from oil spills. See Figure 15-11. Petroleum waste can be made biodegradable by mixing it with soil, nutrients, and microbes that can digest the waste. Other microbes are being used to remove dangerous materials from toxic (poisonous) wastes.

▌**bioremediation**
using bacteria and other organisms to clean up contaminated land and water

Biofuels

▷ What kinds of fuels can agriculture produce?

Biofuels are those made from agricultural products. Biofuels are a source of renewable energy. Materials such as corn, crop wastes, and lumber wastes can all be used to manufacture fuel alcohol. When this alcohol is mixed with gasoline, it is sold in gas stations as gasohol. See Figure 15-12.

Plant and animal wastes can be mixed with bacteria to produce methane gas, which is similar to natural gas and propane. Methane, too, can be used to power vehicles. Some farms sell their waste products to power plants that recycle the wastes to produce electricity.

Figure 15-12 Gasohol is a combination of gasoline and alcohol made from corn and crop waste.

Chapter 15 • Agricultural Biotechnologies **371**

Breathing Easy

NASA has created an artificial ecosystem on the International Space Station. It is also working on systems that could sustain life during space journeys to distant planets. One engineer spent fifteen days sealed in a chamber breathing oxygen produced by wheat plants. All air, food, water, and waste will have to be recycled for such journeys that may take many, many years. Would you want to go along?

Artificial Ecosystems

▷ Why are artificial ecosystems developed?

artificial ecosystem
a human-made, controlled environment built to support humans, plants, or animals

An **artificial ecosystem** is a human-made, controlled environment built to support humans, plants, or animals. It copies some aspects of the natural environment. Astronauts live in an artificial ecosystem that supports their need for air, food, and water. Can you think of others?

Hydroponics

▷ Why is hydroponic farming needed?

hydroponics
growing plants in nutrient solutions without soil

Only about six percent of the earth's surface is suitable for traditional farming in soil. With **hydroponics**, plants are grown in nutrient solutions without soil. See Figure 15-13.

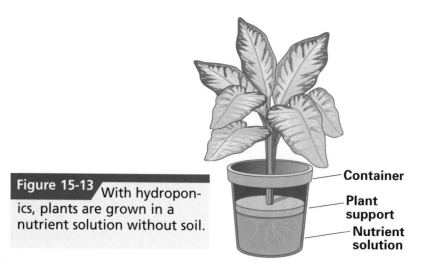

Figure 15-13 With hydroponics, plants are grown in a nutrient solution without soil.

Container

Plant support

Nutrient solution

Figure 15-14 Sprouts grow as the conveyor moves in this hydroponic hot house.

Some hydroponic farms plant their seedlings on a long, wide, slow-moving conveyor system. See Figure 15-14. The system moves at the same speed that the plants grow. When the plants reach the end of the conveyor system, they are ready to be harvested.

Hydroponics re-circulates water and fertilizer, which means the fertilizer used isn't released outside to pollute lakes and streams. Insects can't reach the plants easily. Weather conditions aren't important, so crops can be grown in any climate. All these factors make some people think hydroponics could be used to help relieve world hunger.

Aquaculture

▷ What is grown using aquaculture?

When fish, shellfish, or plants are grown in artificial water ecosystems, the process is called **aquaculture**. The growing pond can be a totally enclosed indoor environment or a pen in a river or ocean. See Figure 15-15 (page 374).

TECH CONNECT

SCIENCE

Which Came First?

Throughout history, the "how" of technology often came before the "why" of science. People used special "seeds" to make bread rise for thousands of years before science discovered the seeds were actually yeast, a type of fungus.

ACTIVITY Obtain a packet of yeast used to make bread and sprinkle it into a bowl of lukewarm water along with a pinch of sugar. Observe what happens over the next twenty minutes.

aquaculture
growing fish, shellfish, or plants in artificial water ecosystems

Tests have shown that fish grown in an aquaculture environment contain lower levels of pollutants in their flesh than fish caught in natural habitats. Predators are also kept out. However, adequate diets must be provided or the animals may be less nutritious.

Agroforestry

▷ What needs does agroforestry meet?

agroforestry
turning forests into controlled environments dedicated to the replacement of trees

Agroforestry is turning forests into controlled environments dedicated to the replacement of trees. Under natural conditions, trees grow very slowly. When clear cutting is done, even if the entire forest is replanted with seedlings, it will be many years before the forest will return to its former state.

Hybridization and genetic engineering can be used to create fast-growing softwood and hardwood trees. New trees are quickly planted to rebuild the stock for future cutting seasons. Biotechnologists have developed a fast-growing tree species that is perfect for papermaking. Paper-mill reforestation programs hope to use these tree saplings to grow enough new trees to replace the ones that are harvested. See Figure 15-16.

Figure 15-16 By planting seedlings, paper mills make sure more trees will grow to replace those they have harvested.

SECTION REVIEW 15C

Recall ▸▸

1. What is bioremediation?
2. What products are produced from pharming?
3. What advantage do biofuels have over fossil fuels?
4. Name three kinds of artificial ecosystems.

Think ▸▸

5. Suggest ways in which the universal systems diagram could be applied to an aquaculture facility used to raise salmon for grocery stores.

Apply ▸▸

6. **Meet the Challenge** Use the Agriculture Technology Challenge guidelines found in the current issue of the *TSA Competitive Events Guide* to develop a project related to biotechnology processes. Enter your completed project in a local, state, or national TSA event.

Summary Activity

For each numbered blank, pick the answer from the list on the right that makes the most sense in the entire passage. Write your answer on a separate sheet of paper. No answer will be used more than once.

Agriculture is a(n) __1__ because it is related to living things. It applies the principles of biology to create commercial products and processes

The __2__ produced a revolution in agriculture. Beginning around 1750, people left the countryside to work in factories in the cities. With fewer workers available, farmers began to rely on __3__ to help them prepare the soil, plant, and harvest crops.

Breeding animals is part of agriculture. When animals of two different species or varieties are bred, the resulting animal is called a(n) __4__ .

After seeds have been planted and animals purchased or bred, they must be cared for, or __5__ . They need water and food, and conditions have to be right for their growth. Many farmers have been giving their animals __6__ to protect them from disease. The medicines promote growth. Unfortunately, this has helped create stronger pathogens.

In recent years, biotechnology has developed new breeding processes. The Austrian monk Gregor Mendel discovered the basic principles of __7__ by careful breeding of pea plants. In the past, improving a species of plant or animal was a slow process. Today, it has been sped up with genetic engineering and cloning.

In some cases, genes from one organism are transplanted into another. The result is a(n) __8__ plant or animal. Cloning is a process that produces an identical copy of a plant or animal. It, too, has caused controversy.

Some crops, animals, and microbes have been genetically modified to produce drugs. This process is called __9__ . __10__ is the use of bacteria and other organisms to clean up contaminated land and water. Biofuels are those made from agricultural products. Agricultural products can be a source of __11__ energy.

Artificial __12__ include hydroponic, aquaculture, and agroforestry farms. The environments create the best conditions for growth of plants or animals.

Answer List

- antibiotics
- bioremediation
- biotechnology
- ecosystems
- genetics
- hybrid
- Industrial Revolution
- machinery
- maintained
- pharming
- renewable
- transgenic

Comprehension Check

1. What is monoculture farming?
2. What is fertilizer?
3. Why is cloning of animals not done on a large scale?
4. What is biosynthesis?
5. What is gasohol?

Critical Thinking

1. **Evaluate.** Organically grown foods are usually more expensive than those conventionally grown. Do you think they are worth the extra money? Explain.
2. **Infer.** Why do you think skill is important in farming?
3. **Compare.** List the advantages and disadvantages of crop rotation versus the use of fertilizers. Which method would you recommend to a farmer today? Why?
4. **Produce.** Create a poster or display giving information about cloning. Include your recommendations as to whether or not cloning should be regulated and why.
5. **Connect.** How do you think courses in social studies would benefit someone working in the field of genetic engineering?

Visual Engineering

Fertilizing. Every year during the planting season, fertilizer manufacturers advertise their products. What is in these fertilizers, and how do they improve the ground? Why are some fertilizers good only for helping plants grow, and others are also good for killing weeds and undesirable plants? Along with a classmate, do a little research to find out how fertilizers and pesticides work. Identify any safety considerations, and what precautions should be taken with chemicals used in the farm or lawn environment. Share your findings with the class.

TECHNOLOGY CHALLENGE ACTIVITY

Constructing a Hydroponic System

Equipment and Materials

- one 6-inch diameter plastic pipe 2 feet long
- two ¼-inch-thick sheets of plastic, 6½-inches square
- one 4-foot length of ⅜-inch flexible tubing
- two tubing/container adaptors with rubber washers and nuts
- two ⅝-inch clamps
- 32-ounce plastic pail
- drill set
- electric drill
- 2¼-inch hole saw
- marking pen
- drill press
- all-purpose plastic cement
- hot glue gun
- hot glue
- pH testing kit
- baking soda
- vinegar
- sterilized garden sand
- fine plastic screening
- plant food
- four small potted plants
- wood blocks
- 18" x 10" x ¾" plywood sheet

Background

To grow plants, you need seeds or young plants, water, sunlight, air, nutrients, and a way to hold the plants in place. Hydroponics is a good method of farming for those regions of the world where rainfall is scarce, soil is poor or scarce, insects are difficult to control, or farmland is in short supply. In hydroponic greenhouses, plants grow faster, take up less space, and produce larger crops. What's more, their water and nutrients are recycled.

Goal

For this activity, you will build and operate a hydroponic system used to grow plants. See Figure A.

Criteria and Constraints

▶▶ You must keep a daily record of solution mixtures and any experiments made on your plants.

▶▶ You must monitor your system daily to be sure it has enough water and to check for leaks.

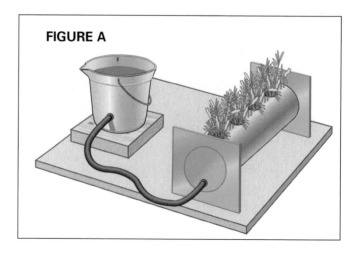

FIGURE A

Design Procedure

1. Tap water contains chlorine. Fill your pail with water and let it stand for two days prior to use to allow the chlorine to escape.
2. Stand the pipe on end in the center of the plastic sheet.
3. Draw a circle at the center of the plastic sheet using the pipe as a template. See Figure B.
4. Mark the location for the tubing adaptor. See Figure B.
5. Drill a hole for the tubing adaptor through the plastic sheet and the pail. See Figure C.

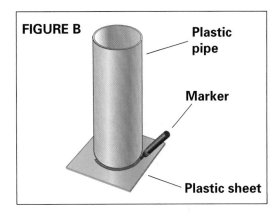

FIGURE B

Plastic pipe

Marker

Plastic sheet

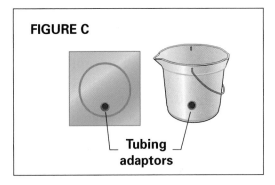

FIGURE C

Tubing adaptors

SAFETY

Reminder

In this activity, you will be using tools and machines. Be sure to always follow appropriate safety procedures and rules. Remember, safety is an attitude that you must develop and maintain at all times.

6. Slip one tubing adaptor through the plastic sheet and one through the pail. Make certain that you place rubber washers on both sides of the fitting to prevent water leaks.
7. Tighten the nuts to hold the tubing adaptors in place.
8. On the inside of your plastic sheet, place a circle of hot glue around the tubing adaptor.

TECHNOLOGY CHALLENGE ACTIVITY

Constructing a Hydroponic System (Continued)

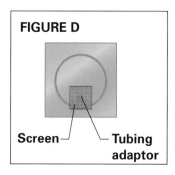

FIGURE D

Screen — — Tubing adaptor

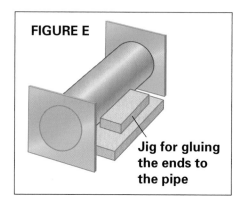

FIGURE E

Jig for gluing the ends to the pipe

9 Press a piece of screening into the hot glue. See Figure D. This protector will prevent sand from escaping when you remove extra nutrients from your planter.

10 Place the pipe on a ¾-inch piece of plywood so that its ends are raised above the work table. For best results, use stop blocks to keep your pipe from rolling. See Figure E.

11 The plastic sheets that you will attach to the ends of the pipe will seal the ends so that the pipe can hold water and also serve as your stand. Check the alignment of the parts before applying glue. Ask another student to help you glue these parts together.

12 Coat both pipe ends with plastic glue. Attach both plastic pieces so that they are straight and even. Make certain that one edge of the plastic is resting squarely on the table so that your finished planter won't rock.

13 Hold or clamp the parts together until the glue sets. Carefully set the project aside to dry.

14 Mark the location for your plant holes. See Figure F.

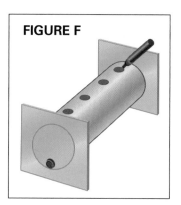

FIGURE F

FIGURE G

Drill

Hole saw

⑮ Drill through the plastic pipe using the hole saw mounted in the drill press. See Figure G.

⑯ Attach flexible tubing to the planter and the pail. Use small hose clamps to lock tubing to adapters.

⑰ Fill the planter with water to check for leaks, and repair where necessary.

⑱ Partially fill the planter with sand.

⑲ Remove plants from pots and carefully wash away soil.

⑳ Carefully place your plants into the planter, filling in sand where needed.

㉑ Test the pH of the water you have set aside. It should read 6 or 7 for best plant growth. Add a few drops of baking soda if pH is low or a few drops of vinegar if pH is high. Then retest your solution.

㉒ Partially fill the pail with a plant-food-and-water mixture. Place the pail higher than the planter so that the mixture fills the planter. Remember that it is better to have too little solution and have to add more than to flood the room.

㉓ Lower the pail so that the extra solution drains from the planter.

㉔ Check daily to see that the sand is moist. Add water to the pail to replace water lost from evaporation. Keep a record of how often water must be added.

㉕ Refer to a reference book about hydroponics to learn how to care for your plants. Devise a series of experiments involving nutrient mixtures, exposure to sunlight, and other variables that affect plant growth. Keep a record of your experiments and the results.

Evaluating the Results

1. Which nutrient mixtures were most effective in promoting growth?
2. If you were to repeat this activity, what would you do differently and why?

Factories
From Cotton Spinners to Computers

Manufacturing has changed from the days of the crude cotton spinners to today's computer-run factories. It is still changing.

 1

The factory system began in Great Britain **during the 1700s** when engine-driven machinery began to be used to produce goods. **In 1790,** Samuel Slater brought the plans for a machine used to spin cotton to the United States. However, labor was scarce and the growth of other factories was slow at first.

Another important innovation was the use of interchangeable parts. **Until 1798,** parts for many products were created individually. Eli Whitney and Simeon North, makers of small weapons, designed parts that could be used interchangeably on any weapon they produced. Also, their **2** workers were trained to specialize in one particular job. Both methods speeded up production.

Another major improvement in factory methods was the moving assembly line, introduced by Henry Ford **in 1914.** Workers in the assembly line added the parts as the product moved past them on a conveyor. **Investigate:** *With what famous vehicle did Ford revolutionize transportation, and what was different about that vehicle?*

3

4 Then, **in the 1950s**, automation became common, as in this automated restaurant. Manufacturing methods also changed with the use of automation—automatic controls over machines used to make products. Automation relied on early computers. One eventual result of automation was industrial robots that replaced human workers for certain tasks, such as welding and spray-painting.

By the 1970s, however, the machinery and methods used in many U.S. factories had become dated. Use of automation was limited to only certain processes. Meanwhile, factories in Germany and Japan, with more modern technology and labor methods, had begun to out-produce those in the U.S. This was particularly true in the automobile industry.

5

6

Starting **in the 1980s**, some U.S. manufacturers began to change. They introduced new factory systems based on using computers throughout the production process. Factories became more flexible and required fewer workers. Product quality improved as well. These changes will lead one day to the factory of the future. **Investigate:** *Find out how many people are employed in factories today.*

Look around at the objects in your classroom. What do you see? Doors? Tables? Books? Desks? These items, and probably everything else in your classroom, are manufactured products. They are made in factories.

Most of the things you use every day were manufactured in factories. In fact, you may find it difficult to name things you use that were never inside a factory. Manufacturing is essential to our modern way of life. Without factories, we would not have cars, contact lenses, televisions, electric drills, breakfast cereals, books, microwave ovens, or CDs.

Manufactured items are sent to stores, where we can purchase them. We are consumers of manufactured products. Consumers are people who buy and use products. We are consuming products just by using soap, wearing clothes, and reading books. Without consumers, there would be no manufacturers.

Manufacturing is a remarkable process. In this chapter and the next, you'll learn to appreciate our manufacturing systems.

◀ This is a flexible manufacturing system.

The Evolution of Manufacturing

▶ **How have manufacturing systems evolved?**

Objectives

▶▶ **Discuss** how manufacturing evolved.

▶▶ **Explain** the difference between durable and non-durable goods.

▶▶ **Discuss** the importance of assembly lines and division of labor.

Terms to Learn

- **assembly line**
- **craft**
- **division of labor**
- **durable good**
- **Industrial Revolution**
- **non-durable good**
- **scientific management**
- **time and motion study**

Standards

- Core Concepts of Technology
- Relationships & Connections
- Cultural, Social, Economic & Political Effects
- Manufacturing Technologies

▌**durable good**
product that lasts three or more years

▌**non-durable good**
product that lasts less than three years

Suppose someone asked you to be responsible for setting up a factory to make skateboards. That would be a big job for anyone. You would probably have many questions, such as . . .

- What would the skateboards be made of? Wood? Plastic? Another material?
- How many skateboards would be made every day? Ten? One hundred? One thousand?
- Would they be expensive or moderately priced?

These important questions are just a few you might have concerning the project. People asked similar questions many years ago as the United States began developing into the strongest manufacturing nation in the world. In terms of value, about twenty-five percent of all the world's manufactured products come from the U.S.

All those products are either durable or non-durable. The federal government defines **durable goods** as those expected to last three years or longer. Television sets, hammers, and bicycles are all examples of durable goods. **Non-durable goods** are expected to last for less than three years. Calendars and packaged foods are examples of non-durable goods.

Early Manufacturing

▶ How has the evolution of manufacturing changed society?

The products that people used in the past weren't always made in factories. Many years ago, people had to make almost everything they needed themselves. They made their own cloth for clothing. They made their own farming tools, their candles for lighting, their wagons for transportation, their furniture, and even their children's toys. See Figure 16-1. It took a long time to make each item.

Figure 16-1 Many years ago, toys were mostly handmade.

TECH CONNECT
SOCIAL STUDIES

Sales of Durable Goods

In determining the state of the economy, government officials often use the most recent data for the sales of durable goods. Durable goods are usually more expensive than nondurable goods. The government thinks that if people are willing to invest in cars, televisions, and washing machines, they are confident about the economy.

ACTIVITY Find out what kinds of durable goods are currently being manufactured the most. Make a list of the top five. Compare the list to the top five from ten or twenty years ago.

Not everyone could do a good job at making so many different things. Some people could make good wagons but not good tools. Others were better at making cloth. People soon began to specialize in just one type of **craft**. They became blacksmiths, wagon makers, and shoemakers.

Around 1750, great changes started to take place around the world because of new methods of manufacturing. Because the new methods were so important, that period is called the **Industrial Revolution**. A revolution brings about great changes. Goods once produced by hand were now produced by machines in factories. See Figure 16-2.

craft
a skilled occupation, usually done with the hands, such as carpentry or sewing

Industrial Revolution
social and economic changes in Great Britain, Europe, and the United States that began around 1750 and resulted from making products in factories

Figure 16-2 This factory in Rhode Island, built in 1790, is a registered historic site.

scientific management
the system of developing standard ways of doing particular jobs

time and motion study
an investigation into the best ways of doing a job; things such as working conditions, wasted time, and unnecessary movement are considered

TECH CONNECT
LANGUAGE ARTS

Cheaper by the Dozen

Two of Frank and Lillian Gilbreth's twelve children wrote *Cheaper by the Dozen*, a book about growing up in their large family.

ACTIVITY Read *Cheaper by the Dozen*. How did Frank and Lillian Gilbreth use efficiency techniques with their twelve children?

division of labor
system of organizing production by giving separate tasks to separate workers or groups of workers

Cities grew in size when factories drew people from farms to cities. The Industrial Revolution was a great turning point in the history of the world, and it caused many social changes. It changed the U.S. economy from one based on farming to one based on industry.

However, factories were wasteful of workers' efforts. As factory production increased, people began to look for better ways to do things. Frederick Taylor worked for a Philadelphia steel mill in 1880. He was the first person to use a stopwatch to measure how long certain factory operations took to complete. He developed standard ways of doing things, making it easier for workers to manufacture products. His method is called **scientific management** and is an important part of modern factories.

The husband-and-wife team of Frank and Lillian Gilbreth also worked on better ways of doing things. Like Taylor, they worked in the new field of **time and motion study**. Time and motion study is a way of finding the way to do a job in the shortest time with the fewest movements. The Gilbreths improved factory working conditions and developed charts and techniques to measure factory efficiency.

Modern Manufacturing

▶ **How can all factories be different and similar at the same time?**

Early experts provided the foundation for establishing factories to manufacture products. However, each product is different and requires different materials and methods. You would not expect a modern company making breakfast cereal to use the same methods as one making computers. Surprisingly, they do have some characteristics in common.

Many companies use a **division of labor**. This means dividing the work into smaller steps done by certain groups of people. Workers develop specific skills. See Figure 16-3. A person who is good at electrical wiring will wire while someone good at painting will paint. Sometimes a flow chart is drawn. A flow chart is a diagram that shows the different steps in producing the product and how they relate to one another.

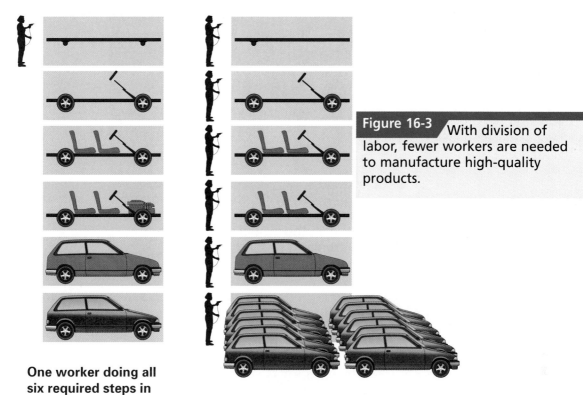

Figure 16-3 With division of labor, fewer workers are needed to manufacture high-quality products.

One worker doing all six required steps in manufacturing a product can make one unit.

Six workers, each specializing in one of six steps, can make twelve units in the same time.

Many companies use **assembly lines** or some type of assembly operation. In an assembly line, the workers often stay in one place and work on the product as it passes by. See Figure 16-4 (page 390). Automobiles and vacuum cleaners are put together this way. Some products are assembled automatically by machines. See again Figure 16-4. For example, most medicines and foods are automatically processed and packaged. Other products don't lend themselves to an assembly operation. Some examples are gasoline, paint, and tissue paper. Can you name some others?

During the manufacturing operation, each company makes many checks on the product's quality. A company making canned soup has to keep the machinery and the food very clean. The product has to taste good and be nutritious. Such companies also have strict governmental regulations they must follow regarding such things as content, safety, and portion size.

assembly line
series of work stations at which individual steps in assembly of a product are carried out as the product is moved along

Figure 16-4 On the left, workers do their part as the product goes by. Above, the product is assembled automatically by the machine.

16A SECTION REVIEW

Recall ▸▸

1. What is the difference between durable goods and non-durable goods?
2. What is the term for work done by blacksmiths, wagon makers, and shoemakers?
3. What was so revolutionary about the Industrial Revolution?
4. What is division of labor?

Think ▸▸

5. Do you think all goods should be durable? Explain.
6. A product can be made more quickly when it moves on an assembly line, rather than the worker moving from product to product. Why do you think this is true?

Apply ▸▸

7. **Time and Motion** Try a simplified version of a time and motion study. Write down, step by step, what you do to get ready for school in the morning. Keep a record of the time it takes. Compare your list with a classmate's list. Which one is more efficient? Explain.
8. **Flow Chart** Create a flow chart that describes the activities that take place in your home to prepare a meal. Show the work done by various individuals and how different activities intersect.

Organizing a Manufacturing System

How is a manufacturing system organized?

SECTION 16B

Objectives

▸▸ **Describe** manufacturing systems using the universal systems model.

▸▸ **Explain** the concept of added value.

▸▸ **Describe** several manufacturing tools and processes.

Terms to Learn

- added value
- abrading
- resin

Standards

- Core Concepts of Technology
- Information & Communication Technologies
- Manufacturing Technologies

Every manufacturing system can be understood by comparing it to the systems model. It requires inputs and uses processes. It produces outputs, and it responds to feedback.

Inputs

What inputs are needed by manufacturing systems?

All manufacturing companies use the same system inputs to produce their products. They use people, materials, tools and machines, energy, information, capital, and time.

People

What are some jobs people do in manufacturing?

Some companies, like General Motors (GM), the largest manufacturer in the United States, employ hundreds of thousands of people at many different plants. GM makes automobiles, automobile parts, trucks, buses, and locomotives at many factories. See Figure 16-5.

Figure 16-5 Some manufacturing plants require a large workforce. What are some of the jobs done in a large plant?

Large companies have many engineers and technologists. Some design the products the companies make and are called design engineers or designers/drafters. Others decide the best way to manufacture the products and are called production engineers. Quality control engineers and technicians inspect the product to be sure that it's well made. Many people operate manufacturing machinery, while others set up or adjust the machines. After the product is made, still other people distribute and sell it.

Production Materials

▷ **What are some of the different materials used to manufacture a product?**

The materials a company uses to make its products are called production materials, or engineering materials. They are different from raw materials. Raw materials are materials in their natural state. They include iron ore, trees, and raw cotton. Most production materials have already been processed to some degree. Trees, for example, have already been cut into lumber and dried.

Materials are chosen for their different properties. Plastic used to make balloons, for example, has to be soft and stretchy.

Companies stay in business by adding value to production materials. **Added value** can sometimes be simple. Take nail making, for example. A nail is a one-piece product made from a roll of strong steel wire. A nail-making machine puts a point at one end and flattens the other end. See Figure 16-6. The machine can make about ten nails a second. Companies add value to steel wire by changing it into nails.

Other companies combine several different materials or make a product with many different parts. An ordinary flashlight has ten parts: plastic base, top, lens, switch, lamp (bulb) holder, spring, two flat pieces of metal, lamp reflector, and metal lamp conductor. See Figure 16-7. The two batteries and bulb are purchased from other companies.

Some companies make products that have hundreds, or even thousands of parts. You probably have some complicated products right in your own home. Your refrigerator is made from many different materials. Can you think of others?

Tools and Machines

▷ **How do people use tools and machines?**

Tools and machines help people do work. A computer is used by writers. Proper equipment is used by space-walking astronauts and fire fighters. Your pencil helps you complete your homework, such as math problems.

added value
the increase in how much a material is worth after it has been processed into a finished product or a part for a finished product

Figure 16-6 The steel wire used to make these nails is more valuable now.

Figure 16-7 A flashlight has parts made from different materials. All the parts have to fit well and work together.

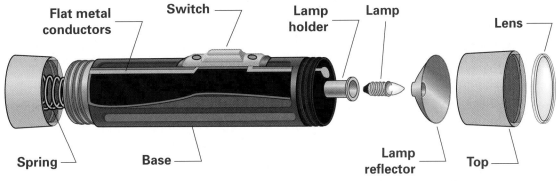

Flat metal conductors — Switch — Lamp holder — Lamp — Lens — Spring — Base — Lamp reflector — Top

In manufacturing, tools and machines change the shape of materials, cut or chip pieces from them, and fasten them to other materials. Hand tools are those that use the power of your hand or arm. A hammer is a hand tool, and so are pliers, screwdrivers, and wrenches. Most machine adjustments are made with hand tools such as wrenches and screwdrivers. See Figure 16-8.

Machine tools are bolted to the floor and operated by electric motors. They bend, cut, drill, grind, and hammer materials, among other things. Machine tools are some of the most important tools used in a factory. Some are as large as a room. Others are smaller, such as a drill press. Machine tools are used for such jobs as drilling holes in jet engines, bending steel for car doors, and cutting screw threads for the jack that may one day help you change a tire.

A portable power tool uses a small motor for power. It can usually be held in your hands. Electric drills and electric saws are portable power tools. See Figure 16-9. A table saw for cutting wood is an example of a power tool that can't be held in your hands or moved easily. People use power tools for such jobs as grinding sharp edges from bicycle frames, assembling electronic parts, and sanding flat surfaces on wooden furniture.

Figure 16-8 This industrial electrician is testing one of the circuits in a large machine. Maintenance of these machines is important in order to keep the factory running smoothly.

Other Inputs

▷ What other inputs do factories need?

As you know, energy is needed to run machines, heat buildings, and provide light. Capital is needed to build factory buildings, buy production materials, and pay workers. Time, of course, is needed to do all the jobs.

Another important input is information. Why do you think a factory needs information? For one thing, information in the form of drawings and instructions is needed by workers in order to make the products. What other kinds of information might be needed?

Processes

▷ What are some manufacturing processes?

The ways a factory adds value to its production materials are called processes. A process is a way of doing something. It involves a number of steps or operations.

Almost everything that takes place during manufacturing involves processes. Someone has to design and

Figure 16-9 Portable power tools are also important for certain kinds of manufacturing operations.

develop the product, organize and manage all the jobs that take place, make the product, distribute and sell the product, and service or make any repairs to products that are returned. In this section we'll talk about some of the processes that take place during production. To review the basics, see Chapter 3, "Processes, Tools, and Materials of Technology."

Changing Size and Shape

▷ How are size and shape changed?

Processing often involves changing the shape or size of a material. The shape or size of wood, metal, and plastic is often changed by separating or forming processes, such as cutting, bending, or casting.

Hand cutting of production materials is done with a knife, saw, and other hand tools. It can also be done with a portable power tool, such as an electric drill, or a large machine like a metal shear. Some cutting is now done with powerful jets of water or lasers.

Bending means to shape a material, usually into a curve or angle. It can be done by hand, as when people make woven wooden baskets. Large machines, called presses and brakes, are often used in factories to bend large pieces of metal.

Casting means to shape by pouring into a mold. See Figure 16-10. You cast ice cubes by freezing water in an ice cube tray. Companies cast plastic and metal. Many plastic toys are cast, and small gears are often made from cast metal.

Figure 16-10 Molten metal is poured into a casting to form a part for the finished product.

abrading
changing the shape of a material by rubbing off small pieces, such as with sandpaper

resin
(REZZ in)
a chemical compound used to make plastics

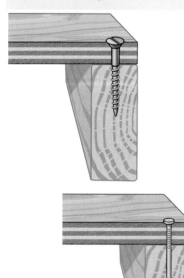

Figure 16-11 The two most common fasteners for wood are nails and screws. If you add glue to the joint, it will be even stronger. Which of the joints shown do you think is stronger?

Cutting, bending, and casting aren't the only ways to shape materials. For example, forging is hammering metal into shape after heating it. Some wrought iron furniture is forged. Extruding squeezes metal or plastic through an opening. It's much like squeezing paint from a tube. Many tubes and bars are extruded, as well as railroad rails. **Abrading** means to scrape or rub off small pieces. Filing, sanding, and grinding remove material by abrading it. Wood furniture is sanded to produce a smooth finish.

Chemical Processing

▶ How are chemicals used in processing?

Another way to process a material is by using chemicals. Food manufacturers do that when they make the biscuits sold in the frozen food department of your grocery store. Among other ingredients, they mix flour, butter, salt (NaCl), sugar ($C_{12}H_{22}0_{11}$), baking soda (NaH[CO_3]) and water (H_20). They combine everything required, shape it, and heat it in an oven. The delicious biscuits that come out are partly the result of chemical processes. When a material's properties are changed in this way, the process is called conditioning.

Chemicals are used for etching the circuit boards in your computer. Chemical acids cut pathways into the surface of a board. The electrical signals follow these pathways.

Plastics are often processed from chemicals. The basic material is called a **resin**. Other ingredients are added to produce special properties, such as strength.

Chemicals are sometimes used to accurately remove metal from special parts, like those in a jet engine. A flowing chemical solution finishes the surface, leaving it very smooth.

Fastening

▶ What are some common fasteners?

Fastening is combining two or more parts together. Furniture, for example, is fastened together with nails, screws, and glue. Nails and screws are mechanical fasteners. Many companies use screws because they have more holding strength than nails. See Figure 16-11. Glue is an adhesive, which is a chemical fastener.

Plastic parts can be fastened with snaps. A small extension on one part snaps into a hole in another and holds everything together. This method allows the parts to be easily taken apart and recycled. Plastic packaging is often made this way.

Plastic and metal parts can also be fastened by melting. Some plastic CD cases are heated and melted together. No mechanical fasteners or adhesives are necessary. Many metal parts of a car body are welded (melted) together with high heat.

Outputs and Feedback

▶ **What outputs and feedback are part of a manufacturing system?**

The products themselves are the main outputs of a manufacturing system. Can you think of others? Providing jobs for a community is one. Producing waste or unwanted by-products might be another.

Feedback can also come in many forms. High demand for the product is part of feedback, and so is low demand. The manufacturer must review what went right or wrong to determine if changes need to be made. Can you think what changes might be needed if a product is *too* successful?

SECTION REVIEW 16B

Recall ▶▶

1. What is the difference between a design engineer and a production engineer?
2. How are production materials different from raw materials?
3. What is added value?
4. Name two kinds of mechanical fasteners.

Think ▶▶

5. Think of a common manufactured item, such as a bike. Use the systems model to describe the system used to make the item.

6. Why do you think it is better to weld a car body together instead of using mechanical fasteners?

Apply ▶▶

7. **Research** Collect samples of different mechanical fasteners. Mount them in a display. Label each one and tell what it is used for.

Objectives

▶▶ **Describe** general steps in setting up and running a small factory.

▶▶ **Explain** the function of market research, quality assurance, and just-in-time delivery.

Terms to Learn

- **commission**
- **just-in-time delivery**
- **market research**
- **marketing**
- **profit**
- **quality assurance**
- **supplier**

Standards

- Core Concepts of Technology
- Attributes of Design
- Engineering Design
- Troubleshooting & Problem Solving
- Energy & Power Technologies
- Manufacturing Technologies

market research
the process of getting people's opinions about a product so that a company knows what changes to make or whether to sell the product

Running Your Own Factory

▶ How could a teen start a factory?

Someday you might come up with a good idea for a new invention or product. You might also think of an innovation for an existing product. What would you do about it?

Suppose you have an idea for a new board game for two to four players. In addition to the board, the game calls for a colored button for each player and cards with questions. You think you might like to manufacture and sell it yourself.

Market Research

▷ How can you find out what products people will purchase?

1. Using your creative imagination, you would plan the game and then sketch what you have in mind. See Figure 16-12. By putting your idea on paper you can show others what you are thinking. The drawing is part of your design.

2. You would want opinions from other people, so you talk to your friends and teachers. Each one tells you how he or she thinks you can improve your idea. This is market research. **Market research** is a way to find out what people will purchase.

Figure 16-12 You might sketch out your game on a big piece of paper and then get input from your teachers and friends.

3. You would make a prototype of the game and test it with some friends. After some changes, you might be satisfied that the game works.

Manufacturing the Product

▶ Who are the suppliers in a manufacturing process?

4. Suppose it's almost summer and you think you'll have enough time to make the game yourself. You plan to manufacture it in your home. You feel that this personal touch might appeal to people who like hand-made crafts.

5. Suppose you can draw and color the playing surface on white paper, but you can't make the heavy cardboard back. You want the playing board to fold up as in other board games. You talk with the owner of a picture framing shop, and he says that he can make 100 cardboard backs for $1.20 each.

   A friend of yours has a computer printer that can print on lightweight cardboard. She agrees to print the game's question cards for $1.50 per set. She also agrees to print the instructions.

   Both the picture framing shop and your friend are your suppliers. A **supplier** is a person or company that provides something for your manufactured product.

   Of course you also want to sell the game. You check with four stores in your town. Three owners agree to offer your game for sale.

6. The framing shop requires an order of at least 100 cardboard game backs. So you make out a budget based on manufacturing 100 games. See Figure 16-13.

100 cardboard game backs, $1.20 each	$120
100 packs of question cards, $1.50 per pack	150
400 colored buttons, $0.10 each	40
100 plastic bags to package games, $0.25 each	25
Colored pens, paper on which to draw the playing surface, rubber cement, and other items	65
TOTAL	**$400**

Your parents go with you to the bank and help you borrow $400 for the project. That $400 becomes your capital.

▌**supplier**
person or company that provides one or more of the materials or parts for a manufactured product

Figure 16-13 You would carefully plan out the costs to produce and manufacture your game, being very specific.

just-in-time delivery
a method of scheduling the arrival of materials at the time they are needed so that storage is not necessary

Figure 16-14 All of the materials would be brought together during production of your game.

quality assurance
the process of inspecting products to make sure they meet all standards that have been set

7. You ask your two suppliers to send you ten game backs and ten sets of question cards every week for ten weeks. You don't have room to store much at home, so you want to receive those items just in time to do the work. **Just-in-time (JIT) delivery** is used by many manufacturers. With JIT, they don't need extra warehouse space or workers to manage items.

8. Before you draw the playing surface, you get your production materials together. This includes the white paper, pens, ruler, and other items necessary to draw the playing surface. See Figure 16-14. Now you go into production and draw the playing surface for ten boards. You have added value to the materials.

9. Malfunctions will affect the quality and enjoyment of your game. You carefully inspect each playing surface you finish. You want to be sure that you did a good job. This inspection is called **quality assurance** or quality control. It means you will meet the requirements, or standards, set for the games. You glue the ten playing surfaces to cardboard backs from the framing shop. You check the glued paper to be sure that you did it well. This is also quality assurance. Quality assurance is one of the most important parts of manufacturing. It enables you to use information about your system to change the system as needed.

10. Now it's time to assemble the parts of the game. You ask a friend to help. You place four different-colored buttons in a small plastic bag. You add a pack of question cards. Then you pass the bag to your friend. Your friend places the game board, instructions, and smaller plastic bag into a larger plastic bag. This is like working on an assembly line in a factory.

ETHICS in ACTION

Good Business

When selling a product, your reputation is important. If you advertise something about your product that isn't true, your customers may not want to do business with you in the future. Keeping ethics in mind while manufacturing, marketing, and selling any product is good for you, your product, and those you do business with.

ACTIVITY Survey five people. Ask if they have ever purchased a product they were not satisfied with. What was the reason behind their complaint? Did they take any action against the manufacturer?

Marketing and Sales

▶ What is involved in marketing your product?

11. Next, you have to plan the **marketing** for your game. Just putting it out for sale isn't enough. You have to tell people about it. They have to know what it is, where to buy it, and what it costs. It's also a good idea to tell them why they might want to buy it—that it's fun to play, that it will tickle their brain cells, that it's hand crafted. One way to do this would be to create flyers using your computer. Then you could print them and post them where they'll be seen.

12. You drop off two at each of the stores that agreed to sell your game. The store owners will charge you one dollar for each game they sell. This is their commission. A **commission** is payment based on sales.

13. You price the games at ten dollars each. How much profit will you make? **Profit** is the money left over when all your bills are paid. Don't forget that the bank loan, too, must be repaid. If you sell all of the games, how much money will you make?

marketing
telling potential customers about products and services in such a way as to make them eager to buy

commission
a payment made to a sales-person or agent for business they have done

profit
money left over after all the bills for making a product are paid

16C SECTION REVIEW

Recall ▸▸

1. What is market research?
2. What do the letters JIT mean?
3. What is a supplier?
4. Describe the purpose of quality assurance.

Think ▸▸

5. Name some items on a bicycle that probably came from the company's suppliers.
6. Name some ways that a sawmill adds value to trees and a furniture company adds value to lumber.

Apply ▸▸

7. **Maintenance** Suppose several customers returned the board games discussed in this section. They report such problems as missing cards or buttons, a loose playing surface, and instructions that are hard to follow. Outline a maintenance system for inspecting and servicing these products so they will function properly.

Summary Activity

For each numbered blank, pick the answer from the list on the right that makes the most sense in the entire passage. Write your answer on a separate sheet of paper. No answer will be used more than once.

In terms of value, about twenty-five percent of all the world's manufactured products come from the United States. Factories make two kinds of goods: durable and __1__ goods. Before there were factories, people specialized in __2__. Then came the __3__ and factories boomed.

In factories, __4__ materials are converted into finished products. The workers develop special skills partly because the companies use a(n) __5__ of labor. The workers put everything together using a(n) __6__.

To start a company or make a new product, a person first needs an idea and creates a(n) __7__. The final product will be made from production materials and parts provided by __8__. When everything is assembled, __9__ will have been added by the manufacturing process.

Factories use tools, materials, and processes. Tools are used to change the __10__ of materials or hold them together. A hammer is a __11__ tool, and an electric drill is a(n) __12__ tool. __13__ is what is used to combine two or more materials together. __14__ are sometimes used to remove metal from special parts or to etch circuit boards in computers.

__15__ is a way to find out what people will purchase. Quality __16__ is a method for being sure products meet quality standards.

Answer List

- assembly line
- assurance
- chemicals
- crafts
- design
- division
- fastening
- hand
- Industrial Revolution
- power
- market research
- non-durable
- production
- shape and size
- suppliers
- value

Comprehension Check

1. Describe general steps in setting up and running a small factory.
2. What is a craft?
3. What is a time-and-motion study?
4. List at least three processes used to change a material's size or shape.
5. What is profit?

Critical Thinking

1. **Organize.** Use the systems model to place the following steps for manufacturing cookies in order: packaging the cookies, a customer eats the cookies, hiring workers, making the cookies, hearing complaints from customers.
2. **Compose.** Write an advertisement for a product you've purchased based on your own reasons for buying it.
3. **Judge.** Complete quality assurance for an X-ray machine, a cereal, and a basketball. Which aspects of each product would be most important to check and why?
4. **Compose.** Many young children worked in factories during the Industrial Revolution. Research and write a paragraph on the child labor laws in your state.
5. **Distinguish.** Determine which of the following products would be best put together on an assembly line: a custom-made guitar, roller skates, a cruise ship. Give your reasons.

Visual Engineering

Toying with Technology. Some antique toys have become very valuable. Certain museums contain just children's toys. Imagine if you had been a small child 150 years ago. What kinds of toys would you have played with? Think of one example. How do you think that toy would have been manufactured without the use of electricity and power tools? How would you make the same toy today?

TECHNOLOGY CHALLENGE ACTIVITY

Manufacturing Bookstands

Equipment and Materials

- three pieces of wood, each about 1 inch thick, 6 inches wide, and 18 inches long
- six pieces of wood, each about 1 inch thick, 6 inches wide, and 8 inches long
- twelve ⅜-inch-diameter dowels, 1½ inches long
- ruler
- pencil
- glue
- hammer
- hand drill with ⅜-inch drill bit
- fine sandpaper
- clear finish in an aerosol spray container

! SAFETY

Reminder
In this activity, you will be using tools and machines, as well as an aerosol varnish. Be sure to always follow appropriate safety procedures and rules. Remember, safety is an attitude that you must develop and maintain at all times.

Background

The manufacturing process includes designing, developing, making, and servicing the product. This involves many steps, so being organized is important. In fact, it may be the most important key to running an efficient system.

Goal

For this activity, you and two teammates will set up a production system to manufacture adjustable bookstands. See Figure A. When you are finished, each of you will have a bookstand to use.

Criteria and Constraints

▶▶ For at least part of your operation, you should use an assembly line and division of labor. At the end of the project, you must turn in a flow chart showing how the materials moved along the assembly line.

▶▶ You should establish a method of controlling quality and servicing products that do not meet quality standards. At the end of the project, you should turn in a brief paragraph describing the process you used.

▶▶ The first bookstand you produce should be considered a prototype. It should be studied for ways in which to improve your methods. Keep a record of any changes you make.

▶▶ At the end of the project, you should recycle or dispose of any waste materials so as to use resources wisely.

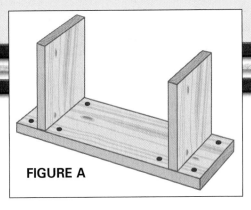

FIGURE A

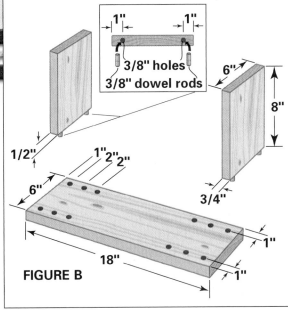

FIGURE B

Design Procedure

The following steps are only for making the bookstands. You must determine your own steps for setting up the system.

❶ Look at the assembly drawing in Figure B to get an idea of how to construct your bookstands.

❷ Establish your method of controlling quality and servicing products that do not meet quality standards.

❸ Start with one of the 8-inch-long pieces of wood. Drill two ⅜-inch-diameter holes, 1 inch deep, in one end. The center of each hole should be 1 inch from the edge. Repeat for the other five bookstand ends.

❹ Place a small amount of glue on the end of a 1½-inch-long dowel rod. Using the hammer, tap the glued end all the way into one of the holes. Repeat for the other eleven dowel rods and holes.

❺ Drill twelve ⅜-inch diameter holes in the bottom of the 18-inch-long pieces of wood. See Figure B for their location. Drill each hole just a little deeper than ½ inch— say ⅝ inch. Repeat for the other two book-stand bottoms.

❻ Wait for the glue to dry on the ends of the dowels. Then see how well the dowel rods in the bookstand ends fit the holes in the bottom. Find the best combination of ends and bottoms.

❼ Use sandpaper to smooth all the surfaces and edges. Using proper ventilation, spray the bookstand with one or two coats of clear finish.

❽ Write a brief paragraph describing the quality-control process you used. Create your assembly line flow chart.

❾ Dispose of waste properly.

Evaluating the Results

1. The bookstand ends probably fit into some holes better than others. That's normal. No matter how hard we try, it's quite difficult to make two parts exactly the same. What could you have done to improve the quality of your adjustable bookstand?

2. Suppose that you wanted to organize your own factory to make and sell thirty adjustable book-stands. In making that many, what would you do differently?

SUPERSTARS OF
MANUFACTURING

INVESTIGATION ////

Jacquard's punched cards are a good example of a technology developed for one purpose that was eventually used for other purposes. Think of a common household item and make a list of at least three other purposes it could be adapted for.

Joseph Marie Jacquard
(1752-1834)

Joseph Jacquard's home country of France was known for its finely woven patterned silk. In 1804, Jacquard invented a punched card system to help make weaving those complicated patterns less difficult. As the cloth was woven, metal rods dropped through the holes in the punched cards to lift or drop different colored threads. By 1812, France had 11,000 Jacquard looms. The first computers used punched cards based on the ones Jacquard created for his loom.

W. Edwards Deming
(1900-1993)

After most factories in Japan were destroyed during World War II, W. Edwards Deming went there to help the Japanese evaluate and improve the quality of their manufactured products. As a result of his success, everyone was soon seeking Deming's advice. American automobile manufacturers and many other companies began to use his methods.

Ernesto Blanco   (b. 1922)

Born in Cuba, Ernesto Blanco emigrated to the United States in 1949 where he earned a degree in mechanical engineering. Today, he teaches engineering design, and he is best known for his many inventions that aid the handicapped. His first major invention was the stair-climbing wheelchair with retractable, spring-loaded spokes. He also created an electric Braille typewriter (shown here) that is still widely used.

Marion Donovan
(1917-1998)

As she was growing up, Marion Donovan spent much of her free time in her parents' factory. After starting her own family, her early experience paid off. At the time, baby diapers were made of cloth, usually leaked, and had to be washed. In 1946, using a series of shower curtains, Donovan invented a leakproof, reusable diaper cover. She then went on to invent the first disposable diaper made of absorbent paper. Today, disposable diapers are almost the only type used.

Jan Matzeliger   **(1852-1889)**

Jan Matzeliger emigrated to the U.S. from Suriname, in South America, when he was 21. He got a job making shoes and eventually moved to Lynn, Massachusetts, where half the shoes in the country were made. At the time, much of the work was done by hand. Matzeliger invented a shoe-stitching machine that automatically assembled shoes five times faster than doing it by hand. Because of decreased labor costs, the price of shoes came down and almost everyone could afford to buy factory-made shoes.

George C. Devol, Jr.   **(1900-1994)**
Joseph Engleberger   **(b. 1925)**

George C. Devol, Jr. patented the first robot in 1954. In 1958, Joseph Engleberger became president of Unimation, a company formed to develop and promote Devol's robots. Their first robot, Unimate I, was built in 1961. Using a combination of computer processing, electricity, and hydraulics, Unimate could respond to 200 commands.

CHAPTER 17

MANUFACTURING IN THE 21st CENTURY

In the United States, there are about 405,000 factories. The U.S. produces more manufactured products than any other country in the world. Each year . . .

- Over 180,000 patents are granted.
- Over $3 billion worth of cell phones are manufactured.
- Over 13 million motor vehicles are built.

Advances in technology have enabled American factories to produce large quantities of goods quickly.

New technologies and products usually result from the demands, values, and interests of businesses and industries, as well as consumers. New materials and processes are constantly being developed and used. For example, twenty years ago most factories made little use of computers. Today, most factories use them in some way.

In this chapter you will learn about how new products are developed. You will also learn about advances in manufacturing. Finally, you will find out how even the way products are sold is changing.

Robotic arms spray paint a new car on the assembly line.

Objectives

▶▶ **Explain** the purpose of research and development.

▶▶ **Discuss** how products are designed.

▶▶ **Describe** a virtual factory.

Terms to Learn

- R&D
- virtual factory

Standards

- Characteristics & Scope of Technology
- Core Concepts of Technology
- Attributes of Design
- Troubleshooting & Problem Solving
- Design Process
- Manufacturing Technologies

R&D

research and development; manufacturing department that searches for and develops new products and processes

Preparing for Manufacturing

▶ How do companies prepare for manufacturing?

Before manufacturing can begin, decisions must be made. What product will be manufactured? How will it be designed? How will the factory be set up to make it?

Research and Development

▶ Where do new products and advances in manufacturing come from?

A company's **R&D** (research and development) department is where invention and innovation take place. That is where new products and methods are usually born. Research looks for the new ideas. Development uses the research to create the new products and processes. The people who work in this department are often creative. See Figure 17-1. They enjoy trying new things and considering all ideas. They often solve technological problems by experimenting.

Manufacturers may study products already on the market and try to improve on them. They may also look at what scientists are investigating and find ways to turn the new information into products people can use. Very often, new products and processes are developed to solve problems or fulfill needs.

Suppose, for example, that the research group at Z. Z. Zipper Company is told that people might be interested in buying a zippered folding boat. They look for a new soft plastic that could be made into a leakproof zipper. The development group makes sure that the new plastic is

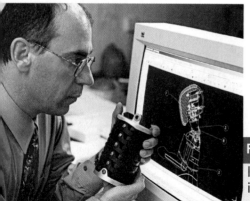

Figure 17-1 Research engineers look for new products or new ideas for old products.

strong enough. They choose a specific nylon fabric for the boat. The research group and the development group work together closely.

Product Design

▷ What are some of the jobs of the engineering design department?

The design department works closely with the R&D department. Design engineers help decide what requirements a product should meet, such as how big it should be. They identify the criteria and constraints involving the color, the material, the shape, and everything else that is part of a new product's design. See Figure 17-2. They also determine what trade-offs may be necessary among competing values. Values include such things as cost, availability of materials, desirability of product features, and waste. For example, can the product be a little smaller to save money or to conserve material supplies?

The engineers at Z. Z. Zipper obtain the R&D department's information. They have to answer an assortment of questions:

- What should be the length of the boat?
- The nylon fabric will wrap around a frame, much like an umbrella. Should the frame be wood or aluminum?
- Should the boat frame fold up or come completely apart for storage?
- Should the nylon fabric be in two pieces or three pieces?

In today's factories, the engineers draw their plans with computer-aided design (CAD) equipment. The plans are stored in the computer's memory and are later used by the production department.

Planning the Factory

▷ In what ways does product design affect factory design?

After a new product has been developed and designed, the factory itself must be planned. What kinds of machines will be needed to make it? How many products will be made, and how many workers will be needed?

Some products are very complicated to assemble. A new automobile model, for example, might have 15,000 to

Figure 17-2 A car prototype like this one will help the design engineers make decisions about the proportions and shape of the new car.

TECH CONNECT

SOCIAL STUDIES

Developing Countries

Developing countries are those that might have ample natural resources but are not prepared to take advantage of them. Unlike the industrialized nations, they usually do not have many factories.

ACTIVITY Obtain a world map and pick a continent. List all the countries on that continent. Indicate which you think are industrialized and which are developing. Look up one of those countries in a library resource to see if you are right about it.

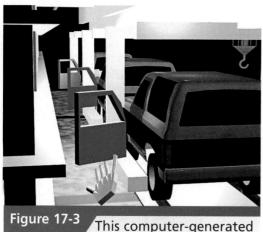

Figure 17-3 This computer-generated virtual factory allows the manufacturing engineers to try various layouts.

20,000 parts. The parts include everything from tiny screws to large door panels. It is sometimes difficult to figure out the order in which everything should be put together.

That's where the virtual factory comes in. A **virtual factory** is a three-dimensional image that appears only on a computer screen. See Figure 17-3. It is not real. Virtual factory software allows manufacturers to "try out" different layouts. Animation lets them observe the work flow before they actually set up the factory or change the process. It is just one of the ways computers have changed manufacturing.

virtual factory
a three-dimensional computer model of a factory

17A SECTION REVIEW

Recall ▸▸

1. What is the name for a factory that appears only on a computer screen?
2. What is the difference between research and development?
3. Name four factors involving a product's design that are the responsibility of the design department.

Think ▸▸

4. Suppose you are a design engineer working with outdoor furniture. Think of at least four questions you might have to answer about your product.

5. Suppose you are responsible for setting up assembly lines to make bicycles. What are some ways in which you could organize the work?

Apply ▸▸

6. **Construction** Use plastic straws to construct a model of the frame for a folding boat. See if you can figure out some problems the manufacturer could avoid. Document your solutions with drawings or written explanations.
7. **Innovation** How innovative can you be? Recycle a plastic milk container into a useful item.

Producing the Product

▶What goes on during production in a modern factory?

The production department is responsible for actually making the company's product. See Figure 17-4. It must be a quality product and manufactured to a specific schedule. A **schedule** is a plan for what must be done by a certain time.

For example, at the Z. Z. Zipper Company, the production engineers decide that one supplier will provide nylon fabric cut to the correct size. Another supplier will manufacture the aluminum frame. The Z. Z. Zipper Company will make the zipper, attach it to the nylon fabric, assemble all the parts, and put them into a box. They must decide which methods to use. Because Z. Z. Zipper is a modern factory, computers will play an important part.

CNC, CAM, and CIM

▶ How are computers used in modern factories?

You already know that CAD stands for computer-aided design or drafting. CNC, CAM, and CIM represent production activities that use computers.

Objectives

▶▶ **Explain** the differences among CNC, CAM, and CIM.

▶▶ **Discuss** the use of industrial robots and e-manufacturing.

▶▶ **Discuss** the roles of quality assurance and safety in modern manufacturing.

▾ Terms to Learn

- CAM
- CIM
- CNC
- e-manufacturing
- NIOSH
- OSHA
- robotics
- schedule
- standard
- troubleshooting

Standards

- Core Concepts of Technology
- Relationships & Connections
- Troubleshooting & Problem Solving
- Design Process
- Information & Communication Technologies
- Manufacturing Technologies

▾ schedule
a plan that includes the work to be done and when it should be finished

Figure 17-4 This aircraft manufacturing company has a clean, pleasant, and well-lit production department.

CNC

computerized numerical control; machine tool operation controlled by commands from a computer

CAM

computer-aided manufacturing; a system that uses computers to coordinate the machinery in a factory

CIM

computer-integrated manufacturing; the use of one computer system to control the design, manufacturing, and business functions of a company

With **CNC**, or computerized numerical control, machine tools operate by commands from a computer. The operator types in the instruction and the machine does the work. With **CAM**, or computer-aided manufacturing, machine tool operators program computers to operate all the machinery. The parts being manufactured automatically go from one machine to another. See Figure 17-5.

With **CIM**, computer-integrated manufacturing, all the computers in the company are linked together, or integrated. Design and production departments can communicate instantly. The purchasing department can tell just-in-time (JIT) suppliers when to deliver production materials. See Figure 17-6. The marketing department can plan when to start advertising products. Management can direct the entire company from one location. CIM is a company-wide process and is more difficult to start up than CAM.

Many companies now do what they call lean manufacturing. Lean manufacturing—also called smart manufacturing and common sense manufacturing—is efficient manufacturing. Lean manufacturers use just-in-time delivery from suppliers. Materials come from suppliers just when they are needed. The company makes only what it can sell quickly, and it keeps only a few finished products in the warehouse.

Figure 17-5 This CAM operator helps program where all the parts for this automobile need to be in the assembly line so that the car will be manufactured efficiently.

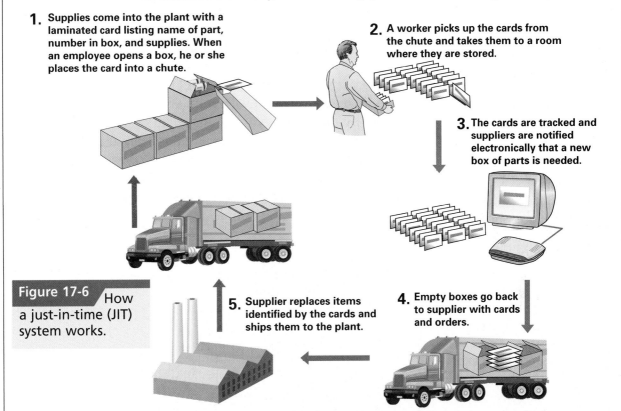

JUST-IN-TIME DELIVERY SYSTEM

Toyota's just-in-time delivery concept calls for the manufacture and delivery of only what is needed, when it is needed. The purpose is to enhance efficiency, reduce storage, and increase savings.

1. Supplies come into the plant with a laminated card listing name of part, number in box, and supplies. When an employee opens a box, he or she places the card into a chute.

2. A worker picks up the cards from the chute and takes them to a room where they are stored.

3. The cards are tracked and suppliers are notified electronically that a new box of parts is needed.

4. Empty boxes go back to supplier with cards and orders.

5. Supplier replaces items identified by the cards and ships them to the plant.

Figure 17-6 How a just-in-time (JIT) system works.

Quality Assurance

▷ Can computers help with quality assurance?

Companies build quality into their products during manufacture. They don't just inspect the final product and throw away all the bad ones. That would be expensive and wasteful.

Products are inspected many times during the manufacturing process. See Figure 17-7. If a problem is found, the operators can quickly adjust the machine tools. Good quality assurance means that there are fewer parts that have to be thrown away. It also means that the product is the best the company can make.

Figure 17-7 Inspection is an important part of quality control. This motorcycle must pass several check points before it leaves the factory.

troubleshooting
a problem-solving method used to identify a malfunction in a system

standard
a rule or guideline for making a product

robotics
the technology involved in building and using industrial robots

Many companies use computers and special instruments to help control quality. One type of computer-controlled machine measures parts and compares the sizes with the dimensions given on plans for the product. Microscopes and X-ray machines may be used as well.

If too many products are defective, a problem may exist in the system. **Troubleshooting** is the method used to identify the malfunction. The problem is studied and different processes are tested until the cause is found.

Some products must be made according to standards set by the government or a special organization. A **standard** is a rule or guideline. Standards are used so that certain products conform to a particular size, shape, or level of quality. The American National Standards Institute (ANSI) publishes standards for such things as screws and other fasteners. Standards provide another way to measure quality.

Robots

▷ Are industrial robots like the robots seen on television and in movies?

Robotics is the technology of industrial robots. Industrial robots were invented in the 1960s and are being used more and more in manufacturing. They take their instructions from a computer that has been programmed by a worker. See Figure 17-8.

Figure 17-8 Many robots are used in the production of this car.

Industrial robots are not the walking and talking type you may have seen on television or in movies. An industrial robot usually has one mechanical arm and is classified as a machine tool. The end of the robot's arm might have a gripper to grasp items and move them from place to place.

The end of the robot's arm might also have a tool or welding tip on it. The robot uses electricity to heat metal until it is soft enough to weld together. There are many uses for robots, but welding and painting are the two most common. Robots can be easily programmed to do other tasks.

Robots do jobs that are dangerous, boring, or otherwise unpleasant for people. Although they are expensive, companies have found them to be very useful.

E-Manufacturing

▷ What is e-manufacturing?

Manufacturers use electronic information in many different ways. For instance, they use it to design parts on a computer, send e-mails, and process paychecks. The term **e-manufacturing** means using electronic information in the manufacture of a product. See Figure 17-9. In one type of e-manufacturing, all machine tools are linked to the Internet. After using a password, a person can connect directly to the machine with a laptop computer, a personal digital assistant (PDA), a cell phone, or other devices. In e-manufacturing, an operator can be anywhere. The Internet provides the link to control the machine.

Almost all manufacturers have suppliers. Some have hundreds. With e-manufacturing, a company can stay in immediate contact with them. The company can manage different suppliers of the same product and keep inventories up to date using the Internet.

LOOK TO THE FUTURE

Machines That Think

As you know, artificial intelligence is the use of computers to solve problems and make decisions ordinarily handled by humans. Artificial intelligence may one day be used in manufacturing. For example, AI can sift through huge amounts of information looking for patterns in the way customers select products. This could save a company time and money by predicting just what customers want to buy.

e-manufacturing
using electronic information in the manufacture of a product

Figure 17-9 Almost all communication is done electronically in the manufacturing environment.

Isaac Asimov wrote science fiction for young people and adults. He created the "Three Laws of Robotics" to govern robot behavior.

ACTIVITY Research Isaac Asimov's three laws. Do you think they apply to industrial robots? Explain.

OSHA

Occupational Safety and Health Administration; the government agency that sets safety rules and checks to make sure the rules are being followed

NIOSH

National Institute of Occupational Safety and Health; the agency that approves the use of protective equipment, such as safety glasses

Safety

▷ How are factories made safer?

Nothing is more important in a factory than the safety of the people who work there. Safety means freedom from injury or any danger of injury.

Eye protection must be worn when a person uses any tool, even a hand tool. A hammer hitting metal might chip a piece from the hammer or metal workpiece. A face shield or safety glasses protect a person's eyes. Since 1947, the Wise Owl Program has recognized employees and students who saved their sight by wearing eye protection. The program is sponsored by the National Society to Prevent Blindness.

Other types of protective equipment include hard hats, earplugs, gloves, and safety shoes with steel toes under the leather. When the air contains dust or vapors, people wear special filters over their nose and mouth to protect their lungs.

When robots are used in a factory, they require special safety devices. Some robots are placed behind high fences, and when the gate is open the robot's power switches off. Another safety device is a carpet that has switches inserted in it. When someone steps on the carpet, the switches turn the power off.

Many factory safety rules are required by federal or state laws. **OSHA**, the Occupational Safety and Health Administration, establishes safety rules and checks up on companies. **NIOSH**, the National Institute of Occupational Safety and Health, approves protection equipment such as hard hats and safety glasses. See Figure 17-10.

Companies look for workers with good safety sense. Do you have good safety sense? Just follow five simple rules:

- Make sure you follow safe procedures.

- Wear the proper safety equipment.

- Learn how to use a tool before you start. Check with your teacher for instructions.

- Inspect tools regularly. If the tools are worn or damaged, they should be immediately repaired or replaced.

- Work slowly and always keep safety in mind.

Figure 17-10 All industrial safety equipment must be approved by NIOSH, the National Institute of Occupational Safety and Health.

SECTION REVIEW 17B

Recall ▸▸

1. What are the two most common tasks industrial robots do?
2. What does the term e-manufacturing mean?
3. What do CNC, CAM, and CIM stand for?
4. What do OSHA and NIOSH stand for?

Think ▸▸

5. Why do you think an industrial robot is classified as a machine tool?
6. Companies sometimes claim that "Quality is built into our products." What do you think they mean?

Apply ▸▸

7. **Design and Test** Design and build your own protective container for shipping an egg. To test it, drop the container holding a fresh egg from a height of 6 to 10 feet onto the floor. Some recommended rules: (a) The container must be no larger than 4 x 4 x 4 inches. (b) Everyone cleans up his or her own mess. (c) The egg will be put into the container before the test, so the container cannot be built around the egg. (d) Every container must have solid sides; no fair just wrapping it with foam rubber. (e) To promote creative packaging techniques, no padding should be used.

Objectives

▶▶ **Explain** the purpose of the marketing department.

▶▶ **List** several forms of advertising.

▶▶ **Explain** the difference between a wholesaler and a retailer.

Terms to Learn

• advertising
• retailer
• wholesaler

Standards

■ Core Concepts of Technology

■ Design Process

■ Information & Communication Technologies

■ Manufacturing Technologies

advertising
making a public announcement that a product is available

Selling the Product

▶ **Have there been advances in the way products are marketed?**

The final steps in manufacturing are handled by the marketing department. As you know, marketing is telling potential customers about the company's products and services in such a way as to make them eager to buy.

Marketing can be done in two main ways. A company usually advertises its products. It may also hire salespeople to tell customers about the products directly.

Advertising

▶ **How is advertising done?**

Advertising is making a public announcement that your product is available. You have probably seen thousands of advertisements during your lifetime. They appear on television, are heard on the radio, and are seen in newspapers and magazines. Advertising can help create demand for a product.

In recent years, one form of advertising has changed. Outdoor advertising used to be done primarily on billboards—big signs along a roadway. Today, outdoor advertising is all around. See Figure 17-11. You can see it on the sides of trucks and other vehicles and on clothing. Do you have a pair of designer jeans with the designer's name on them? If so, you're a walking advertisement for that designer's products.

Another big change in advertising has occurred because of the Internet. It's become a new medium in which companies can talk about their products. Some companies have their own Web sites. Others insert ads on related sites. Still others send e-mails to potential customers.

Figure 17-11 Have you ever seen advertising like this? What was being advertised?

Assertiveness

People often trust the judgment of assertive people and believe in their capabilities.

When you present yourself, your abilities, and your ideas with confidence, you are showing assertiveness. Some people confuse assertiveness with arrogance, but an arrogant person is full of self-importance and contempt for others. Assertive people show respect for other people and their abilities. To become assertive, you need to:

- ✔ **Share your opinions and ideas in a friendly way.**
- ✔ **Stay informed so that those opinions and ideas are worth sharing.**
- ✔ **Make an effort to know people in authority better and ask for their advice.**

Sales

▷ **What is the difference between direct and indirect sales?**

wholesaler
the merchant who buys large quantities from a manufacturer and sells smaller quantities to retailers

retailer
a merchant who buys products from a wholesaler and sells them to consumers

Salespeople are responsible for closing the deal with a customer. Some sales are made directly to the person or people who will use the products. At other times, the product is sold to a store of some kind.

For example, remember the Z. Z. Zipper Company? The production department assembled all the boat parts and put them into a strong cardboard box. Twenty packed boats went to the company's warehouse every day. To sell the boats, the marketing department contacted the B. B. Boat Sales Company.

B. B. Boat is a wholesaler. A **wholesaler** is a company that purchases large numbers of products from a manufacturer and then sells smaller numbers to retailers. A **retailer** is a company or store that sells products to consumers. See Figure 17-12. Examples of retailers are Target and Wal-Mart. Moving products from the manufacturer to the consumer is called distribution.

Figure 17-12 Working in a retail store requires good communication skills. What else would be important?

The Internet has also changed the way selling is done. Many stores and manufacturers now have Web sites where you can order products. Other sites, like Amazon.com, do all their business on the Internet. They do not have a store that you can visit.

Some manufacturers have begun to work directly with the customer starting at the design stage. Customers can order custom-made products over the Internet. All processes are computerized, and many extra steps can be eliminated. One example is a company that sells a new kind of music CD. Customers can order copies of just those songs they're interested in instead of an entire album. The next time you surf the Internet, see if you can find other examples.

Other common sales techniques are using the mail to send information to possible customers and calling on the telephone. While telemarketing is popular with sellers, many people dislike being called at home. The U.S. government has made efforts to limit such sales calls. Sellers who annoy buyers may not make many sales.

SECTION REVIEW 17C

Recall ▸▸

1. What is the job of the marketing department?
2. What is the difference between a wholesaler and a retailer?

Think ▸▸

3. What products might be difficult to sell successfully over the Internet? Why?
4. Why do you think a manufacturer might want to sell directly to a customer?

Apply ▸▸

5. **Advertising** Select a product and create an advertisement for it using the medium of your choice.
6. **Research** Interview a salesperson about the best way to sell a product. Share your findings with the class.

Summary Activity

For each numbered blank, pick the answer from the list on the right that makes the most sense in the entire passage. Write your answer on a separate sheet of paper. No answer will be used more than once.

The United States is the world's largest producer of __1__ and has about 405,000 factories. Many use industrial __2__, which have one mechanical arm. Others use __3__, which is using electronic information in the manufacture of a product. For example, all machine tools can be linked to the Internet. The operator can be anywhere, as long as he or she has access to the Internet.

In a company organization, the __4__ and development department looks at new ideas and creates new products. This is where invention and __5__ take place. After the company decides to make a new product, __6__ engineers decide how it will look. After all the testing is finished, the __7__ department decides how to manufacture the product.

With __8__, machine tools operate by commands from a computer. The operator types in the instruction and the machine does the work.

When computers are programmed to operate all the machinery, __9__ is being used. The parts being manufactured automatically go from one machine to another.

CIM stands for computer-__10__ manufacturing. The computer linkage can also tell the company's __11__ suppliers when to deliver production materials. The completed product is inspected by the __12__ assurance department.

The marketing department determines how to __13__ the product. A(n) __14__ sells to retailers. Some manufacturers sell directly to consumers on the phone, through the mail, and over the __15__.

Answer List

- advertise
- CAM
- CNC
- design
- e-manufacturing
- innovation
- integrated
- Internet
- just-in-time
- manufactured goods
- production
- quality
- research
- robots
- wholesaler

Comprehension Check

1. List at least three forms of advertising.
2. How many patents are issued each year in the U.S.?
3. What does R&D stand for?
4. What is troubleshooting?
5. What is distribution?

Critical Thinking

1. **Appraise.** Many consumers object to tele-marketers. Manufacturers say that tele-marketing helps them sell products. Do some research on how telemarketing should be regulated, if at all.

2. **Criticize.** Count the number of commercials and the number of minutes devoted to advertising in a half hour of TV. In your opinion, is this an appropriate amount of time or not? Explain.

3. **Decide.** The federal government oversees workplace safety. Do you think this is necessary, or should companies be allowed to establish their own safety rules? Explain.

4. **Infer.** Why might a background in science be important to a designer of automobiles?

5. **Propose.** Suppose you are responsible for making a pizza using an assembly line. Make a sketch showing the various work-stations and label them according to the work done.

Visual Engineering

The Importance of Quality. It is obvious that quality control is very important if a manufacturer wants to produce a quality product. How many times have you bought something and taken it home and found out that it didn't work? It makes you question whether you want to buy that same name brand again, doesn't it? Quality control is important in almost any product you buy or sell. What do you think would be important to inspect on this motorcycle before it leaves the manufacturer? Think of a simple product that you might buy and list what should be inspected on it before it is sold.

TECHNOLOGY CHALLENGE ACTIVITY

Testing Products

Equipment and Materials

- flashlight
- six batteries that fit your flashlight: two of brand A, two of brand B, and two of brand C; make sure they all have the same ratings printed on the outside (for example, "Size D Alkaline" or "Size AA Heavy Duty")
- one or two plastic grocery bags
- clock
- calculator (optional)
- computer (optional)

! SAFETY

Reminder
In this activity, you will be using tools and machines. Be sure to always follow appropriate safety procedures and rules. Remember, safety is an attitude that you must develop and maintain at all times.

Background

A product manufactured by one company is usually a little different from the same product manufactured by another company. Even though they all try to meet the same standards, sometimes one company makes a better product than the others. An evaluation of different products like cars, door locks, and toothpaste is published in magazines such as *Consumer Reports*.

Goal

For this activity you will choose several brands of a single product and test them. You will record the test results and determine which product is the best buy.

Criteria and Constraints

▸▸ The procedure in this activity is for testing flashlight batteries. However, you may choose a different product with your teacher's permission.

▸▸ You must determine in advance what qualities you will test the product for. If the product comes with an instruction manual, you must read it to learn the correct way to use the product.

▸▸ You must keep a record of the test results.

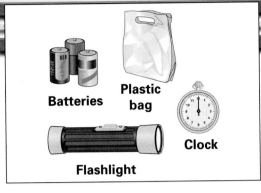

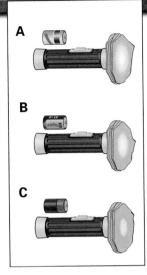

FIGURE A

FIGURE B

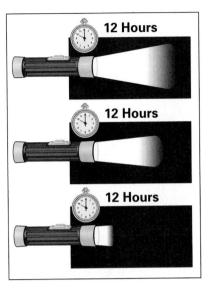

FIGURE C

Design Procedure

1. Record the cost of each brand. An expensive set of batteries might last only a little longer than a less expensive set. Use your own judgment to determine the importance of cost. Neatly write down all your information. See Figure A.

2. Determine your test procedure. Your first test might be to see which set of batteries produces the strongest light. One way to do this is with white or light brown plastic bags from your local grocery store. See Figure B. See how many layers of plastic it takes to completely block the light from a flashlight that has brand A batteries inside. Keep folding the plastic over and over. It's not unusual to use twenty or more layers. Repeat for brands B and C. Neatly write down all your information.

3. Perhaps another test would be to see how long it takes to wear out the batteries. See Figure C. Put two new brand A batteries in the flashlight and turn it on. Check the time. Keep the flashlight on until it no longer puts out much useful light. Discard the batteries and repeat for brands B and C. Neatly write down all your information.

Evaluating the Results

1. Based on your testing, which brand provided the most value? Why do you think so?

2. What other tests could you have conducted on the batteries? For example, how would you have decided which brand is best in very cold weather?

3. Did you have any trouble reading the information you wrote down? How important was it to clearly write down all the information from the test?

Construction
Great Achievements in Design

All great construction projects require careful planning and engineering.

1 Egypt's three great Pyramids at Gizeh were built nearly **4,500 years ago**. That of the pharaoh Cheops, the largest of the three, contains 2,300,000 finished stone blocks, each averaging two and a half tons. The mystery of how the blocks were put in place has never been solved. The mathematics the builders used was surprisingly accurate.

The world's longest structure and one of the largest construction projects ever undertaken is the Great Wall of China. Built to keep out invaders, it was put together entirely by hand with earth and stones held in wood frames. Construction began during the Qin dynasty (221-206 B.C.E.), and the wall was given its present form during the Ming dynasty (1368-1644). **2** The main wall is about 1,500 miles long. **Investigate:** *Find a list of the Seven Wonders of the Ancient World. How many are extraordinary examples of construction? Why do you think this is so?*

Until 1971, the tallest building in the world was the Empire State Building in New York City. Completed **in 1931**, after only 16 months of construction work, the building measured 1,250 feet in height and had 102 stories. **In 1950**, its height was increased to 1,472 feet by the addition of television antennas.
Investigate: *Which skyscraper has the most inhabitable floors and where is it located?* **3**

The Biltmore Estate in the mountains outside Asheville, North Carolina, was designed to rival the great mansions of Europe. Created for George Washington Vanderbilt III **in 1895**, the house is the largest private residence in the United States, with 250 rooms, 34 master bedrooms, three kitchens, 43 bathrooms, 65 fireplaces, a bowling alley, a gymnasium, and an indoor swimming pool.

4

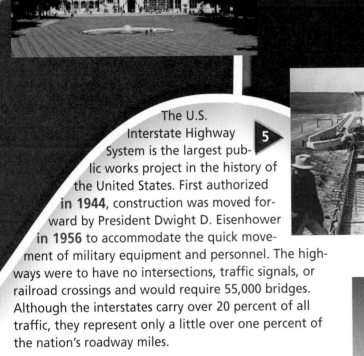

5 The U.S. Interstate Highway System is the largest public works project in the history of the United States. First authorized **in 1944**, construction was moved forward by President Dwight D. Eisenhower **in 1956** to accommodate the quick movement of military equipment and personnel. The highways were to have no intersections, traffic signals, or railroad crossings and would require 55,000 bridges. Although the interstates carry over 20 percent of all traffic, they represent only a little over one percent of the nation's roadway miles.

6 The geodesic dome is one of the lightest, strongest, and most cost-effective structures ever devised. First developed by R. Buckminster Fuller **after World War II**, the domes are assembled from interlocking polygons, usually triangles. The geodesic dome is able to enclose more space without internal supports than any other structure. Perhaps the best known is the Epcot Center dome at Walt Disney World in Florida.

Suppose that it's raining outside. The rain is falling really hard. The wind is howling. It's pitch dark. You're not worried because you know you can stay warm and dry inside a building. Perhaps you're inside a house or apartment building. You might also be inside a school, store, restaurant, bus terminal, or hospital. All of the buildings we use every day for shelter were assembled by people who work in the construction industry.

Another important part of construction is building roads, bridges, and tunnels. Cars and trucks travel swiftly from city to city because people have constructed smooth, level roads. Strong bridges allow trains, trucks, and cars to cross gorges, rivers, and large bodies of water. Tunnels allow us to travel underground. Other large construction projects include such things as dams, airport runways, and space stations.

The Jumeirah Beach Resort in Dubai, United Arab Emirates, is a unique sail-shaped building and is the world's tallest hotel.

The Evolution of Construction

▷ What types of construction have all civilizations needed?

Objectives

▶▶ **Discuss** how different construction systems evolved.

▶▶ **Name** some important structures.

Terms to Learn

- skyscraper
- stick construction
- surfacing

Standards

- Characteristics & Scope of Technology
- Cultural, Social, Economic & Political Effects
- Role of Society
- Influence on History
- Design Process
- Construction Technologies

Ancient civilizations were interested in the same types of construction projects as we are today. They used construction to solve problems. They wanted comfortable buildings, smooth roads, and safe bridges. Technology helped them build these things.

Construction systems have existed for as long as people have built shelters, bridges, and other structures. Some structures are temporary. These include such things as buildings used to house county fairs and dams built to divert water around a construction site. Other structures, such as homes and airports, are permanent.

Construction falls into three basic types: residential, commercial/industrial, and civil. See Figure 18-1.

Figure 18-1 The three types of construction are commercial/industrial, residential, and civil.

Residential construction is done to build places where people live, including both apartment buildings and single-family homes. Commercial/industrial construction methods are used to build stores, shopping malls, and factories, as well as churches and other houses of worship. Civil construction is done to create large structures for public use. Hospitals and schools are considered civil construction projects, as are roadways, bridges, tunnels, and dams. Most civil construction is paid for with tax money collected by the government.

Structures are built using many different processes and procedures. All structures rest on some kind of foundation (base), and all need to be maintained so that they continue to function.

Buildings

▷ What kinds of shelters did early people construct?

Ordinary people in the past lived in houses that look much like those of today. Natural materials such as wood and stone were used in their construction.

During the American colonial period, many trees grew in the eastern half of the United States. Most early houses were made from wood. Log houses were quite popular because they could be easily put together. See Figure 18-2. Log houses, however, were wasteful because so much wood was used for one house. Their construction also required a great deal of strength to position the heavy logs.

Figure 18-2 Log cabins were easy to build and the materials were easy to find.

Figure 18-3 Stick construction is usually done with 2 x 4 or 2 x 6 lumber.

stick construction
method of building construction in which lightweight pieces of wood or steel are used

skyscraper
a very tall building

A new type of house construction started to appear in the 1840s. Instead of being made from logs or large wooden beams, these new houses were made of light-weight pieces of wood. This method was soon called **stick construction**. See Figure 18-3. The frames went up quickly and provided both safe and strong dwellings. People soon began to prefer this type of house. New wants and needs were created. Almost all modern houses are still built this way. Even **skyscrapers** can be built with a type of stick construction using steel.

The first skyscraper was the 1885 Home Insurance Building in Chicago. At ten stories high, it was not as tall as some other buildings. However, the Home Insurance Building was the first to use a metal frame as a basic part of its design. The outside walls were connected to the metal frame. The walls did not support the building as in log houses.

Today's tallest skyscraper is Taipei 101 in Taipei, Taiwan. It is 1,676 feet tall. See Figure 18-4. Modern skyscrapers are made with concrete and steel.

Roads

▷ How were early roads constructed?

Most ancient roads were little more than dirt paths. The great road builders of the past were the Romans, who started building improved roads about 300 B.C.E. They first dug a wide trench a foot or more deep. Then they filled it

Figure 18-4 Taipei 101 is the tallest skyscraper in the world at 1,676 feet. What is the tallest building in the United States? How tall is it?

Figure 18-5 This road was built by the Romans in Petra, Jordan.

with rocks and topped it off with flat stones. The roads were higher in the center so that rainwater would drain off and be carried away by ditches along the sides. The Romans built over 50,000 miles of these roads, and some are still used today. See Figure 18-5.

Roads can more easily support heavy loads if the roads are covered with a durable material. This is called **surfacing** a road. The Romans used flat stones but some early American roads were surfaced with logs or planks laid crosswise. Both types of road were bumpy and had to be traveled very slowly. Wagons could be damaged by too much shaking. Horses' hooves sometimes became stuck between the logs or planks. The first section of the National Road in the eastern United States, now U.S. Highway 40 or Interstate 70, was made from logs.

Around 1800, George McAdam, a Briton, developed a method for easily making a smooth road surface. He used tar, which came from crude oil, heated it, and spread it over a thick layer of crushed rocks. It is still a common way

surfacing
applying a material to a roadway that provides a smooth surface

Figure 18-6 Road workers lay down asphalt for a highway.

of surfacing roads, driveways, and parking lots. See Figure 18-6. Americans commonly call the material asphalt or blacktop. The British call it macadam or use the brand name Tarmac.

Modern highways are often surfaced with concrete. The world's longest national highway is the Trans-Canada Highway, which spans 4,860 miles.

Bridges

▶ When were the first metal bridges made?

The Romans gave us not only long-lasting roads, but also strong, well-designed bridges. They developed the arch—wedge-shaped stones locked in a curve. The arch distributed weight sideways as well as down. See Figure 18-7. Like their roads, some Roman bridges have lasted for centuries.

During the late 1700s, improvements in metal manufacturing greatly reduced the price of iron. That's when people started to make bridges out of iron. One hundred years later, bridge builders started to use steel, a much stronger material. The Eads Bridge, the first major all-steel bridge, was built across the Mississippi River at St. Louis in 1874. It appeared so delicate that twenty-seven of the country's leading engineers said it would quickly collapse or be washed away by a flood. They were wrong. The Eads Bridge has safely spanned the most powerful and frequently flooded river in the U.S. for more than 100 years. It was named for its designer and chief engineer, James Buchanan Eads.

Modern bridges are still made of steel and supported by concrete. The modern bridge with the longest distance between supports is the Akashi Kaikyo Bridge, near Kobe, Japan.

Tunnels

▷ **What is the purpose of a tunnel?**

Bridges take people over obstacles, while tunnels take them through or under them. Tunnels are less noticeable than bridges and can be less inviting. They often make us think of mysterious caves. See Figure 18-8.

The first major tunnel in the United States was constructed through the Hoosac Mountains in western Massachusetts. The tunnel is almost five miles long, took over twenty years to complete, and opened in 1875. It is still used today.

The world's longest tunnel is the Seikan in Japan, which is over thirty-three miles long. More than fourteen of those miles are under water.

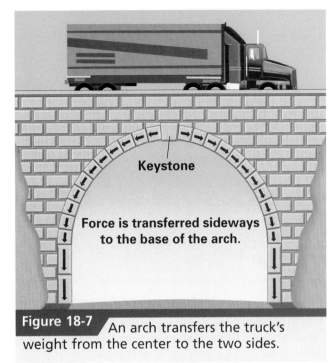

Keystone

Force is transferred sideways to the base of the arch.

Figure 18-7 An arch transfers the truck's weight from the center to the two sides.

Other Construction

▷ **What is the world's most unusual construction project?**

Let's look at some types of construction that we may not see every day. Dams, canals, and space stations are examples of other kinds of construction.

Dams are used to divert the flow of water or to cause it to form a pool. Sometimes this is done to prevent floods. Huge earthen dams were built during ancient times on the Nile River to help control flooding. The Romans built dams in Italy out of brick and stone. Modern dams are much larger and built based on scientific and engineering principles. Dams often improve water supplies and the economy of the area; however, these needs can compete with needs of the environment. Animal habitats can be changed or destroyed, and areas downstream can become too dry. One of the best-known dams in the United States is the Hoover Dam on the Colorado River. When the Three Gorges Dam in China is completed, it will be the world's largest.

Figure 18-8 The Bay Bridge and Tunnel connect the cities of San Francisco and Oakland, California.

Canals are human-made waterways. Early peoples used small canals to bring water to crops. The Panama Canal was built by American engineers to connect the Atlantic and Pacific Oceans. See Figure 18-9. It was first opened to ships in 1914. Ships no longer had to make the long, dangerous journey around the tip of South America. Trade and travel increased as a result.

The world's most advanced and unusual construction project is taking place in outer space. The International Space Station, built by the United States and fifteen other nations over a period of years, circles the globe every ninety minutes. Rotating teams of astronauts live there for weeks at a time.

Figure 18-9 The Panama Canal connects the Pacific and Atlantic Oceans. It was completed in 1914. Which American president was instrumental in building the canal?

18A SECTION REVIEW

Recall ▶▶

1. Why were the log houses built during America's colonial period so wasteful?
2. What new type of house construction began in the 1840s?
3. What was the first skyscraper and in what city was it built?
4. What was the name and location of the first major all-steel bridge in the U.S.?

Think ▶▶

5. Why do you think skyscrapers are popular for large cities?

6. Suppose you have been asked to build a dam or a tunnel. Name at least three problems you would have to overcome.

Apply ▶▶

7. **Design** Make a model of a teepee or other early Native American dwelling. Tell why you think this design was used.
8. **Construction** Build an arch from wooden blocks or foam plastic. Test its strength by placing various weights on top.

Design Requirements

> ▶ How are construction systems like other systems?

Like other technological systems, construction systems include inputs, processes, and outputs. Inputs include the seven resources—people, materials, tools and machines, information, energy, capital, and time. Processes include designing the structure and putting it together. Outputs include the structure itself.

Malfunctions of any part of a construction system may affect the way the system works and the quality of the outputs. For example, if steel of poor quality is used to build a bridge, the bridge may collapse. Most construction systems also include feedback. Maintenance, which involves inspecting and servicing the bridge, often results from feedback. With maintenance, the bridge continues to function properly, and its life is extended.

Many requirements are placed on the development of construction projects to ensure safety and long life. Other requirements are based on the purpose of the structure. Buildings, for example, have many different uses. That is partly why they look like they do and are constructed with different materials. One example would be the floor in your school classroom. It has to be strong enough to safely support you and your classmates. The floor in a single-family home has to support a smaller load. So your classroom floor might be concrete, but the floor in a house might be made of wood. Can you think of other examples?

Although there is no perfect design, many requirements are met during the creative design process. Some requirements involve materials, building codes, and safety considerations.

Objectives

▶▶ **Explain** the purpose of several construction materials.

▶▶ **Discuss** building codes and safety.

Terms to Learn

- **building code**
- **composite**
- **concrete**
- **conservative design**
- **curtain wall**
- **lumber**

Standards

- Core Concepts of Technology
- Cultural, Social, Economic & Political Effects
- Role of Society
- Influence on History
- Troubleshooting & Problem Solving
- Design Process
- Construction Technologies

Materials

> ▷ What kinds of materials are used in construction?

Before scientific knowledge was developed, early peoples built their homes and other structures from whatever materials they had available. Today, scientists work to develop new and better materials for construction projects.

lumber
pieces of wood cut from logs into convenient shapes

Wood

▷ What is wood used for?

Trees growing in the forest or on commercial tree farms produce wood in its natural state. People cut the wood into **lumber**. See Figure 18-10. Most single-family homes are made from lumber because it is readily available and easy to work with. Lumber is also less costly than some other materials.

Wood is a renewable resource. After trees are cut down, new ones are planted in their place. However, trees take many years to grow, so it is important to stretch our supply of wood. For this reason, crooked trees, sawdust, and other wood wastes are used to create engineered wood materials, such as plywood and hardboard.

One of wood's disadvantages is inconsistent quality. Pieces of lumber cut from the same tree might not be equally strong. Insect damage, knots, and other problems cause these differences. Also, if wood is not treated or protected with paint or other coverings, it can rot and weaken the structure. Engineered wood materials usually do not have these disadvantages. Other substances, such as strong glues, are added to the wood and improve its qualities. Some engineered wood beams, for example, are stronger than beams of natural wood.

Wood is usually fine for houses and smaller buildings. However, we often use different materials to build schools, malls, factories, theaters, and other structures used by the public.

Steel

▷ **Why is steel popular?**

Metal has been used to make structures for many years. Steel is the most useful because it is very strong. It is a mixture of iron and small amounts of carbon. Steel is primarily used to provide support. Skyscrapers, for example, are built with a steel framework. Then floors and interior walls are added. The outside walls are hung on the metal frame. They do not help support the building. These exterior walls are called **curtain walls**. They can be made of many different materials, even glass. See Figure 18-11.

Commercial buildings are sometimes made with steel supports and painted steel walls. Many suspension bridges are made of steel. When constructing bridges over water and deep gorges, it is easier to work with steel than other materials.

curtain wall
exterior wall on a skyscraper that does not help support the building

LOOK ᵀᴼ ᵀʰᵉ FUTURE

Houses of Steel

Although steel has been used in constructing large buildings for over 100 years, it is seldom used for houses. This may change. The demand for high-quality lumber has reduced the supply available for framing houses, and builders are beginning to use more steel. The cost of framing a house with steel is about the same as framing it with wood. Although the framing methods are similar, different tools are used and parts are often assembled at a factory. The pieces of steel are held together with screws, welds, or pressure. Is there a house of steel in your future?

Figure 18-11 This triangular office building has glass curtain walls.

Figure 18-12 These workers are screeding the concrete for a driveway. What does screeding mean?

concrete
a mixture of cement, sand, stones, and water that hardens into a construction material

Figure 18-13 Steel rebar makes concrete much stronger.

Concrete

▷ **What is concrete made from?**

A mixture of dry cement, sand, stones, and water, **concrete** is the most adaptable, common, and useful construction material. It is used for large buildings, highways, dams, bridges, and the foundations of houses. See Figure 18-12. Nearly every structure contains some concrete. Some people call concrete *cement,* but cement is just one item in a concrete mixture. It is the binder that holds everything together.

Concrete begins as a pasty mixture that resembles ordinary mud. It is poured into forms that hold it in place while it hardens. These forms give the concrete its final shape and allow it to be used for many purposes. Steel reinforcing rods or other steel shapes are sometimes placed in the wet concrete to help strengthen it after it hardens. See Figure 18-13.

In building construction, concrete is used for such things as supporting columns, foundations, walls, and floors. Many homeowners purchase bags of the dry mixture to use for such things as walkways. Concrete is even used for sewer pipes and birdbaths.

Other Materials

⏵ What other materials are used?

Many construction methods use different materials for specific applications. Asphalt, for example, is an important surfacing material made from crude oil and other substances. It is used mainly for road construction and repair, driveways, and walkways.

Some bridges are made of composite materials. A **composite** is a combination of two or more materials, such as fiberglass and carbon. Composite materials are much lighter than concrete and do not have to be protected from the weather as steel does. See Figure 18-14. Composites are used for many other construction products, such as bathtubs and roofing materials.

composite
a material made from two or more other materials

TECH CONNECT
SCIENCE ⟩

Like Peas in a Pod
Sometimes innovative technologies are created when old materials are used in new ways. A new kind of temporary pod-like shelter for use in natural disasters has been created out of paperboard covered with plastic.

ACTIVITY Make a sketch of your own idea for a temporary shelter. Suggest at least three materials for the shelter based on properties that would make it portable, sturdy, wind-resistant, and waterproof.

Figure 18-14 This bridge is made of composite materials in a factory and then shipped to the site where it will be used.

In addition to composites, all kinds of other materials are used in construction. Some of them include masonry (bricks and stone), window glass, vinyl exterior wall coverings, fiberglass insulation, copper wire, and plastic water pipes. They are often part of a structure's subsystems.

Regional Requirements

▷ Are all buildings built the same way?

The many regions across the United States have different types of terrain and weather. Structures designed for one region might not be the best choice for another region. For example, flatter roofs commonly appear on houses in the south, where there is little snow. See Figure 18-15. However, people in the northeast have to be concerned about the weight of snow on their roofs in the winter. If flat roofs were used there, the snow would not slide off easily, and the roofs might collapse. The people in California and some other states have to be ready for earthquakes. People in some central states receive more tornadoes. Other regions have to be ready for floods and rising water, excessive summer heat, or landslides.

ETHICS in ACTION

Being Neighborly

New technologies can bring about the need for new laws. Many homes are now heated with solar energy from the sun. What if a construction company decides to build a tall apartment building next to such a home, blocking the sunlight? Does the construction company have a right to do that? Several states have already enacted laws resolving such disputes over solar power.

ACTIVITY How do you think such disputes should be resolved? Find out if your state has passed any laws in recent years that affect the use of alternative energy sources.

Building Codes

▶ What are building codes?

Local governments have established rules regulating the types of structures that can be constructed in their region. These rules are part of the information inputs used by construction systems. They are called **building codes**.

Construction specialists agree that there are certain general standards that must be met with each structure. For example, commercial buildings must have enough exits in case of an emergency. Houses must have safe water supplies. Bridges and tunnels must be able to support heavy loads. These are just a few of the many objectives of building codes.

Building codes can vary by state, county, or city. They often are modified as new materials and methods of construction are developed. Before construction begins, builders contact a local inspection office at city hall or the county courthouse. They are expected to follow the regulations that have been published in the building code books.

building code
rule used to control how structures are built

Construction Planner

Large construction company is looking for someone to evaluate locations for small shopping centers in this state. Must have working knowledge of the construction business and good personal skills. Neatness and good personal grooming are important.

Personal Grooming

Always remember that first impressions can make lasting impressions.

First impressions really do count, especially during a job interview. The first thing a potential employer notices is a candidate's appearance. A poor appearance can be so distracting that good qualities are overlooked. Begin now to "dress for success" by checking yourself in the mirror before you leave for school each day. Remember to check the following:

✔ **Are your clothes and body clean?**
✔ **Is your hair combed?**
✔ **Are your clothes businesslike and appropriate?**
✔ **Do your grooming and appearance send the right message?**

Your appearance will not only affect how others see you, but how you see yourself. Personal grooming skills will make you more qualified for any task.

TECH CONNECT
SOCIAL STUDIES

Locating the Codes

Books containing building codes are usually kept at city hall or the county courthouse. Different books hold the code for different subsystems.

ACTIVITY Go to the local office where the building codes are located. Ask to see one about foundations, electrical subsystems, sanitation requirements, or any other that the office has available. Note the year when the code became official, how many different sections the code has, and how many total pages are in the document.

Figure 18-16 Houses on stilts are common along rivers, lakes, and oceans.

For example, a builder might plan to construct a tall building in a town's historic district. That might not be permitted because people want to preserve the historic appearance of the area. Someone might hope to build a house near a lake, river, or seashore. The building codes might permit it but only if the house is built on strong poles to keep it well above floods. See Figure 18-16.

Building codes are designed to protect people's safety and the environment. That is why they also cover such things as electricity, plumbing, and energy use. Every aspect of construction is covered. That is one reason why there is a lot of paperwork associated with building a new structure.

Safety

▷ How safe and strong are our major structures?

Any major structural failure usually appears on the national news. That is because it is such a rare event. There are probably some buildings in your town that were built over 100 years ago and are still safely used today. Buildings, bridges, and all other structures are designed to support much more weight than could possibly be placed on them. Engineers call this **conservative design**.

conservative design
design strategy used to be sure structures can bear more weight than required under normal conditions

Of course, things happen that nobody can predict. For those situations, engineers design escape routes and safety devices. All tall buildings have elevators, but they also have stairways in case electricity is cut off to the elevators. They have sprinklers to put out small fires and keep them from spreading. See Figure 18-17. Entrance doors and emergency exits all open outward to allow people to quickly leave the building. When accidents or other problems do occur, engineers take steps to troubleshoot, which involves studying the problems and preventing them from happening again. These kinds of controls change the system. New innovative technologies have sometimes resulted.

Safety is also important on a construction site while the structure is being built. Workers must keep the area cleared of waste and other materials that could cause people to trip or fall. Holes dug for pipes and foundations must be clearly marked and protected. What other hazards can you think of?

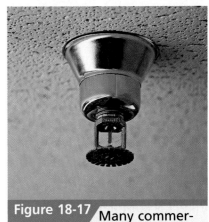

Figure 18-17 Many commercial buildings and some homes have sprinkler systems to protect against fires.

SECTION REVIEW 18B

Recall »

1. Name the four ingredients of concrete.
2. What is a composite material? Give an example.
3. Certain standards must be met when constructing a building. What are these standards called?
4. All bridges use a conservative design. What does this mean?

Think »

5. Some all-metal commercial buildings have walls made of aluminum, but most are made of steel. Why do you think steel is more popular?

6. Suppose you were asked to design a house built on the side of a hill and a house built in an area that has tornadoes. What differences in design would you include? Why?

Apply »

7. **Construction** Make a composite material by gluing a small section of window screening between two 3-inch by 5-inch index cards. Evaluate the product for strength.
8. **Research** Observe work at a construction site in your community. What is being done? What kinds of materials are being used? What safety precautions are in place? Report your findings to the class.

Summary Activity

For each numbered blank, pick the answer from the list on the right that makes the most sense in the entire passage. Write your answer on a separate sheet of paper. No answer will be used more than once.

Construction is carried out at a building site that __1__ as the project is completed. The first __2__ was the Home Insurance Building in Chicago. It was the first to use a(n) __3__ as part of its basic design. Roads are also included in construction projects. The first really great road builders were the __4__. They were the first to make roads higher at the center to drain off water. The British developed a material called macadam for surfacing roads that is still used today. In the U.S., we call the material __5__. The first major all-steel bridge was built across the Mississippi River at the city of St. Louis. The first major American tunnel was the __6__ Tunnel in western Massachusetts.

Many structures are built of __7__, which is wood cut into useful shapes. Insect damage and __8__ can cause wood to have inconsistent quality. Tall buildings are built with a steel framework. The outside walls are hung on the framework and are called __9__ walls. Concrete is also a construction material. It is a mixture of __10__, sand, stones, and water. Because it can take on any shape, concrete is the most __11__ material used in construction.

Structures built in different regions have different __12__. For example, people in the northeast have to be concerned about __13__. Building __14__ vary by regions and are the responsibility of local governments. They are designed to protect the environment and people's __15__. For example, engineers have developed __16__ in case there is a fire.

Answer List

- adaptable
- asphalt
- cement
- changes
- codes
- curtain
- Hoosac
- knots
- lumber
- metal frame
- requirements
- Romans
- safety
- skyscraper
- snow loads
- sprinkler systems

Comprehension Check

1. How did road building evolve?
2. When were bridges first made from iron?
3. What is the purpose of a tunnel?
4. Which structures are most often built with wood?
5. What is the purpose of a building code?

Critical Thinking

1. **Infer.** If you were a worker on a construction site for a house, what kinds of safety hazards would you be mindful of?
2. **Hypothesize.** How do you think the discovery of the arch affected the design of later structures?
3. **Connect.** How do you think the improved Roman roads affected Roman society?
4. **Decide.** Building roads and dams often affects the nearby environment. Trees may be cut down, areas may be flooded, or animals may be driven away. Do you think this is a fair tradeoff? Explain.
5. **Classify.** Rate the building materials discussed in this chapter from most important to least important. Give reasons for your decisions.

Visual Engineering

Constructing an Arch. The arch was probably first used for building bridges. It is still seen in cathedrals and other large buildings. Arches are found in the windows of many older structures in the United States. One of the most famous monuments in the U.S. is the St. Louis Arch. If you spend enough time planning its construction, an arch is not hard to build. With your team, do some research on how arches are designed and built. Then carefully plan how to make an arch out of small wooden building blocks. Use CAD software to lay out the design and calculate the angles of the blocks. Finally, construct the arch and any support pieces.

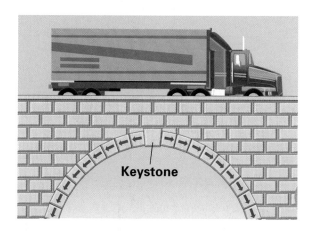

Keystone

TECHNOLOGY CHALLENGE ACTIVITY

Testing the Strength of a Column

Equipment and Materials

- four 3- by 5-inch index cards
- one or two light-duty rubber bands
- light-duty plastic cup
- small weights, such as metal washers or bolts
- scale for weighing the cup and weights
- plastic tub

Reminder
Be sure to always follow appropriate safety procedures and rules. Remember, safety is an attitude that you must develop and maintain at all times.

Background

Observe the concrete columns supporting highway bridges and overpasses, particularly on an interstate highway. Some columns are cylinders, and others have square corners, like boxes. They might be wide or they might be narrow. Their shape depends upon the load that the bridge or overpass is expected to carry. Column shape affects strength.

Goal

For this activity, you will fold index cards into different shapes to discover how much of a load each one can carry before it collapses.

Criteria and Constraints

▶▶ Paper folds must be sharp and uniform.
▶▶ You must keep a record of the maximum loads carried by each of the columns. You will observe the influence that each shape has on the column's load-carrying ability.

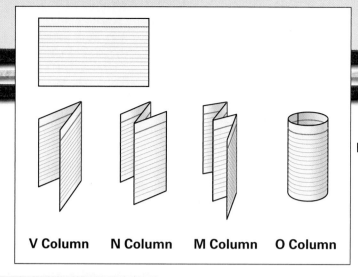

V Column   N Column   M Column   O Column

FIGURE A

FIGURE B

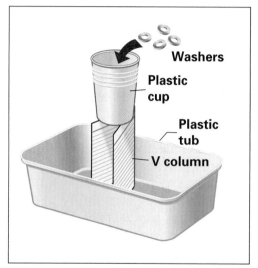

Washers

Plastic cup

Plastic tub

V column

Design Procedure

① Review the drawing of column shapes in Figure A. Let's call them V (one fold), N (two folds), M (three folds), and O (tube). The folded cards will look like those letters when viewed from the top.

② Weigh the plastic cup on the scale. Then weigh ten washers or bolts. After you know what ten of them weigh, you will be able to calculate how much 20, 35, or any other number will weigh. Do not mix types of weights. Use only washers or only bolts.

③ Fold one card into a V and place it in the plastic tub. Carefully set the plastic cup on top of the folded card. If necessary, use a light-duty rubber band to help your V column keep its shape.

④ Slowly and carefully, add weights to the cup as shown in Figure B. Eventually, the column will collapse and scatter the washers or bolts inside the tub. Collect and count them, then calculate the maximum load supported by the V column. Although it will be a small amount, don't forget to add in the weight of the plastic cup.

⑤ Repeat the procedure for the N column and the M column.

⑥ Now make the O column and use the rubber band to help it hold its shape. Repeat the procedure.

Evaluating the Results

1. Of the V, N, and M columns, which one carried the largest load? Why do you think it turned out that way?
2. Did the O column carry a smaller or larger load than the folded columns?
3. What conclusions can you draw about column design from this experiment?
4. What do you think would happen if you made all the columns a little bit shorter? Would they hold more or less load? Why?

SUPERSTARS OF CONSTRUCTION

INVESTIGATION ////

Every structure requires a design. Suppose you are an architect who wants to use Fuller's geodesic dome as part of a house or other building. How would you use the space under the dome? Make sketches of possible uses for your geodesic dome and show where you would place furniture and other items.

Thomas Telford
(1757-1834)

If any single person can be credited with the design for our modern highways, it is probably Thomas Telford. He used the principles established by the ancient Romans to construct durable highways in his native Scotland. Telford dug down a foot or more into the soil and hand-laid a foundation of large stones. Then he added smaller stones to make a crowned surface. He was also responsible for developing the system of subcontracting jobs to other firms, which he used when building many canals, bridges, and ship docks. Telford was also the first contractor to provide carefully written instructions that others were expected to follow.

Frank Lloyd Wright
(1869-1959)

Frank Lloyd Wright developed a uniquely American architecture, called Prairie Style. His buildings tended to blend with nature, and he preferred natural forms and materials like wood and stone. He developed the ranch-style house, a single-story house with long, low lines that appealed to many people. Altogether, Wright designed about 1,000 structures in America and around the world. His productivity was far greater than any other 20th century architect.

Othmar Amman
(1879-1965)

After earning a degree in civil engineering in his native Switzerland, Othmar Amman came to the United States in 1904 and took a job with a bridge-building firm. After many successful projects, he was named chief engineer for the Port of New York in 1927. In following years he designed three of the world's longest suspension bridges: the George Washington Bridge in New York, San Francisco's Golden Gate Bridge, and the Verrazano-Narrows Bridge.

R. Buckminster Fuller

(1895-1983)

R. Buckminster Fuller believed that human creativity was limitless and that technological progress could give all people richer lives. An inventor, designer, and philosopher, Fuller is perhaps most famous for inventing the geodesic dome, which he perfected in 1947. The dome was the result of a quest for an affordable housing design that could be built using mass-production techniques. Fuller decided on the geodesic dome because it encloses a greater volume with less material than any other form. He also designed concert halls, banks, auditoriums, and other structures. Fuller held more than 2,000 patents and wrote more than 20 books.

Kenzo Tange

(b. 1913)

Japanese architect Kenzo Tange has designed many important structures in Japan and around the world. Some of his designs use elements from Japanese culture. One office building was formed from interlocking concrete beams similar to the wood beams found in Japanese temples. A 1960 design for apartment buildings was based on temple-shaped clusters connected by transportation links. In one of his most innovative designs, he eliminated entire floors of a building in order to create roof terraces or spaces for future expansion. Although Tange has taught architecture at MIT, Harvard, and Yale, his only completed structure in the United States is the expansion of the Minneapolis Art Museum.

Patricia Billings

(b. 1926)

Artist Patricia Billings is responsible for inventing Geobond®—one of the most revolutionary materials in construction history. After several of her plaster sculptures broke, Billings set out to create a sturdier type of plaster. Eight years and dozens of experiments later, she tried mixing a special cement additive with gypsum and concrete. The resulting plaster was almost unbreakable. Then a scientist friend realized that Billings's new material was also incredibly resistant to heat. Billings returned to her lab, and in eight more years she had created Geobond. Geobond products are so heat resistant they refuse to burn even after being heated with a 2,000° F flame for as long as four hours. Geobond is the first workable replacement for asbestos.

Look at your arm. You can see your skin—the outer covering of your arm—but you don't see the bone that is part of its internal support. You see only what's on the outside, not what's inside. A house is a lot like that.

The outside of a house may be covered with such materials as brick or wood. The walls inside may be covered with paneling, wallpaper, or paint. The floors may be covered with carpeting, tile, or other materials. All these coverings hide the inner structure of the house. The inner structure supports the walls, floors, ceilings, and roof. That structure has to be strong enough to support the loads the house must carry but lightweight enough to be easy to construct. Houses should also be easy to maintain.

Single-family houses make up about two-thirds of the households in the United States. This chapter is about the different kinds of housing and how a house is built.

Habitat for Humanity enables young people to help others build their own home.

Objectives

▶▶ **Identify** different residential dwellings.

▶▶ **Explain** the differences between manufactured houses and site-built houses.

Terms to Learn

- building site
- geodesic dome
- residential building

Standards

- Use & Maintenance
- Construction Technologies

residential building
(rezz ih DEN shul)
building in which people live

Homes for People

⊳ **In what kinds of housing do people live?**

Buildings are of three basic types: commercial, public, and residential. Commercial buildings are those used by businesses, such as stores. Public buildings, such as a city hall, are for public use. **Residential buildings** are the ones people live in. They include multiple-family housing and single-family homes.

Multiple-Family Housing

⊳ What is multiple-family housing?

The cost of land in some communities is quite high. Large buildings that can house many families require a smaller area of land per person than single-family homes. They include apartments, townhouses, and condominiums. See Figure 19-1.

The federal government estimates that, in the United States, buildings with ten or more housing units in them provide about 13 million units all together. They are often built using techniques similar to those for commercial buildings. Those methods will be discussed in Chapter 20, "Heavy Construction."

Figure 19-1 In large cities, many people live in high-rise apartment buildings.

Single-Family Housing

▶ What kind of construction is used for most single-family homes?

A single-family house is free standing and designed for the members of one family. The property it is built on is called the **building site**, or construction site.

Stick-built construction is used for most single-family homes. Workers assemble pieces of lumber resembling sticks over a foundation (base) of some kind. However, other styles and methods are also used.

Log Houses

▶ How are log houses sold?

Many people associate log houses with the American pioneers. However, they were originally designed for the forests of Switzerland, Germany, and Scandinavia. Immigrants from those countries brought this construction knowledge with them to the United States.

Their links to the past and pleasing appearance have made log homes popular again. Today, log houses are usually sold in kit form and the owner often hires a builder to assemble the house. See Figure 19-2.

building site
location for construction of a building

TECH CONNECT

SOCIAL STUDIES

This Old House

Many communities seek to preserve old buildings of historical value. Historical societies, museums, and community Web sites often feature them.

ACTIVITY Identify such an old house in your community. Find out when it was built and who the owners have been. Share your information with interested neighbors and classmates.

Figure 19-2 Log homes have become popular again. This one is being built from a kit.

Geodesic Domes

▶ How much material is needed to build a geodesic dome compared to other buildings?

geodesic dome
spherical structure the surface of which is formed by triangles or other polygons

Figure 19-3 How would you like to live in a house constructed from a geodesic dome?

The **geodesic dome** was invented in the late 1940s by R. Buckminster Fuller (1895-1983). The simplest version is a sphere (or portion of a sphere), the surface of which is formed by many polygons. The dome is self-supporting and does not require inside walls to help hold it up. Some people have used the design to build unusual looking houses. Geodesic domes are also sold in kit form. See Figure 19-3. A geodesic dome requires only about three percent of the material needed for a traditional building of similar size.

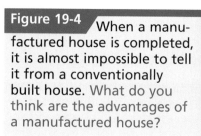

Figure 19-4 When a manufactured house is completed, it is almost impossible to tell it from a conventionally built house. What do you think are the advantages of a manufactured house?

Manufactured Houses

▶ Where are manufactured houses made?

Most stick-built houses are built on the site where they will be located. Manufactured houses, or parts of these houses, are made in factories. They are also called prefabricated houses. Factory production of houses uses less energy, less raw material, and less labor than building them on the site. Factories also can make more houses in less time.

It is sometimes difficult to tell a manufactured house from a site-built house. The houses are often made in two halves. The two halves are delivered by trucks to the site where a concrete foundation has already been installed. Large cranes position the parts on the foundation, and everything is bolted together. See Figure 19-4.

Mobile homes and trailers are also manufactured houses. They have their own wheels and can be easily moved. Including all types, there are about nine million manufactured homes in the United States.

TECH CONNECT
LANGUAGE ARTS

What Are the Benefits?

Advertisers usually focus on the ways in which a customer could benefit from their product. In selling a house, for example, they might tell potential buyers with children that the house is close to schools.

ACTIVITY Suppose you wanted to sell your house or the house of someone you know. What benefits does the house offer? Write an advertisement explaining the benefits.

SECTION REVIEW 19A

Recall ▸▸

1. Name the three basic types of buildings.
2. Name three examples of multiple-family housing.
3. What advantages do manufactured houses have over site-built houses?

Think ▸▸

4. Name three advantages and three disadvantages of living in an apartment as compared to a single-family home.
5. How might you be able to tell the difference between a manufactured and a site-built house?

6. Geodesic dome houses are less popular than other house designs. Why do you think this is so?

Apply ▸▸

7. **Research** Use the Internet to find the cost of a log house kit and a geodesic dome kit. Which of the two designs would you prefer? Why?

Objectives

▶▶ **Describe** how a building site is chosen.

▶▶ **Explain** how a house is assembled.

Terms to Learn

- drywall
- footing
- foundation
- gable roof
- insulation
- mortar
- sheathing
- stud
- subfloor
- truss

Standards

- Assessment
- Construction Technologies

House Construction

▶ **How difficult is it to build a house?**

Constructing a house can't be too difficult. Until fairly recently, many people built their own. During the 1800s, there were no construction companies in many regions of the United States. Pioneer farmers had to construct shelters for themselves and their animals. Neighbors helped, and everyone worked until the walls were up and the roof nailed down.

Those early amateur house builders had no power tools, few metal nails, and no printed plans. In spite of those difficulties, many houses they built are still standing. The basic construction techniques they used are still used today. See Figure 19-5.

Choosing a Location

▷ **Can you build a house anywhere?**

All sorts of locations for houses are possible. A house can be built in a busy city neighborhood or along a quiet country road. It can be built on the side of a hill or on a sandy beach.

Some building sites are better than others. Before choosing one, there are some questions that need to be answered.

- Are electricity lines close by? How about city water, natural gas for heating, and telephone lines? If city sewers are not available, the house must have its own sewage treatment system.

- Will the house face south? Houses facing south are sometimes less expensive to heat than houses facing north. Can you figure out why?

Figure 19-5 Can you imagine the time and labor it must have taken 200 years ago to build a house like this using posts and beams?

- How much will property taxes cost?
- Most construction sites require moving dirt. How much dirt will have to be moved? Houses on hillsides are more expensive to build than those on level land.

Selecting a House Design

▶ Why do we have different house designs?

Selecting a house design depends on several factors in addition to the personal style preferences of the builder or owner. These factors include such things as building laws and codes, cost, and climate. Other factors may include the way in which the house will function. For example, will it be used for a growing family or a retired couple?

The basic design of a house should fit in well with its surroundings. A log house might look out of place in a city neighborhood of brick houses. However, it might look fine along a tree-lined country road. Sometimes local laws restrict the type of houses that can be constructed in certain areas. These laws protect the home owners already there. These home owners can be sure that the neighborhood will always look about the same.

After choosing the general design, the next step is to determine the size of the house. House size is calculated by its floor area. The smallest houses have about 1,000 square feet of living area. An average-size house might have 1,200 to 1,500 square feet. However, houses can be many different sizes.

Plans are drawn and the contractor is hired. Construction can then begin.

Figure 19-6 Concrete footings like this support the foundation of a house. If the footing and the foundation are weak and develop cracks, the entire house might be structurally unsafe.

The Foundation

▶ How is a foundation constructed?

Buildings usually contain a variety of subsystems. The **foundation** is the part of the house that rests on the ground and supports the structure. The foundation starts with a trench dug where the outside walls will be located. Its depth depends on the local building code. Concrete is poured into the trench to create **footings**. See Figure 19-6.

foundation
the part of a house that rests on the ground and supports the upper structure

footing
the bottom part of a foundation, made of hardened concrete and located under the foundation wall

mortar
a mixture similar to concrete used to fasten concrete blocks or bricks together

Figure 19-7 Many foundation walls are made using concrete blocks. Mortar is spread between the blocks to hold them in place.

The foundation also usually includes a low wall. The size of the wall depends on whether a full basement or only a crawl space is desired. The foundation wall is made from poured concrete or precast concrete blocks that have large holes inside to reduce their weight. The blocks are carefully positioned on the footing and fastened together with mortar. See Figure 19-7. **Mortar** is similar to concrete, but it doesn't have stones in it.

The Basic Structure

▶ **What holds up the house?**

Wooden supports in the walls, floor, and roof make up the inner structure of a house. The lumber used is larger than that used anywhere else because it has to be strong enough to hold the structure's weight.

The Floor

▶ **How is a floor attached to a foundation?**

After the foundation is completed, the floor is the next step. Bolts are embedded in the top of a poured foundation wall. In a concrete block wall, holes in the blocks are filled with mortar, and the bolts are placed into the mortar while it is wet. After the mortar hardens, a piece of lumber is bolted to the foundation. Floor supports are nailed to this lumber, which is called a sill plate. See Figure 19-8.

The floor is supported underneath by floor joists. Joists are boards that extend from the front of the house to the rear. When the foundation area is large, a wooden or steel center beam, called a girder, is installed in the middle.

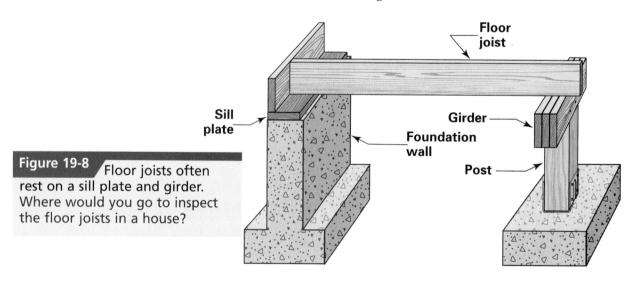

Figure 19-8 Floor joists often rest on a sill plate and girder. Where would you go to inspect the floor joists in a house?

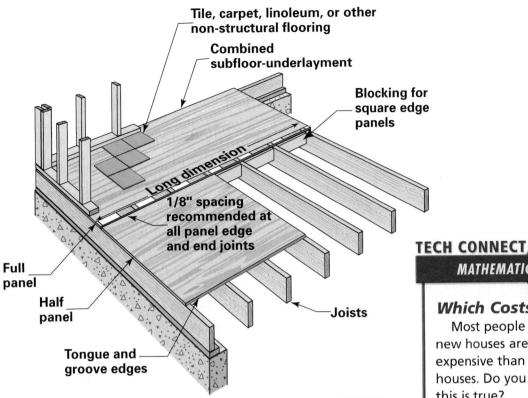

Tile, carpet, linoleum, or other non-structural flooring

Combined subfloor-underlayment

Blocking for square edge panels

Long dimension

1/8" spacing recommended at all panel edge and end joints

Full panel

Half panel

Tongue and groove edges

Joists

Figure 19-9 The subfloor is made from plywood that is laid over the floor joists. Tile, carpeting, and linoleum go on top of the subfloor or the finished floor.

A **subfloor** is nailed to the floor joists. The subfloor is usually made of plywood sheets, which go down quickly. See Figure 19-9. During construction, the subfloor takes a great deal of abuse. Tools are dropped on it, nails are accidentally driven into it, and paint is spilled. The subfloor absorbs that wear. Later, a finished floor is attached to it.

Walls

▶ **Why are walls put together on the floor?**

With a solid subfloor on which to work, constructing the walls is the next step. The vertical wall supports are called **studs**.

The wall framework is put together on the subfloor, with openings left for windows and doors. It is much easier to construct a wall on a flat floor, where it can be easily supported. When the framework is finished, it's raised into position by two or more people and nailed to the floor. See Figure 19-10 (page 464).

Figure 19-10 (page 464).

TECH CONNECT

MATHEMATICS

Which Costs More?

Most people think that new houses are more expensive than older houses. Do you think this is true?

ACTIVITY Look at the homes-for-sale section in your local newspaper. Find ads that list the number of square feet in the house as well as the price. Divide the price by the square footage to find the cost per square foot. Compare older homes for sale with new homes in similar neighborhoods. Which is more expensive per square foot?

subfloor
the first layer of flooring, usually made of plywood

studs
pieces of lumber, usually 2 x 4s, used for the framework of walls

As each wall is built, it's nailed at the corners to those walls that are already put up. The walls are strengthened by ceiling joists connecting the top of the front wall to the top of the back wall. After the frame is in place, a layer of **sheathing**, often made of plywood, is added to enclose the structure on the outside.

Inside walls dividing the house into rooms are made much like the outside walls. The frames are assembled on the floor, raised, and nailed in place.

The Roof

▶ What is a roof truss?

The roof must keep water out when it rains. It also must be strong enough to support heavy snow loads in cold climates. Most roofs have two sloping sides that meet at the center board, or ridge. This style is called a **gable roof**. The shape makes a strong and secure cover for the house.

The gable roof is constructed by building a sloped frame. Long pieces of lumber called rafters are nailed to the top of the walls and meet at a peak. They make up the basic roof structure. See Figure 19-11. The ridge board at the peak acts like a central support for the rafters. The rafters are covered with sheets of lumber called roof decking.

Many roofs are made from prefabricated triangular frameworks called **trusses**. Trucks deliver trusses, which

Figure 19-10 After a wall has been nailed together on the floor, it is raised into position and braced to keep it straight. What size lumber is normally used in a framed wall?

sheathing
a layer of material between house framing and the outer covering

gable roof
a roof with two sloping sides that meet at the ridge and form a triangular shape at either end

truss
prefabricated triangular framework that supports a roof

Figure 19-11 These are the basic components of a framed gable roof.

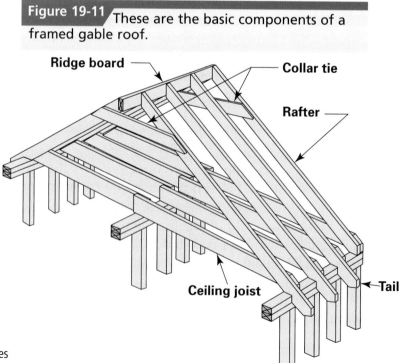

Ridge board — Collar tie

Rafter

Ceiling joist — Tail

are placed on top of the walls by a small crane. See Figure 19-12. After being nailed in place, trusses are covered with roof decking. The major advantage of trusses is that the roof can be made much more quickly, sometimes in about half a day.

Wall and Roof Coverings

▷ What are some siding and roofing materials?

The external appearance of a house is what gives it character. Some houses just seem to look better than others. It might be because the builders or owners carefully picked siding and roofing material. Siding covers the walls. Roofing covers the roof.

Many different kinds of siding material can be fastened to the outside walls of a house. Wood, plastic, metal, and brick are only a few. They are attached to the sheathing.

On most houses, asphalt shingles are nailed to the roof decking over a layer of heavy asphalt paper called underlayment. See Figure 19-13. The two layers provide a seal against rain and snow. The first row of shingles is applied near the edge of the roof, and the following rows overlap. This method makes sure that water runs off the roof instead of under the shingles.

After the windows and doors are installed, the outside is almost completed. A person walking by might think the house could be occupied.

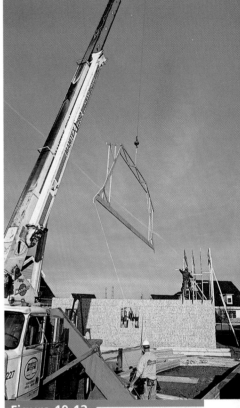

Figure 19-12 Trusses can save time in building a roof. A crane like this one lifts the trusses in place.

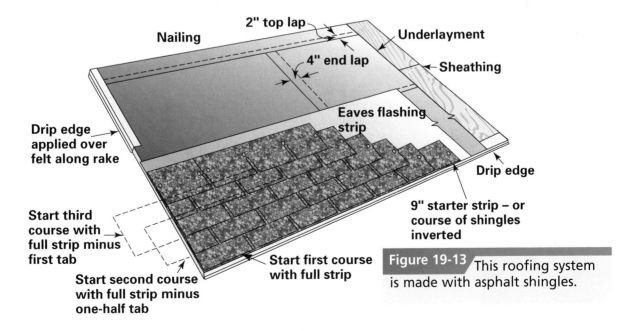

Figure 19-13 This roofing system is made with asphalt shingles.

Labels in figure:
- 2" top lap
- Nailing
- Underlayment
- 4" end lap
- Sheathing
- Eaves flashing strip
- Drip edge applied over felt along rake
- Drip edge
- Start third course with full strip minus first tab
- Start second course with full strip minus one-half tab
- Start first course with full strip
- 9" starter strip – or course of shingles inverted

Figure 19-14 A worker must always use the proper safety equipment when installing insulation.

insulation
material used to keep heat or cold from leaving or entering a building

Figure 19-15 It is very important that electrical wiring be done correctly and safely. There is no margin for error when working with electricity.

The Interior

▷ **How is the interior of a house completed?**

The interior of a house is completed after the exterior is finished. Workers have the advantage of being protected from bad weather. They install such things as insulation, utilities, and interior wall coverings.

Insulation

▷ **What does insulation look like, and how does it work?**

Insulation is like a blanket tucked inside the walls and ceilings. The more you have and the better it's tucked in, the warmer you will be in the winter. The most popular insulation is a fluffy type of fiberglass. It's made in long rolls wide enough to fit between floor joists, ceiling joists, and wall studs. See Figure 19-14.

The fluffy fiberglass roll sometimes has waterproof paper attached to one side. The paper usually faces the inside of the house and is a moisture barrier. It keeps moisture away from the wooden framing materials. Moisture would cause the wood to become wet and perhaps rot.

Cracks always appear in a wood house during its construction. Those cracks can let in cold air. Caulking is used to seal them up. Caulking looks a little like toothpaste and is applied with a special tool called a caulking gun.

Utilities

▷ **What types of utilities are used in most houses?**

Buildings contain a variety of subsystems that are commonly referred to as utilities. When construction workers refer to utilities, they mean electricity, natural gas, water, and sewage disposal. These services are provided by businesses called public utilities. Electricity is used for heating, lighting, cooking, air conditioning, and other purposes. Natural gas is used for heating and cooking. Water is used for drinking and cleaning. Sewage disposal removes wastes from the house. These utility subsystems are installed by subcontractors who specialize in each type.

Electric Systems Electricians install all the wiring, outlets, and light fixtures in the house. See Figure 19-15. Electricity comes from the public utility company and enters the house through a heavy insulated wire called a service entrance cable. A meter located outside measures how much electrical power is used.

Heating and Cooling Systems Heating and cooling systems require such equipment as furnaces, air conditioners, pipes, and ducts. In forced-air heating systems, for example, a fan blows heated air from the furnace through metal ducts to vents in each room.

Heating systems are usually powered by natural gas, electricity, or fuel oil. Air conditioning is usually powered by electricity.

Water and Sewage Systems These systems are installed by plumbing subcontractors. These subcontractors are responsible for the network of pipes that carry water to the kitchen, laundry room, and bathrooms and the pipes that carry wastes out of the house.

Water and sewage services are usually provided by the same public utility company. Pressurized water is delivered to the house through underground pipes. Inside the house, pipes branch out to the places where water is used.

The main water shutoff valve is in the water line where it enters the house. You should know where it's located. Suppose you accidentally break the water line under a sink. If you quickly find and turn off the main water valve, no harm will be done. Otherwise, the water could do quite a bit of damage.

Sewer pipes are much larger than water lines and are not pressurized. These lines are installed to carry the waste from the house.

LOOK TO THE FUTURE

Self-Sustaining Homes

In an experiment, over 1,800 houses in a California housing development have been equipped with solar panels that help turn solar energy into electricity. This electricity is used for heating and other power needs within each house. Any electricity not used each day is then sold to a local utility company. The money earned is used to pay for power that must be purchased from the utility company on cloudy days. This net-zero technology—technology that pays for itself—may one day be used to provide electricity to homes in many regions of the United States.

drywall
the inside covering of walls and ceilings, made from plaster and sturdy paper

Inside Wall Coverings

▶ **How are inside walls covered?**

The walls inside the house must also be covered. Workers use sheets of drywall to cover both walls and ceiling. **Drywall** is made from plaster sandwiched between two layers of sturdy paper. See Figure 19-16. Even if a wall will be covered with paneling, drywall is frequently installed first. The heavy drywall sheets are nailed or screwed to the walls and ceiling. A plaster-like material called spackling fills in the cracks between the sheets and covers the nail heads. The walls may then be painted or covered with paneling or wallpaper.

Figure 19-16 Drywall is normally screwed to the wall and ceilings with a special screw gun. Stilts are used when working on high ceilings.

CAREERS

Architectural Interns

Architects need three trainees. Responsibilities include checking house designs and construction documents, such as cost analyses. Successful applicants must have some drawing ability and a basic knowledge of drafting and design software. Good interpersonal skills are a must.

Interpersonal Skills

Using good interpersonal skills will help you form strong relationships at work and help the overall team.

Interpersonal skills are those you use in dealing with other people. Why are they important in the workplace? They are important because for most jobs people must interact to get the work done. If someone is unfriendly or grouchy all the time, people may want to avoid him or her. The work doesn't flow as smoothly. Practice your interpersonal skills now so they'll be ready when you need them! Here are a few tips:

✔ **Be courteous and use good manners.**
✔ **Learn people's names and speak to them by name.**
✔ **Learn to listen. Don't interrupt or insist on talking only about yourself.**
✔ **Don't gossip. Gossiping wastes time and hurts other people.**
✔ **Be dependable and trustworthy.**
✔ **Say please and thank you.**

Finishing Touches

▶ **What are the last jobs completed during house construction?**

After the walls and ceilings are finished, the floor is completed. It is usually done last to prevent damage to it while other work is going on. Various materials may be used to finish a floor, including wood, stone, tile, and carpeting.

Trim is then installed around the edges of the floors, around doors and windows, and sometimes where ceilings and walls meet. See Figure 19-17. Trims are usually made of wood or plastic.

Interior doors are then hung and kitchen and bathroom cabinets are added. Shelves and countertops are also installed at this time.

Landscaping—planting trees, grass, and bushes—is usually done while the interior of the house is being finished. After all the work is completed, the contractor checks to be sure everything has been done properly. The house is then ready to be turned over to the owner.

Figure 19-17 This trim carpenter is installing baseboard. A nail gun instead of a hammer is often used for this kind of work.

SECTION REVIEW 19B

Recall ▸▸

1. Why are walls put together on the subfloor?
2. What are trusses?
3. Why do roof shingles overlap each other?
4. What is the purpose of a moisture barrier on fiberglass insulation?

Think ▸▸

5. Some floors require 2 x 10 lumber, but a wall requires only 2 x 4 lumber. Why do you think this is true?
6. What do you think are some advantages of wood siding over plastic or metal siding?

Apply ▸▸

7. **Experiment** Conduct a test of different types of insulation. Paint three or more boxes black on the outside. Leave one box empty; line the walls of each remaining box with a different type of insulation. Put thermometers inside each box and place the boxes in the sun. Record the results and analyze your findings.

CHAPTER 19 REVIEW

Summary Activity

For each numbered blank, pick the answer from the list on the right that makes the most sense in the entire passage. Write your answer on a separate sheet of paper. No answer will be used more than once.

Apartments, townhouses, and condominiums house many families. Most houses are built at a specific location or __1__. Parts for manufactured houses are prefabricated in a(n) __2__. Log houses are usually sold in __3__ and the owner hires a builder. Another type of house looks like a sphere (or portion of a sphere) and is called a(n) __4__. Houses that face __5__ are sometimes less expensive to heat than houses facing north.

The foundation walls of a house are made from poured concrete or __6__. Bolts connect the foundation to the __7__. The floor is __8__ by joists that rest on the foundation. The __9__ absorbs wear during construction and supports a finished floor.

The walls are nailed together on the floor and raised into place. Ceiling __10__ help hold them in position. The roof structure can be made with rafters or __11__. The roof is covered with shingles. The first row of shingles is applied at the __12__ of the roof.

Fiberglass insulation has waterproof __13__ attached to one side, which acts as a moisture barrier. When cracks in a house appear during construction, they are sealed with __14__. Inside walls are covered with drywall, which is made from __15__ sandwiched between two layers of sturdy paper.

The utilities installed in a new house can include electricity, natural gas, and __16__. Pipes, wires, and ductwork for utilities are usually installed by __17__ who specialize in it. The finished floor, trim, cabinets, and __18__ are done last.

Answer List

- caulking
- concrete blocks
- edge
- factory
- geodesic dome
- joists
- kit form
- landscaping
- paper
- plaster
- sill plate
- site
- south
- subcontractors
- subfloor
- supported
- trusses
- water

Comprehension Check

1. Describe a geodesic dome.
2. List at least two questions that must be answered when choosing a building location.
3. How is the size of a house calculated?
4. What is sheathing?
5. What is the purpose of insulation?

Critical Thinking

1. **Compare.** Suppose you were designing two homes, one for a family with two young children and one for a retired couple. How would your designs differ and why?
2. **Design.** Make a rough sketch of a home you would like to own someday. What three features would be most important to you and why?
3. **Organize.** Put the following tasks in order and describe your reasoning: installing ceiling light fixtures, installing carpeting, and painting walls and ceiling.
4. **Infer.** Most early homes in the U.S. had two stories. Today, many houses are one-story ranch houses. Why do you think the ranch-style house became so popular?
5. **Propose.** What safety factors should be considered when working in the room in number 3 above?

Visual Engineering

Framing a Wall. The first thing you need to know in order to frame a wall is the names for its different parts. The easiest way to find all the names is to look in a carpentry textbook. Terms like stud, cripple, header, and trimmer are just a few of those that apply to walls. Make a drawing of a typical framed wall that has a door and window in it. Label all the structural parts. Find out the best way to nail such a wall together. Is there a correct procedure? How can you be sure the wall will be square and straight?

TECHNOLOGY CHALLENGE ACTIVITY

Building a Model House

Equipment and Materials

- posterboard
- white glue
- marking pens
- waxed paper
- razor knife
- wooden cutting surface
- house floor plan

! SAFETY

Reminder

In this activity, you will be using cutting tools. Be sure to always follow appropriate safety procedures and rules. Remember, safety is an attitude that you must develop and maintain at all times.

Background Many different house designs exist because people have different wants and needs. Each has his or her own opinion of what makes a desirable house. When a new design is being considered, a model is sometimes made. A model helps people visualize what a building will look like.

Goal For this activity, you will make a model of a house that you find interesting or attractive.

Criteria and Constraints

▶▶ Your model should be made of posterboard.

▶▶ You should choose a fairly modern house design rather than one of historical interest, which might be too complex to make.

▶▶ You should use a scale of one-half inch equals one foot.

▶▶ You must submit your sketches along with your model.

▶▶ You must label the rooms and any special features of the house.

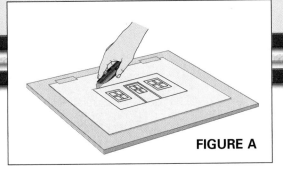

FIGURE A

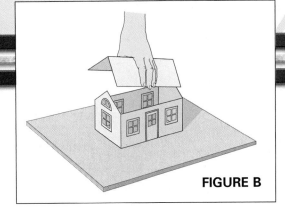

FIGURE B

Design Procedure

1. Do some research on house plans. Sunday editions of newspapers often show plans for a house being featured that week. Magazines or books on house design also include plans.

2. Select a house design that interests you. A single-story house will be easier to construct than one with two stories.

3. Make sketches of your model's floor plan and outer walls. Use a scale of one-half inch equals one foot. If a dimension shown on the original plans is 10 feet, for example, it will be 5 inches on your model.

4. Transfer your drawings to the posterboard. Use the razor knife to cut out the front wall. See Figure A. Draw in the windows and doors with your marker. Also, draw in any desired details such as shutters.

5. Repeat for the other three walls. It is better to do one wall at a time. Cutting two or more walls and then folding them might cause alignment problems if your measurements, folds, or cuts are a little off.

6. Place a small section of waxed paper onto your work surface to protect it. Hold the front wall against one of the side walls and glue the corners together along the inside of the walls. Repeat for all the other walls.

7. After the glue has dried, glue the four walls onto another piece of posterboard. This will help hold the model together by providing a solid foundation.

8. Sketch the entire roof onto posterboard. Cut it out and fold it in the middle along the ridge line. Place it on the walls, but do not glue it. The fold will keep it in place. See Figure B.

9. Remove the roof and add some interior walls. Place labels in the various rooms to identify them.

10. Add a porch (large or small) if the design does not already have one (optional).

11. Add a deck if the design does not already have one (optional).

12. Add some landscaping features like bushes and trees cut from posterboard.

13. Label the features of your house that you especially like.

Evaluating the Results

1. Write a paragraph explaining why you chose this house design.

2. Discuss as a class whether or not the models helped you visualize the house better than just a drawing.

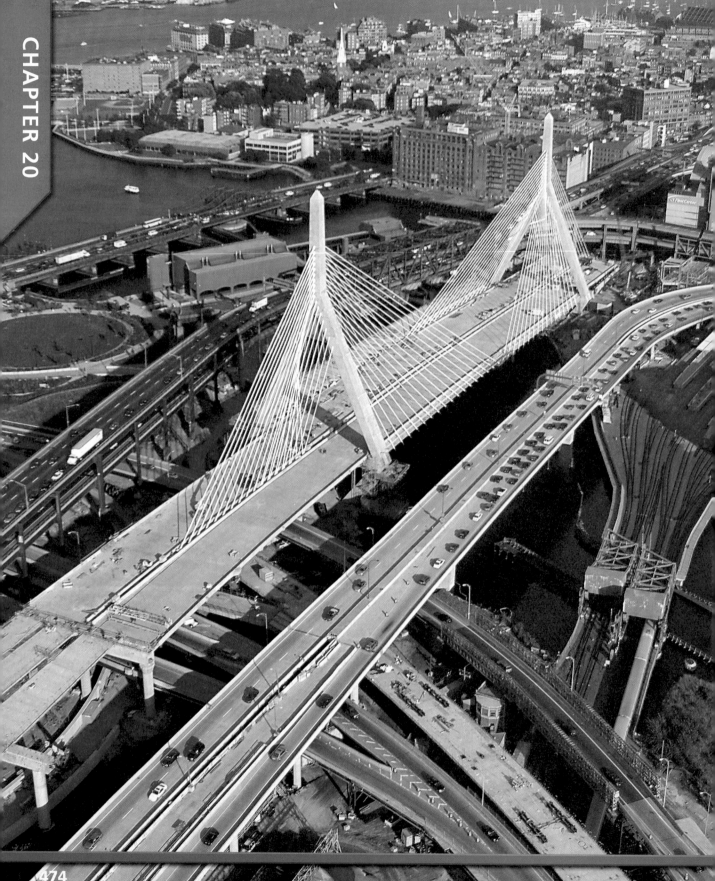

Y ou may have seen some of the impressive structures built with heavy construction methods. They include the tall skyscrapers in large cities. They also include civil construction projects, such as massive concrete dams, deep tunnels, and tall monuments.

Heavy construction refers to large projects that require special equipment and involve many workers with different skills. The projects usually take years to complete. The designs for these structures come from the efforts of many qualified engineers and architects who focus on safety. Each project has different needs. Architects and engineers who work in heavy construction often specialize.

 This construction project is taking place in downtown Boston, Massachusetts.

Large Buildings

▶ **How are large buildings different from houses?**

Objectives

▶▶ **Explain** how the construction of large buildings differs from that of houses.

▶▶ **Describe** basic methods used in building skyscrapers.

Terms to Learn

- crane
- excavation
- pier
- pile

Standards

- Core Concepts of Technology
- Relationships & Connections
- Design Process
- Use & Maintenance
- Construction Technologies

Some large buildings, called skyscrapers, are very tall and do not cover much land area. They are usually built in large cities where land is scarce and very expensive. Other large buildings spread out over a great deal of land but are not very tall. They may be built where land is inexpensive or where it has been set aside for public use.

Skyscrapers

▶ **What is the tallest skyscraper in the United States?**

Of the five tallest buildings in the United States, New York City has two. They are the Empire State Building (1,250 feet) and the Chrysler Building (1,046 feet). Chicago has three: the Amoco Building (1,136 feet), the John Hancock Center (1,127 feet) and the Sears Tower (1,454 feet). The Sears Tower is the tallest building in the United States. See Figure 20-1. The world's tallest skyscraper is Taipei 101 in Taipei, Taiwan (1,676 feet). However, the Sears Tower still holds the record for the highest floor in a skyscraper where people can live. It is habitable up to 110 stories, but Taipei 101 is habitable only to 101 stories.

Figure 20-1 The Sears Tower rises high above the Chicago skyline.

Foundation and Framework

▶ **How are skyscrapers made?**

The foundation of a skyscraper must support a huge load. A deep **excavation** is dug, and **piles** made of wood, concrete, or steel are driven deep into the soil until they hit solid rock. Rock gives the building the proper support. Foundation walls are then built and extra support is added in the form of strong **piers**, or columns, made of reinforced concrete.

The inner framework of a skyscraper is usually made of steel. The structure must be strong and rigid. It is put together floor by floor. As each floor is completed, large **cranes** lift the steel members to the next floor level. Workers weld, bolt, or rivet the members in place. Then the curtain walls are lifted into place and attached. As you know, these curtain walls are merely hung on the structure and provide little support.

It is not unusual for people to be concerned about their safety in tall buildings. Modern construction procedures, building codes, and strict rules cover fire protection and ways of escape. Tall buildings have restrictions and controls about the use of fire-resistant construction materials. Even lightning conductor systems are built into modern tall buildings. See Figure 20-2. All skyscrapers must sustain strong winds without swaying. In areas that have earthquakes or violent storms, special supports are used.

TECH CONNECT
MATHEMATICS

Scraping the Sky

The United States has many tall skyscrapers and the top five are not all that different in height.

ACTIVITY Draw scaled representations of the top five skyscrapers. Use this scale: 1 inch equals 100 feet. Label your drawings.

excavation
a large hole

pile
a large shaft driven deep into the soil to support a structure

pier
a concrete column used to add support to a foundation

crane
large machine used to lift heavy loads by means of a hook attached to cables

Figure 20-2 Lightning has struck the Empire State Building in New York City many times.

Other Large Buildings

▶ Why might other large buildings differ in construction?

Some buildings are spread out over large areas of ground. Your school may be one such structure. Like skyscrapers, they are often steel-framed buildings. However, the shape and appearance of these large buildings are usually quite different. Your school, for example, would probably have different outer walls than would a grocery store or a place for religious worship. Like skyscrapers, all these structures are carefully designed by engineers and architects.

Very wide buildings, such as auditoriums, cannot usually use column-and-beam framing. Large areas may then be supported by built-up girders, members that are joined together. Trusses and arches can also be used. See Figure 20-3.

Figure 20-3 Trusses like the ones in the ceiling can span long distances. This exhibit hall has few posts for support.

20A SECTION REVIEW

Recall ▶▶

1. What city is home to the tallest skyscraper in the United States and what is the name of that skyscraper?
2. How is the foundation for a skyscraper different from that for a house?
3. Why is it possible for skyscrapers to have walls of glass?

Think ▶▶

4. Into which category or categories of buildings do you think skyscrapers fit: residential, commercial/industrial, or public? Explain.

5. Does your community have buildings over twenty stories tall? Why do you think this type of building was chosen??

Apply ▶▶

6. **Construction** Build a model of a well-known skyscraper.
7. **Research** Visit the Web site of a well-known skyscraper, such as the Sears Tower. Report any interesting construction details to the class.

Roadway Projects

> ▷ What are some typical roadway projects?

Roadway projects include streets, roads, highways, bridges, and tunnels. Although many bridges are built across rivers, they also cross deep valleys and other roadways. Tunnels are often built to extend highways or railways through an obstacle rather than over or around it.

Roads and Highways

> ▷ What is below the surface of a highway?

Did you know that there are millions of miles of roadway in the United States? That includes everything from rural roads to city streets to interstate highways. The interstate highway system in the United States was started in 1952. It is over 40,000 miles in length and follows routes that connect about ninety percent of all major cities. Federal, state, and local governments spend billions of dollars every year to maintain existing roadways and construct new ones.

All paved roads are made in three layers. See Figure 20-4. The **subgrade** is the natural soil along the roadway. If it's not level or firm enough, heavy machines scrape and pack the soil. Next comes the base. A common base is sand or gravel. It provides support and keeps water from collecting underneath, which could freeze and break the pavement. Finally, the surface material is added. The surface is smooth and higher at the middle to drain off water.

Objectives

▸▸ **Explain** why asphalt and concrete are preferred materials for roadways.

▸▸ **Describe** the ways in which bridges are supported.

▸▸ **Describe** methods used to build tunnels.

Terms to Learn

- abutment
- cable-stayed bridge
- cantilever bridge
- girder bridge
- shield
- span
- subgrade
- suspension bridge
- truss bridge

Standards

- Core Concepts of Technology
- Design Process
- Use & Maintenance
- Information & Communication Technologies

▌**subgrade**
the layer of soil beneath a roadway

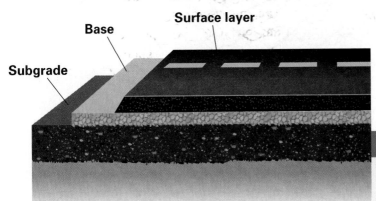

Base

Surface layer

Subgrade

Figure 20-4 You have probably seen highways being constructed in your area.

Figure 20-5 Concrete with steel reinforcement bars embedded in it is used in the construction of busy interchanges like this one.

The surface material chosen depends on the type of traffic that will use the road. Highways, which are main roads, are commonly surfaced with asphalt or concrete. Asphalt is a brownish-black, flexible material made from crude oil and other substances. Concrete is often preferred because it is easy to mix, doesn't have to be heated like asphalt, and dries to a hard, durable surface. Concrete can be strengthened with steel bars or steel mesh placed into it when it is wet. That is why it is used where the heaviest traffic is expected. See Figure 20-5. Airport runways and interstate highways are often surfaced with strengthened concrete that is about ten inches thick.

Asphalt is often used for less important roads. It is also used to repair worn or damaged concrete roads. Asphalt is flexible and sticks better to old concrete than fresh concrete does. Many interstate highways have been repaired with asphalt or resurfaced over the existing concrete base.

Bridges

▷ What are the different kinds of bridges?

The foundation of a bridge consists of the roadway and everything below it. At the ends of a bridge where it meets the land, **abutments** support both the bridge and the earth. If the bridge is long, piers may support the roadway between the abutments. The distance between supports is called a **span**. (Sometimes the entire length of a bridge is also referred to as the span.) If the earth beneath the bridge is not stable, piles may also be used.

abutment
a structure that supports the end of a bridge or dam

span
distance between bridge supports; the entire length of a bridge

Figure 20-6 The Golden Gate Bridge in San Francisco, California, is one of the most famous in the world. It was built in 1937. What type of bridge is it?

Figure 20-7 This beautiful cable-stayed bridge is the Sunshine Skyway Bridge in St. Petersburg, Florida.

Bridges are among the most beautiful human-made structures. The Golden Gate Bridge in San Francisco is an example of a beautiful suspension bridge. **Suspension bridges** hang from large cables and are used to cross wide spans. When it opened in 1937, the Golden Gate Bridge had the world's largest span (4,200 feet) and the highest supporting towers (746 feet). See Figure 20-6.

Another type of suspension bridge is the **cable-stayed bridge**. Inclined cables, called stays, connect the roadway to tall support towers. See Figure 20-7. Cable-stayed bridges are cheaper and easier to construct than traditional suspension bridges.

A **truss bridge** is one held together with steel beams. The beams are fastened together in the shape of many triangles. The Eads Bridge, discussed on page 436, is a **cantilever bridge** strengthened with trusses. A cantilever is a self-supporting beam that is fastened to the ground at one end. Two cantilevers meet in the middle to make the bridge. See Figure 20-8 (page 482). This strong design resists high winds.

A **girder bridge**, or beam bridge, has a simple structure. The roadway rests on girders laid across the span. Girder bridges are frequently supported by piers partway along the span. Many interstate highway bridges use this design.

suspension bridge
a bridge that hangs from large cables across a wide span

cable-stayed bridge
a bridge supported by inclined cables connected to towers

truss bridge
a bridge made from steel beams fastened together in triangular shapes

cantilever bridge
(CAN-tih-lee-vur)
a bridge made of beams fastened to the ground at only the ends; the beams meet in the middle of the bridge

girder bridge
a bridge made of girders that rest on the ground on either side of the span

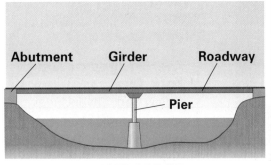

GIRDER BRIDGE

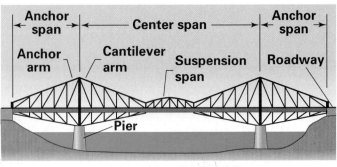

CANTILEVER BRIDGE

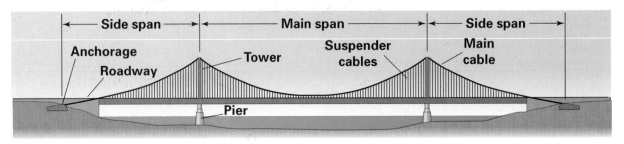

SUSPENSION BRIDGE

Figure 20-8 Bridges can be constructed in several ways.

shield
metal tube that fits inside a tunnel to support the walls

Figure 20-9 The Eisenhower Tunnel in Colorado is 1.7 miles long. It is one of the highest tunnels in the world.

Tunnels

▶ How are tunnels made?

Bridges go over natural barriers, but tunnels go under or through them. Tunnels are constructed through mountains and under water. Modern tunneling methods involve blasting with explosives and drilling with huge machines. A large metal tube called a **shield** fits inside the tunnel as it is drilled. For immersed tunnels, pre-built sections are brought to the site and sunk into an excavation. Then workers connect the sections together.

At 1.7 miles, the Eisenhower Memorial Tunnel in Colorado is the longest highway tunnel in the United States. See Figure 20-9. At an altitude of 11,000 feet, it is among the highest tunnels of its type. The Fort McHenry Tunnel in Baltimore, Maryland, is a part of Interstate 95. It is an immersed tunnel and was not constructed like most others. Workers first dug an undersea trench to accommodate sealed twin-tube steel and concrete sections. The sections were floated over the trench and workers pumped in additional concrete to make

Leadership

Being a leader can be a very rewarding experience, especially at work.

Leadership is the ability to guide, direct, and influence other people. Good leaders don't necessarily tell others what to do. Sometimes they act as coaches who train and support others. However, good leaders almost always inspire others to do their best. Few people are born with leadership skills. They must usually be learned and developed. Would you like to be a leader someday? Here are some tips for developing your leadership skills:

✔ **Volunteer for the leader's role. Practice taking responsibility.**

✔ **As a leader, explain your decisions; don't expect people to blindly obey.**

✔ **Delegate responsibility. Don't try to do everything yourself. Let others share in decision making and doing important jobs.**

✔ **Encourage everyone on the team to participate.**

✔ **Be supportive. Show appreciation.**

✔ **Share the credit for a job well done.**

Lead Surveyor

Large civil-engineering company is looking for a take-charge person to supervise surveying crews for major road and bridge projects. The successful applicant must be licensed, have at least three years of experience as a surveyor, and have at least one year's experience in a leadership position.

---

each section settle onto the bottom. The tunnel has thirty-two concrete tube sections positioned together to make it almost one and one-half miles long.

One tunnel project stands above all others as being the most expensive private construction project in history. It's the 32-mile-long tunnel under the English Channel between Folkestone, England, and Calais, France. See Figure 20-10 (page 484). It was a joint venture among British and French companies. The project started in 1986, and scheduled train service began in 1994. With 23.6 miles of its length underwater, it's the largest undersea tunnel ever built. Its depth varies from 90 to 480 feet below the bottom of the seabed.

The project involves three tunnels: two for high-speed electric trains and one between them for maintenance. Massive boring machines 27 feet in diameter removed material. Drilling was carried out from both ends of the tunnel at the same time.

Figure 20-10 A train, with its cargo of loaded trucks, leaves the Channel Tunnel on the French side of the English Channel.

20B SECTION REVIEW

Recall ▸▸

1. What two materials are most often used to surface roadways?
2. Why are roadway surfaces higher in the middle?
3. What is the difference between a suspension bridge and a truss bridge?
4. What device is used to prevent tunnel walls from collapsing after they are dug?

Think ▸▸

5. Concrete highways are poured in sections with strips of tar between them. What do you think is the purpose of the tar strips? Why aren't the concrete sections joined tightly together?
6. What special safety precautions do you think would be necessary for workers building a bridge or a tunnel?

Apply ▸▸

7. **Evaluating** What causes roads to crack or develop holes? Observe a section of roadway that has been repaired. Has an entire section been replaced or only holes filled? Why do you think that method was chosen? What materials do you think were used?
8. **Research** Do an Internet search for Web sites featuring famous bridges or tunnels, such as the Golden Gate Bridge, the tilting Gateshead Millennium Bridge, the Seikan Tunnel, or the English Channel Tunnel. Make a model or drawing of the structure, labeling key parts.

Other Structures

What other structures are built using heavy construction?

Heavy construction methods are used to build many other structures. They include such things as dams, monuments, canals, airports, shopping malls, and subway systems. Dams, monuments, and the International Space Station will be covered in this section.

Dams

What is the purpose of a dam?

The Grand Coulee Dam in the state of Washington and the Hoover Dam on the Arizona-Nevada border are among the largest single pieces of concrete in the world. A dam is built across a river to block the flow of water. It is usually done for one of three reasons: to provide a water supply for nearby communities and farms, for flood control, or to provide electrical power.

Objectives

▶▶ **Describe** the three main parts of a dam.

▶▶ **Tell** the purpose of a monument.

▶▶ **Explain** how construction in space is different from that on Earth.

Terms to Learn

- embankment
- outlet works
- spillway

Standards

- Core Concepts of Technology
- Cultural, Social, Economic & Political Effects
- Environmental Effects
- Role of Society
- Assessment

LOOK TO THE FUTURE

Dinosaurs of Construction?

As shoppers' needs change, shopping malls are changing too. Over the next several years, the big enclosed malls anchored with a large department store at each end may gradually disappear. Shoppers and store designers seem to prefer smaller, more open designs. Busy customers can park close to a store and dart into it without having to walk a long distance. Some stores are avoiding malls completely, preferring their own space where they can expand as needed. Do you see such changes beginning to take place in your area?

embankment

the main part of a dam that holds back the water

outlet works

section of a dam with gates that allow water to flow through

spillway

portion of a dam that allows excess water to spill over the dam

Dams may be made in several different shapes. However, most dams have three main parts: the embankment, the outlet works, and the spillway. The **embankment** is the large section that blocks the flow of water. The **outlet works** contains gates that allow a certain amount of water to flow through the dam. When too much water builds up behind the dam, the **spillway** allows the excess to flow around or through it and prevents the dam from breaking.

Hydroelectric dams (see Chapter 4, "Energy and Power for Technology") also include a power station. As water flows through the outlet works, it turns the blades of a turbine, which generates electricity.

The construction of a dam usually takes many years. The Hoover Dam was opened in 1936 after about eight years of work. See Figure 20-11. The project was too large for any one company, so six of them banded together to build the dam. There are countless smaller concrete and earth dams in the United States. An earth dam is made of carefully selected soil that is hauled to the site. See Figure 20-12. Layer upon layer of the soil is compacted with heavy rollers to form a watertight mass.

While dams have many benefits, they also impact the environment. Wildlife habitats are often lost, and there are many trade-offs. The gigantic Three Gorges Dam being built in China is an example. When finished, the dam will be the largest in the world. It will control severe flooding and produce billions of dollars worth of electricity. At the same time, more than fifteen cities, 1,500 towns and villages, and uncounted wildlife habitats will be lost. The river valley will be under almost 200 feet of water.

Figure 20-11 The Hoover Dam was such a huge project when it was built in the early 1930s that six construction companies had to band together to finish it.

Figure 20-12 Small earth dams like this one are often seen on small private lakes.

Monuments

▷ **What is the purpose of a monument?**

While most of the structures discussed in this chapter appeal to basic human needs, monuments appeal to the human spirit. They are designed to honor and remember the past and look forward to the future. The Statue of Liberty, the St. Louis Gateway Arch, and other grand monuments often inspire feelings of patriotism and pride. Some are so huge and unique that they have required more careful design and construction than most traditional structures.

The Statue of Liberty that greets all ships entering New York City's harbor was a gift from France in 1886. It is meant to celebrate the personal freedoms that Americans enjoy. Its internal structure of iron beams resembles the metal cage of a skyscraper. See Figure 20-13. The framework is covered with molded copper sheets. As the copper oxidizes, the statue turns green in color.

The St. Louis Gateway Arch is in the middle of a national park on the banks of the Mississippi River in St. Louis, Missouri. It was built in memory of the Louisiana Purchase of 1803 and the westward expansion that followed. The arch is constructed of stainless steel, and its foundation extends 60 feet into the ground. It was built starting from both ends at the same time. The two halves were connected at the top in 1965, after four years of construction. See Figure 20-14. The Arch is 630 feet tall, which makes it the tallest monument in the United States.

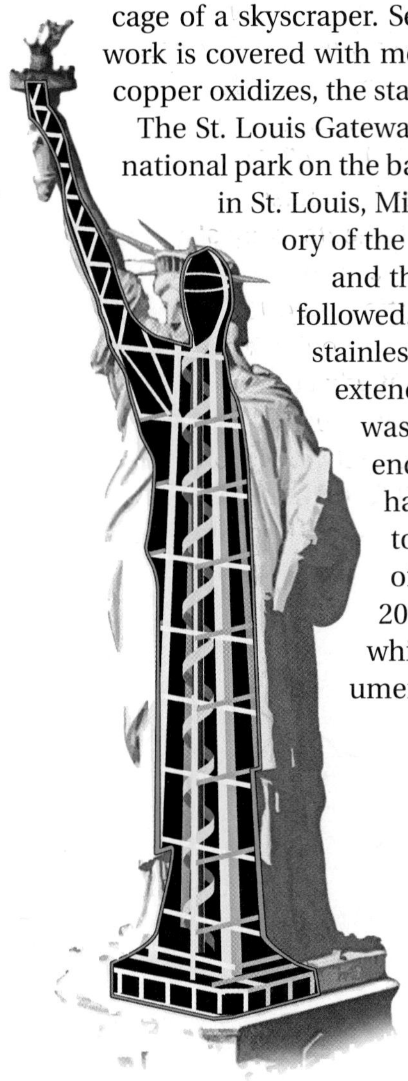

Figure 20-14 The completion of the Gateway Arch in St. Louis, Missouri, took place in 1965.

Figure 20-13 The Statue of Liberty is supported by a strong interior metal frame.

The Lincoln Memorial in Washington, D.C., the Eiffel Tower in Paris, France, and the Taj Mahal in Agra, India, are also examples of monuments.

The International Space Station

▷ How does construction in space differ from construction on Earth?

The United States, Russia, and fourteen other nations are building the International Space Station (ISS) in orbit around the earth. See Figure 20-15. The station is now permanently occupied by rotating crews of three people.

The ISS has several purposes. It serves primarily as a laboratory in which to do experiments. Researchers learn about the effects of very low gravity on certain materials and processes. They are also learning important information about what abilities are needed for people to live and work in space.

Figure 20-15 The International Space Station orbits the earth every ninety minutes.

Since it began in 1998, all assembly of the station has taken place in space. Modules (sections), materials, and equipment must be brought up from Earth in American space shuttles or Russian spacecraft. All maintenance must be done in space as well.

The framework of the ISS is a series of trusses. Modules for such things as living quarters are attached at various points. As the astronauts work, their tools and other equipment must be tied down to prevent them from drifting away. What other difficulties do you think workers in space have?

The ISS has huge solar collectors that draw energy from the sun and recharge its electrical batteries. When the station is completed, the collectors will extend about 350 feet, which is longer than a football field.

SECTION REVIEW 20C

Recall ▸▸

1. Give three reasons why dams are constructed.
2. Of what material is the St. Louis Gateway Arch constructed?
3. What kind of energy is used to power the International Space Station?

Think ▸▸

4. Hydroelectric dams are an environmentally clean way to generate electricity. Yet only about five percent of America's electrical power comes from hydroelectric dams. Why do you think this is so?
5. Monuments can be expensive. Do you think they are worth the investment? Explain.

Apply ▸▸

6. **Research** Do some research on the Great Pyramids of Egypt. Why were they built? How large are they? What materials and construction methods were used? Share your findings with the class.
7. **Survey** Not everyone agrees on the value of the International Space Station. Do a survey of at least twenty people. Do they think the ISS is a worthwhile investment or not? Graph your results.

Summary Activity

For each numbered blank, pick the answer from the list on the right that makes the most sense in the entire passage. Write your answer on a separate sheet of paper. No answer will be used more than once.

Major projects like skyscrapers are called __1__ construction projects. The word *skyscraper* has come to mean a tall building with an interior __2__. The tallest skyscraper in the United States is the __3__. Skyscraper safety is an important concern. Modern construction procedures, __4__, and other forms of legislation lay down strict rules.

All paved roads are made in three layers. The top layer is called the __5__. Concrete is frequently used as the top layer. A highway can be strengthened with steel __6__ or mesh placed in the wet concrete. Many concrete highways are repaired with asphalt. It sticks better and is __7__ than concrete.

One of the most beautiful __8__ bridges in the world is San Francisco's Golden Gate Bridge. Many bridges are strengthened with trusses. A truss is made of steel beams fastened together in the shape of many __9__. Many interstate highway bridges are __10__, or beam, bridges. A cable-stayed bridge uses inclined cables called __11__.

A metal tube called a(n) __12__ fits inside a tunnel as it is constructed. The English Channel Tunnel has a total of three tunnels. Passengers ride in __13__. It connects England with France.

Dams have three main parts: the embankment, the outlet works, and the __14__. Monuments are large structures that appeal to the human spirit. The tallest monument in the United States is the __15__. The International Space Station orbits the earth. Its framework is a series of __16__.

Answer List

- bars
- building codes
- girder
- heavy
- more flexible
- Sears Tower
- shield
- spillway
- St. Louis Gateway Arch
- stays
- support frame
- surface
- suspension
- trains
- triangles
- trusses

Comprehension Check

1. What are piles and piers?
2. What material is used for the inner structure of a skyscraper?
3. Name the three main parts of a dam.
4. About how thick is the concrete used on interstate highways?
5. What is the purpose of a monument?

Critical Thinking

1. **Compare.** Suppose you are an architect who will design a module to be attached to the International Space Station. What would you propose as substitutes for the following: walkway, front door, water faucet?
2. **Decide.** What do you think would be the hardest part of building the International Space Station? Why?
3. **Design.** Sketch a design for a monument to honor someone from your area who achieved something special.
4. **Appraise.** What social and economic impacts do you think building the interstate highway system had on U.S. society at the time?
5. **Decide.** Would you rather live in a single-family house or a large apartment building? Why?

Visual Engineering

The Big Dig. The city of Boston, Massachusetts, has undertaken one of the biggest construction jobs in the city's history. Research this project on the Internet. How do you suppose such a huge project was organized? Along with two or three of your classmates, list concerns that the city probably addressed before starting. Find out how long it will take to complete the project, how it will affect the city during construction, what it will cost, and where the money for it comes from.

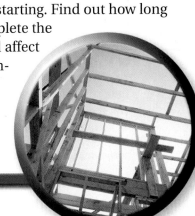

TECHNOLOGY CHALLENGE ACTIVITY

Building a Truss Bridge

Equipment and Materials

- six 36-inch-long pieces of ³⁄₁₆-inch balsa strips
- quick-drying adhesive
- pencil
- paper
- ruler
- cutting blade
- 2- to 3-gallon plastic bucket with handle
- sand
- nylon cord
- small piece of hardwood
- scale
- safety goggles and long-sleeved shirt

! SAFETY

Reminder
In this activity, you will be using cutting tools. Be sure to always follow appropriate safety procedures and rules. Wear safety glasses. Remember, safety is an attitude that you must develop and maintain at all times.

Background

Much of the strength of a bridge comes from its design. Truss bridges were developed during the 1500s. The parts of a truss are arranged in the form of many triangles.

Goal

For this activity, you will build a truss bridge and test it.

Criteria and Constraints

▶▶ Your bridge must measure 18 inches long, 4 to 6 inches tall, and 4 to 6 inches wide. It must be able to accommodate the piece of hardwood used in the test.

▶▶ Your bridge must be constructed from balsa wood. No metal may be used.

▶▶ All joints must be flush. No joints may overlap.

Design Procedure

1. Research truss bridge designs. Make a full-scale drawing of a bridge that is exactly 18 inches long, approximately 4 to 6 inches tall, and approximately 4 to 6 inches wide.
2. Construct the bridge using the ³⁄₁₆-inch balsa strips. To improve stiffness, glue two strips together for the main horizontal supports. See Figure A. All joints must be flush. No overlapping joints are allowed.

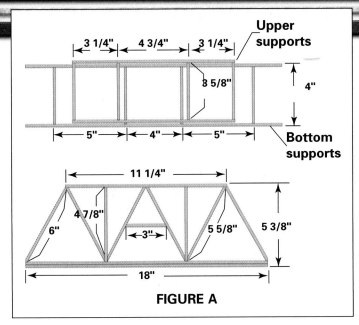

FIGURE A

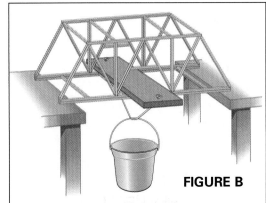

FIGURE B

3 Your design must be able to accommodate the piece of hardwood used for loading the bridge. At the middle of the sides, leave an opening that measures at least 1½ x ½ inches.

4 After the adhesive has completely dried, weigh your bridge.

5 Arrange two tables 16 inches apart, and place the bridge across the gap. Put the hardwood on the bridge and tie the bucket to it. The bottom of the bucket should be about 4 to 6 inches above the floor. See Figure B.

6 Put on your safety glasses and long-sleeved shirt. If possible, ask another student to videotape the testing procedure to document how long your bridge lasts. Slowly pour sand into the bucket until the bridge breaks. It may fail quickly and might scatter broken balsa wood. Make sure the bucket of sand doesn't tip over when the bridge fails and the bucket drops to the floor. Weigh the bucket to see how much load your bridge carried. Record the results.

Evaluating the Results

1. How much weight did your bridge carry in comparison to its own weight? Was it five times more? Ten times more? More than that? Can you see why the design of a bridge can help it carry very heavy loads?

2. Where did your bridge begin to fail? Was it at the center, the edges, or the top? How could you change your design so that your bridge would carry a heavier load?

The Wheel
Rolling through History

The invention of the wheel dramatically changed transportation.

1 The earliest evidence for the invention of the wheel comes from the area between the Tigris and Euphrates rivers in what is now modern Iraq. There, **around 3500** B.C.E., the ancient Sumerians formed wheels from three planks of wood that were bound with leather tires held in place by copper nails. A hole in the middle held an axle.

Use of the wheel for organized warfare created a demand for something that weighed less and that would permit greater speed. **Around 2000** B.C.E., the use of spokes eliminated most of the heavy wood in the wheel's center. Sculptures from Chinese tombs from the **thirteenth century** B.C.E. show the use of spokes on chariot wheels.

2 Investigate: *When and how did use of the wheel spread to the New World (Western Hemisphere)?*

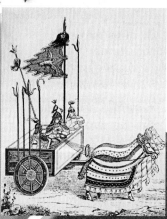

In the 1800s, heavy railroad cars required sturdy metal wheels that ran along a track. Locomotives in the United States used large driving wheels to **3** help increase their speed to about 30 mph. Some of these wheels were over six feet in diameter. Investigate: *Find out why such large wheels are no longer used on locomotives.*

4 The wheels on horse-drawn vehicles were often covered with a solid, rubber tire. However, the solid rubber was hard on roadways and did not produce a comfortable ride. The first pneumatic (air-filled) tire was patented **in 1845** in England by Robert W. Thomson and used on carriages. **In 1888**, John Dunlop of Ireland patented a pneumatic tire for bicycles and formed a company for manufacturing them.

Around 1900, use of the pneumatic tire spread to automobiles. Separate inner tubes held the air, and the outer covering was made from rubber-coated cotton cloth. **In 1954**, the tubeless tire was introduced permitting high speeds by reducing surface friction.

5

Investigate: *The first tires were made from natural rubber. Find out when and why synthetic rubber began to be used in large quantities.*

6

In 2001, Dean Kamen introduced the latest wheeled vehicle, his battery-powered Segway Human Transporter. Its two wheels are made of plastic and surrounded by air-filled tires. What makes the Segway innovative is that each wheel is powered by its own motor and controlled by a computer. Gyroscopes help the rider maintain balance. Kamen believes his invention will eventually replace the automobile in crowded downtown areas.

Try to count all the engines and motors that you and your family operate. Cars and lawn mowers may be the most obvious, but some people also use snowblowers, chain saws, and motorcycles.

We have become dependent on engines and motors. They produce the power we need to carry people and goods. It would be difficult to operate other technologies if we couldn't generate power for transportation. Without power for transportation, how would you get to school or visit far-away relatives or friends?

Do you know the difference between engines and motors? Both convert energy into motion that enables us to do work. Motors usually run on electric energy, and the motion they produce is rotary (circular) motion. Engines use heat energy to produce motion. People often use the terms motors and engines interchangeably. You can say, for example, that a car is powered by an engine or a motor. However, there are no electric engines, only electric motors.

Can you imagine the tremendous power coming from these rocket engines?

Objectives

▶▶ **Explain** how an external combustion engine works.

▶▶ **Describe** the difference between a steam engine and a steam turbine.

Terms to Learn

- **external combustion engine**
- **piston**
- **turbine**

Standards

- Relationships & Connections
- Influence on History
- Design Process
- Energy & Power Technologies
- Information & Communication Technologies

external combustion engine
an engine that burns fuel to create energy; its power source is outside the engine

piston
a plug that slides inside a cylinder in an engine

External Combustion Engines

▶ What are some types of transportation power?

Until about three hundred years ago, the only forms of transportation power that people used came directly from nature. They included wind, flowing water, and muscle power from people and animals. There were no engines of any kind until the first steam engine was built in England in 1712. It was the size of a small building and operated pumps to remove water from coal mines.

Steam Engines

▶ How do steam engines work?

All steam engines are **external combustion engines**. See Figure 21-1. *External* means that the power source is outside the engine. *Combustion* refers to burning. Steam engines use the heat from burning coal or wood to change water into steam. Since the fire is under a boiler, which is outside the engine, the power source is external.

Steam engines have a piston that moves back and forth. A **piston** is a plug that just barely fits inside a closed cylinder. Expanding steam from a boiler pushes on one end of the piston and causes it to move inside the cylinder. See again Figure 21-1. As the piston moves, it closes one valve and opens another. The steam is then sent to the other end of the piston and moves it back to its original position. This back-and-forth movement turns a circular flywheel.

By connecting the spinning flywheel to a vehicle's wheels, steam was used to power land transportation. This is a good example of how technology developed for one setting can be used in another setting. Steam engines were used in some early cars like the Stanley Steamer. They also powered huge locomotives that pulled passengers and cargo. The engines made a loud chuffing noise and produced clouds of steam.

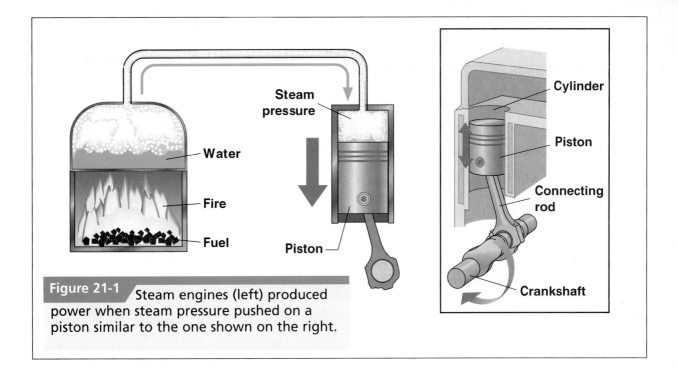

Figure 21-1 Steam engines (left) produced power when steam pressure pushed on a piston similar to the one shown on the right.

Steam engines are not used very much in modern-day America because their efficiency is low and they do not produce much power for their size. People's needs and wants have changed. However, today's technologies can be better understood by studying those used in the past.

Steam Turbines

▷ What are steam turbines?

Steam turbines operate from steam pressure, just like steam engines. That is where the similarity ends, however. As you know, steam engines produce power by means of pistons moving up and down. Steam turbines develop power from spinning disks. The two kinds of power sources are very different.

A **turbine** is a continually spinning disk that resembles a pinwheel. Blow on a pinwheel and it spins. See Figure 21-2. You could call the pinwheel a "breath turbine" because your breath makes it spin. Steam from a boiler spins steam turbines, as shown in Figure 21-3 (page 500).

Steam turbines power oceangoing ships. Ship turbines develop as much as 150,000 horsepower. Turbines are also used in electrical plants to produce electricity.

Figure 21-2 Steam turning a turbine is similar to blown air turning a pinwheel.

turbine
(TUR-bin)
a disk or wheel that changes the energy of moving gases or liquids into rotary motion

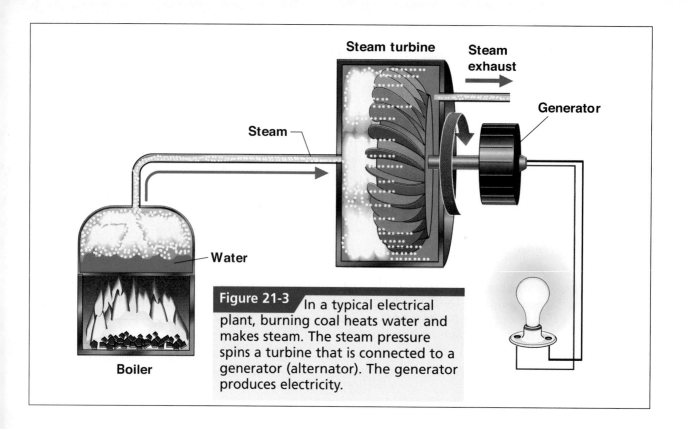

Figure 21-3 In a typical electrical plant, burning coal heats water and makes steam. The steam pressure spins a turbine that is connected to a generator (alternator). The generator produces electricity.

Steam turbine

Steam exhaust

Generator

Steam

Water

Boiler

21A SECTION REVIEW

Recall ▸▸

1. Where and when was the first steam engine built? What was it used for?

2. Why are steam engines called *external* combustion engines?

3. A steam turbine operates differently than a steam engine. Explain the difference.

Think ▸▸

4. What safety problems do you think early steam engines caused?

5. All engines and motors are systems. Identify the input, process, and output in an external combustion engine.

Apply ▸▸

6. Research Look up improvements made on steam engines by James Watt (1736-1819). Make sketches illustrating his changes.

7. Design Create a paddlewheel turbine using Styrofoam® and plastic spoons. Turn it with water from a faucet.

Internal Combustion Engines

⊳**Where does burning take place in an internal combustion engine?**

Did you know that there's a fire inside the engine of your family car? It's what makes the engine operate. You can't see the flames because they're deep inside. This internal (inner) fire means that your family's car is powered by an **internal combustion engine**. Gasoline, diesel, gas turbine, and rocket engines are all internal combustion engines. Although you can see the flames in some jet engines and all rocket engines, the combustion still takes place inside the engine.

Most engines that we use in our daily lives create power from a piston sliding inside a cylinder. Fuel and air are placed inside the cylinder. See Figure 21-4. An electric spark causes this mixture to ignite (catch fire) and build up high pressure in a very short time. The pressure pushes the piston down with great force. Another device, such as a flywheel, changes the downward motion to circular motion. This process produces enough power to cause automobiles to travel on land, boats to move through water, and small airplanes to fly in the air.

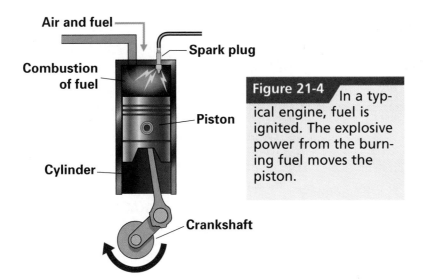

Figure 21-4 In a typical engine, fuel is ignited. The explosive power from the burning fuel moves the piston.

Objectives

▸▸ **Explain** how an internal combustion engine works.

▸▸ **Describe** four-stroke and two-stroke engine cycles.

▸▸ **Explain** the purpose of a crankshaft.

▸▸ **Identify** differences among various engines.

Terms to Learn

- crankshaft
- ignition system
- internal combustion engine
- jet engine
- maintenance
- propellant
- reciprocating motion
- rotary motion
- thrust

Standards

- Core Concepts of Technology
- Relationships & Connections
- Environmental Effects
- Role of Society
- Design Process
- Use & Maintenance
- Energy & Power Technologies
- Transportation Technologies

◢ **internal combustion engine**
an engine that burns its fuel inside

Engine Cycles

▶ Are engine cycles like other types of cycles?

When a set of events happens over and over again, we say that the events go in cycles. The seasons of the year follow a cycle. There are also life cycles, food cycles, and business cycles, just to name a few.

A bicycle has *cycle* as part of its name. In pedaling a bicycle, your legs go up and down, repeating the motions over and over again. One motion of your leg makes a downward stroke. Lifting your leg creates an upward stroke. (A stroke is movement in one direction.) Your legs make two strokes before repeating the same motions. We could say that your bicycle is operated by a two-stroke human power plant. Your legs deliver power to the rear wheel, which is called the driving wheel. Wheels under power are called driving wheels. Other wheels are called trailing wheels.

Four-Stroke Cycles

▶ What is involved in a four-stroke cycle?

Much like your legs pedaling a bicycle, the pistons inside an engine move up and down. The most popular type of engine is the four-stroke cycle engine. The pistons make four strokes before they repeat themselves. See Figure 21-5. These strokes are the intake, compression, power, and exhaust strokes.

Figure 21-5 This is what happens during each stroke of a four-stroke cycle engine.

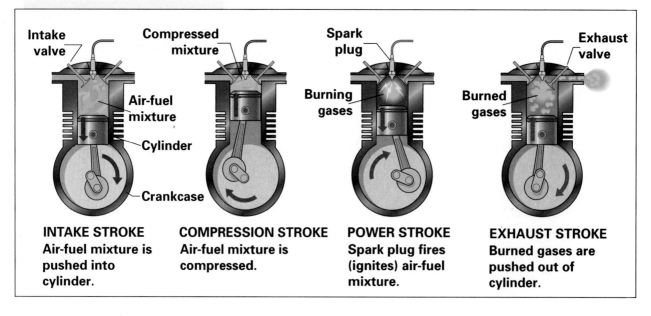

INTAKE STROKE
Air-fuel mixture is pushed into cylinder.

COMPRESSION STROKE
Air-fuel mixture is compressed.

POWER STROKE
Spark plug fires (ignites) air-fuel mixture.

EXHAUST STROKE
Burned gases are pushed out of cylinder.

- **Intake stroke.** The piston moves down, creating a partial vacuum in the cylinder. The intake valve is open and fuel and air flow into the cylinder.
- **Compression stroke.** The valve closes and the piston moves up. It squeezes the air-fuel mixture to about one-eighth of its original volume in the top of the cylinder.
- **Power stroke.** An electric spark from a spark plug ignites the mixture. The fuel and air mixture burns very rapidly and increases the pressure inside the cylinder. This pressure forces the piston down.
- **Exhaust stroke.** The exhaust valve opens. The piston moves up and pushes out the exhaust gases. The four strokes of the cycle begin to repeat.

Most cars have four-stroke cycle engines. When oil is added to the engine for lubrication, it is put in separately. It is not mixed with the fuel.

Two-Stroke Cycles

▶ How does a two-stroke cycle work?

Some small gasoline engines operate with only two strokes. The intake and compression strokes are combined. The power and exhaust strokes are also combined. See Figure 21-6. Such engines operate on a two-stroke cycle. The piston makes two strokes before it begins to repeat itself.

Figure 21-6 This is what happens during each stroke of a two-stroke cycle engine.

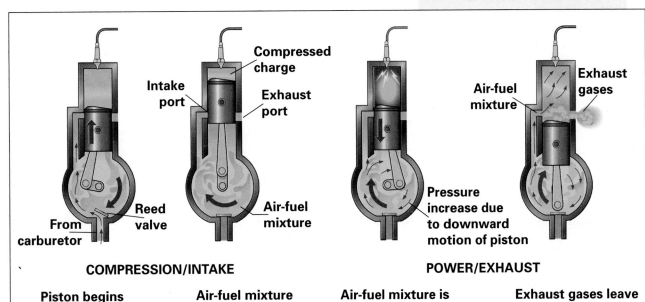

COMPRESSION/INTAKE

Piston begins compression. Air-fuel mixture enters crankcase.

Air-fuel mixture is compressed. Ports are sealed.

POWER/EXHAUST

Air-fuel mixture is ignited. Piston is forced down.

Exhaust gases leave cylinder. Air-fuel mixture enters cylinder.

Figure 21-7 This engine runs very quietly even though there are thousands of explosions going on inside every minute.

These engines power mopeds, string trimmers, chain saws, lawn mowers, and other devices. Two-stroke cycle engines are less efficient and emit more pollutants. In them, the fuel and oil are mixed together.

Gasoline Engines

▷ How do gasoline engines produce power for automobiles?

There are more gasoline engines in the world than any other type. See Figure 21-7. They are used in cars, boats, and small airplanes. They start easily, are inexpensive to make, and can be made in almost any size. Very small gasoline engines are used in lawn trimmers. The largest gasoline engine ever built powered some American military airplanes in the late 1940s. Named the R-4360 Wasp Major, it was more than forty times larger than the engine in most automobiles.

Automobile engines operate with a four-stroke cycle. Many modern engines have four cylinders, but others may have six or eight cylinders. Some 1934 Cadillacs had as many as sixteen cylinders.

The piston in a gasoline engine moves only up and down. This up-and-down, straight-line motion is called **reciprocating motion**. Unless we travel by pogo stick, reciprocating motion cannot be used for transportation. We need a way to convert it to **rotary motion**, or circular

reciprocating motion
(ree-SIP-roh-kay-ting)
up-and-down or back-and-forth motion that occurs in a straight line

rotary motion
circular motion

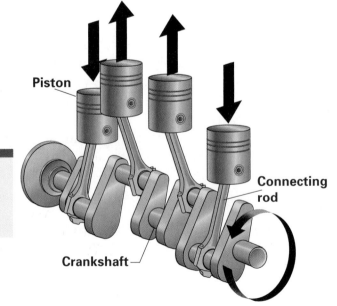

Piston

Connecting rod

Crankshaft

Figure 21-8 A crankshaft is a heavy, strong piece of metal that changes reciprocal motion into rotary motion.

motion. All our transportation methods, except rocket-powered spacecraft, require rotary motion to turn wheels or propellers.

Reciprocating motion is changed to rotary motion by a **crankshaft**. See Figure 21-8. The pedals on your bicycle are also attached to a crankshaft. The crankshaft converts your reciprocating leg motion to the circular motion of the wheels. Your legs move up and down, but the bicycle wheels rotate to move you smoothly forward.

An automobile has a crankshaft inside the engine. Each piston is joined to it by a connecting rod. The rotating crankshaft transfers power to the driving wheels.

Government Regulations

▷ Why are exhaust emissions controlled?

Through the Environmental Protection Agency, the federal government regulates the exhaust emissions and miles per gallon for all cars sold in the United States. These regulations mean that modern gasoline engines have many differences when compared to those made only a few years ago. The emissions must be low to keep harmful pollutants from damaging the environment. The miles per gallon must be high to help ensure that the world's oil supply will last for a long time.

Diesel Engines

▷ How are diesel engines different from gasoline engines?

Diesel engines are best suited for heavy-duty work. Although used in some automobiles, they are more important for powering trucks, buses, locomotives, and ships. They can operate smoothly under heavy loads that would cause a gasoline engine to stall, or stop running. Diesel engines last longer and require less maintenance than gasoline engines. Diesels are too heavy for airplane use.

The internal parts of a diesel engine are much like those inside a gasoline engine. Diesels have pistons, cylinders, and a crankshaft. They come in four-stroke and two-stroke cycle versions. Generally, cars and small trucks use four-stroke cycle engines. Two-stroke cycle engines are used in cross-country semitrailer trucks and locomotives. Two-stroke cycle diesels are a bit more efficient and use less fuel than four-stroke cycle diesels.

▶ **crankshaft**
the part of an engine that changes the reciprocating motion of the pistons to rotary motion that turns the wheels

LOOK TO THE FUTURE

From Space to the Expressway

Technology used in space exploration is often applied to problems here on Earth. Ceramic tiles, made from minerals heated at high temperatures, protect the space shuttle from the heat of reentry. Engineers hope these same ceramics could be used to make the pistons in diesel truck engines, which produce a lot of heat. Engine efficiency and emissions could be improved. Today, diesel truck engines have a lifetime of a million miles or more. With ceramics they may last even longer!

ignition system
the engine system that starts the fuel burning

The major differences between diesel and gasoline engines are in the diesel's fuel system and in the process by which the fuel is ignited. Diesel fuel is similar to kerosene and cannot be easily ignited with a spark plug. The diesel engine uses a different type of **ignition system**. It uses hot air.

The engine's four strokes are the same as those in a gasoline engine. However, there are some operating differences as shown in Figure 21-9.

- **Intake stroke.** Only air enters the cylinder. It is not an air-fuel mixture.

- **Compression stroke.** The air is squeezed to about $\frac{1}{22}$ of its original volume. Squeezing air that much causes its temperature to rise to about 1,000°F. Then diesel fuel is squirted directly into the cylinder. The high air temperature ignites the fuel immediately.

- **Power stroke.** Pressure in the cylinder builds up very quickly and pushes the piston down.

- **Exhaust stroke.** The piston moves up. Burned gases are pushed out of the cylinder.

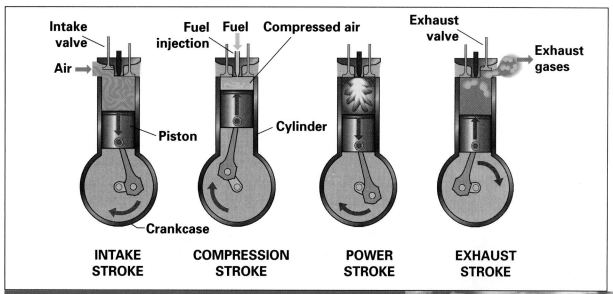

Intake valve — Air → | Fuel injection — Fuel | Compressed air | Exhaust valve — Exhaust gases

Piston | Cylinder

Crankcase

INTAKE STROKE | **COMPRESSION STROKE** | **POWER STROKE** | **EXHAUST STROKE**

Figure 21-9 A diesel engine makes four strokes. Large eighteen-wheelers, like the one shown on the right, use diesel engines to pull heavy loads.

Practically all automobiles use gasoline engines, but some have diesel engines. Why do we have both types? What are the advantages and disadvantages of each? What are the trade-offs? See Table 21-A for a brief comparison for automobile engines of about the same size.

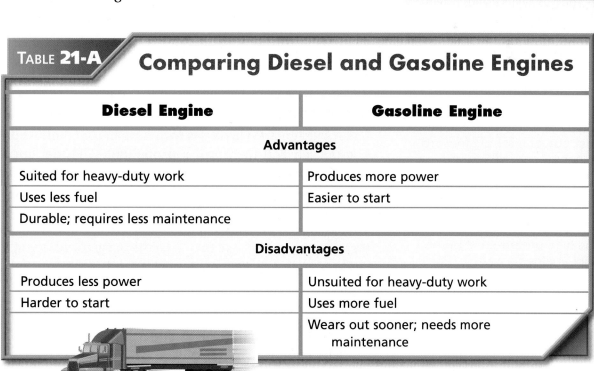

TABLE **21-A** **Comparing Diesel and Gasoline Engines**

Diesel Engine	**Gasoline Engine**
Advantages	
Suited for heavy-duty work	Produces more power
Uses less fuel	Easier to start
Durable; requires less maintenance	
Disadvantages	
Produces less power	Unsuited for heavy-duty work
Harder to start	Uses more fuel
	Wears out sooner; needs more maintenance

Gas Turbine Engines

▷ **Why do commercial airplanes use gas turbine engines instead of gasoline engines?**

Gas turbine engines power many large airplanes, ships, trucks, and some locomotives. They are complicated, but they are the most reliable internal combustion engine. A reliable engine is one that rarely stops working because a part breaks. It requires less **maintenance** (inspection and servicing).

Gas turbines are smaller and lighter than other engines of the same power rating. They also have a long engine life. Their biggest disadvantage is high cost. The gas turbine has parts that spin at high speeds and are kept at high temperatures. Such parts must be carefully made from special materials. This makes the engine very expensive. Some engines used by airplane companies cost over $1 million each.

There are three basic types of gas turbine engines: the turbojet, the turbofan, and the turboshaft, or turboprop. See Figure 21-10. A more common name for a turbojet or turbofan engine is **jet engine**. Jet engines push airplanes through the air with a jet of high-pressure exhaust gas. Have you ever blown up a balloon and then let it go? The pressurized air escapes through the nozzle. The force of the air makes the balloon dart around. This force is known as **thrust**.

Rocket Engines

▷ **Why are rocket engines used for sending satellites into orbit?**

A rocket engine carries its own oxygen for combustion. The oxygen and fuel form a **propellant**. The fuel burns with the oxygen to produce high-speed exhaust gas. The gas rushes out the rear to produce thrust. Jet engines also develop thrust, but they depend on oxygen from the air. There's no air in outer space. Only rocket engines can travel in space, because they take the oxygen with them.

The simplest rocket engines use a solid propellant. Solid propellants do not need a combustion chamber. They are simply ignited, and they burn. However, the combustion cannot be stopped or controlled once it has begun.

The thrust from five powerful rocket engines lifts the space shuttle off the launching pad. Two are solid-propellant rocket engines that are strapped to each side and look like long tubes. See Figure 21-11 (page 510). They are called solid rocket boosters (SRBs). They help boost the space shuttle into orbit. Each one develops 2.6 million pounds of thrust and burns out in about two minutes. Then the SRBs drop into the ocean, where a ship recovers them. They are overhauled and used on other flights.

The three middle engines on the space shuttle are called space shuttle main engines (SSMEs) and are a permanent part of the shuttle. These engines develop lift-off thrust. They do not drop away like the SRBs. They use liquid pro-

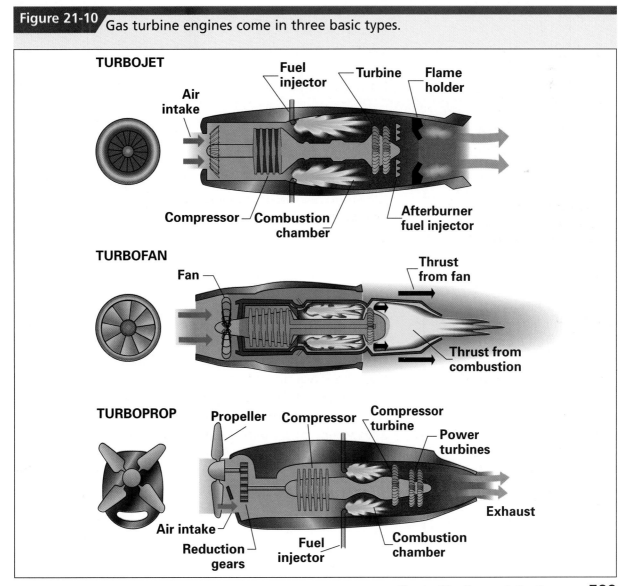

Figure 21-10 Gas turbine engines come in three basic types.

TURBOJET

Air intake — Fuel injector — Turbine — Flame holder

Compressor — Combustion chamber — Afterburner fuel injector

TURBOFAN

Fan — Thrust from fan

Thrust from combustion

TURBOPROP

Propeller — Compressor — Compressor turbine — Power turbines

Air intake — Reduction gears — Fuel injector — Combustion chamber — Exhaust

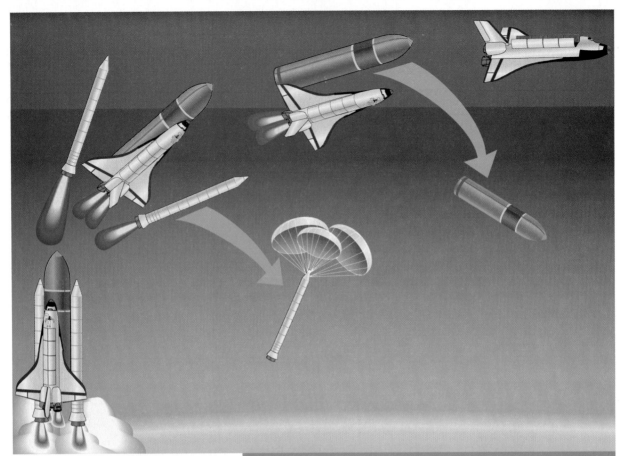

Figure 21-11 During the launch of the space shuttle, the two solid rocket boosters are released first. The liquid propellants are carried in the big center tank that drops away later in the flight. What liquid propellant is used?

TECH CONNECT
SOCIAL STUDIES

Lift Off!

America's *Saturn V* rocket, which put Neil Armstrong and Edwin Aldrin on the moon in 1969, remains the largest and most powerful rocket ever built. It was 363 feet tall compared to the space shuttle's 184 feet.

ACTIVITY Watch the movie *Apollo 13*. Write a paragraph describing the effects of the space program on the families of astronauts.

pellants. The propellants used in the SSMEs are liquid oxygen (LOX) and liquid hydrogen (LH$_2$). The propellants are carried in a large external tank that's covered with insulation. After it's emptied, the tank drops away and is not recovered. See again Figure 21-11. We use both liquid- and solid-propellant rocket engines because each has its advantages and disadvantages. See Table 21-B.

Commercial airplanes aren't powered by rocket engines because rocket engines use so much propellant. They are also much more expensive to make than jet engines.

TABLE **21-B**

Comparing Solid and Liquid Propellants

Solid Propellant	Liquid Propellant
Advantages	
Fuel can be stored for a long time.	Thrust is easy to control.
Engine is not complicated to make.	Combustion can be started, then stopped.
Engine can be made in small sizes.	
Disadvantages	
Thrust is not easy to control.	Fuel cannot be stored for a long time.
Combustion cannot be stopped once started.	Engine is complicated to make.
	Engine cannot be made in small sizes.

SECTION REVIEW 21B

Recall ▸▸

1. What is the difference between a stroke and a cycle?
2. What is used to turn reciprocating motion into rotary motion?
3. What ignites the fuel inside the diesel engine?
4. How is solid propellant different from liquid propellant?

Think ▸▸

5. Name the source of energy, the process, and the load for a four-stroke cycle engine.
6. The huge SRBs on the space shuttle use up all their propellant in only two minutes. Explain why this is desirable.

Apply ▸▸

7. **Construction** Make a crankshaft from cardboard or balsa wood.
8. **Construction** Simulate an engine with a small plastic bottle, baking soda, and vinegar. Place the baking soda and vinegar in the bottle. Place the bottle on the surface of a basin of water, as if it were a boat. The base-acid reaction will form a gas that pushes the bottle across the water's surface.

Objectives

▸▸ **Describe** how electric motors are used to power locomotives.

▸▸ **Explain** how a hybrid automobile is powered.

▸▸ **List** the advantages and disadvantages of hybrid and fuel cell cars.

▼ Terms to Learn

- fuel cell
- hybrid

Standards

- Relationships & Connections
- Cultural, Social, Economic & Political Effects
- Environmental Effects
- Role of Society
- Use & Maintenance
- Energy & Power Technologies

Electric Motors

▶ **Can you name some transportation vehicles that use electric motors?**

Many transportation vehicles use electric motors. Subways and electric trains are two examples. See Figure 21-12. However, other transportation devices like elevators, escalators, and rides in amusement parks are also powered by electric motors. Some devices combine electric motors with other power systems.

Locomotives

▷ **How do electric trains work?**

One type of locomotive uses a diesel-electric drive system. A diesel engine turns a generator to produce electricity. The electricity is sent by wires to electric motors directly connected to the driving wheels. Diesel-electric locomotives can develop over 6,000 horsepower and pull 200 railroad cars.

Figure 21-12 Trains that use electric motors help reduce air pollution and are quieter than diesel-powered trains or buses.

A few locomotives use a gas turbine engine to operate the generator. You can tell if a locomotive uses a gas turbine or diesel engine by listening carefully as a train passes by. Gas turbine engines make a high-pitched sound. Diesel engines make a lower-pitched sound.

Another type of locomotive is all electric. It has no diesel or gas turbine engine. It usually draws its electrical power from an overhead cable. The stainless steel *Acela Express* is a high-speed all-electric locomotive used for intercity passenger travel. It moves along the popular Washington-to-New York-to-Boston corridor at speeds over 150 mph. Its travel time between Washington, D.C., and New York City is under two-and-one-half hours.

Electric Cars

▷ **Will we ever drive cars powered by electricity?**

In response to public concerns, automobile companies have experimented with electrically powered cars in order to find a technology that does not harm the environment. For several years, General Motors leased battery-powered EV1 electric cars to customers in California and some nearby states. The driver pressed an accelerator pedal just as in an engine-powered car. The pedal controlled how much electricity was sent to the motor from the batteries, and the car sped up or slowed down.

Automobile companies continue to experiment with all-electric cars. For those completely dependent on batteries, the biggest problem has been the distance you can drive on a single battery charge. It tends to be around fifty miles, which means you can drive only about twenty-five miles away from a power source.

Hybrid Automobiles

▷ **Can an electric car have a gasoline engine too?**

Although no all-electric car is currently for sale to the general public, the hybrid comes fairly close. A **hybrid** is a combination of different elements. The hybrid car uses the combination of an electric motor and a gasoline engine. See Figure 21-13 (page 514).

The electric motor is used at low speeds around town and limits exhaust emissions. The small gasoline engine is used for higher speeds on the open highway where emissions do not tend to collect. The gasoline engine also

TECH CONNECT
MATHEMATICS

Speedy Metrics

"Kilometers per hour" (kph or km/h) appears on many speedometers. One mile per hour (mph) equals 1.61 kph. A car moving at 60 mph is moving at about 97 kph (60 × 1.61).

ACTIVITY Convert the following mph speeds to kph: 35, 40, 65, 75. Round your answers to the nearest whole number.

▸ **hybrid**
something made from a mixture of different elements

operates a generator that recharges the batteries that power the electric motor. One type of hybrid automobile has a 44 horsepower electric motor and a 70 horsepower gasoline engine.

Fuel Cell Automobiles

▷ What are the advantages and disadvantages of a fuel cell?

fuel cell
a device that generates electricity by converting chemical energy of a fuel, such as hydrogen, into electrical energy

If battery-powered electric automobiles are not practical, fuel cells might be the answer. A **fuel cell** converts hydrogen to electricity. Hydrogen is an unlimited resource. Electricity is then converted to mechanical power. Experimental fuel-cell-powered electric automobiles have gone 280 miles before having to be refueled. Some cities are testing fuel-cell-powered buses. See Figure 21-14.

Fuel cells don't pollute the atmosphere because their only emission is water. They also don't have to be recharged, just refilled with hydrogen. Fuel cells might power a car that you will drive in the future. So far, however, there are none available to the public, partly because they are so expensive. Car manufacturers are working to make fuel cell cars more practical.

Figure 21-14 This bus is powered by a fuel cell.

SECTION REVIEW 21C

Recall »

1. Name five examples of transportation devices that use electric motors.
2. How does a diesel-electric locomotive work?
3. How fast does the *Acela* travel?
4. Why is the hybrid automobile so named?
5. What emission does a fuel cell produce?

Think »

6. Electric trains are sometimes designed to draw power from overhead wires or from a protected high-voltage rail on the ground. They have two electrical pickup locations. Why do you think this is so?

7. All-electric cars are very quiet compared to gas-powered cars. Why do you think this is so?

Apply »

8. **Research** Find out which automobile companies sell hybrid cars. If a dealership is in your community, pick up a brochure. Based on what you read, decide what you like and don't like about hybrid automobiles.

Summary Activity

For each numbered blank, pick the answer from the list on the right that makes the most sense in the entire passage. Write your answer on a separate sheet of paper. No answer will be used more than once.

Much of our useful power comes from engines. The energy source used by an engine is __1__. Engines that use gasoline are found in cars. Rocket engines power space shuttles and __2__ engines are used in jumbo jets.

The world's first engine was a steam engine used to pump __3__ from coal mines. The wood or coal used by steam engines was burned outside the engine. That's why the steam engine is a(n) __4__ combustion engine.

Gasoline engines use a piston that slides up and down inside a(n) __5__. __6__ forces the piston to move, and the piston moves up and down only a few inches. Each upward or downward motion of the piston is a(n) __7__. The up-and-down motion is converted to more useful __8__ motion by a crankshaft. Because the fuel is burned inside the engine, the gasoline engine is called a(n) __9__ combustion engine.

The diesel engine also uses pistons. Its kerosene-like fuel is ignited by __10__. Diesel engines are best suited for __11__. Currently, only a few automobiles are powered by diesel engines.

Gas turbine engines use a jet of high-pressure exhaust gas to push airplanes through the air. They are more expensive than gasoline engines but are also __12__. That's why all large commercial airplanes use gas turbine engines.

Unlike gas turbine engines, rocket engines carry oxygen as well as fuel. That's why only rocket engines can travel __13__. Space shuttles have two solid propellant boosters strapped to each side.

Many trains are powered by electric motors. Some locomotives use diesel engines to turn __14__ that produce electricity. Automotive companies are experimenting with electric cars. The hybrid car runs on an electric motor and a(n) __15__. Fuel cell cars convert hydrogen to electricity. Their emission is water.

Answer List

- cylinder
- external
- fuel
- gasoline engine
- gas turbine
- generators
- heavy work
- hot air
- in space
- internal
- more reliable
- pressure
- rotary
- stroke
- water

Comprehension Check

1. Why is a steam engine considered an external combustion engine?

2. What is the difference between a motor and an engine?

3. What is the purpose of the electric spark used in an internal combustion engine?

4. Which is the most reliable internal combustion engine?

5. What is a fuel cell?

Critical Thinking

1. **Decide.** Consider the automobile engines discussed in this chapter. If you were to buy a car, which type of engine would you prefer and why?

2. **Analyze.** Some people want to convert immediately to electric or fuel cell-powered cars. Others want to wait until the world's supply of oil is completely used up. What are the advantages and disadvantages of each position?

3. **Connect.** How do you think someone who maintains and repairs cars would benefit from language arts skills?

4. **Infer.** What needs and wants do you think inspired people to develop the gasoline engine for automobiles instead of continuing to use the steam engine?

5. **Combine.** Design your own hybrid vehicle using two different power systems. Write a brief paragraph explaining how it works. Your design does not have to be serious.

Visual Engineering

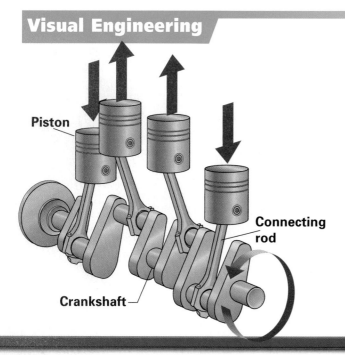

Piston

Connecting rod

Crankshaft

Conversion of Power. The crankshaft in an engine converts the up-and-down motion of the pistons into circular motion. How is the circular motion of the crankshaft used to power the car? Does a front-wheel-drive car handle differently than a real-wheel-drive car? Do some research to find out. Then make a drawing or build a simple model to show how the power from the engine gets to the wheels. Share your findings with the class.

TECHNOLOGY CHALLENGE ACTIVITY

Building a Steam Turbine

Equipment and Materials

- copper tube, 1 inch in diameter and 4 inches long
- two corks that fit tightly into the ends of the copper tube
- aluminum beverage can
- plastic bead
- small nail
- two metal clothes hangers
- propane torch
- scissors
- razor knife
- heavy wire cutters
- file
- safety glasses or goggles
- stop watch

SAFETY

Tool Safety
In this activity, you will be using dangerous tools, such as a propane torch. Be sure to always follow appropriate safety procedures and rules. Remember, safety is an attitude that you must develop and maintain at all times.

Background

In a steam power plant, steam is directed toward a fan-shaped turbine. The pressure from the steam causes the turbine to spin. The rotary motion is used to turn an alternator that generates electricity for use in your home and school. Turbines in steam power plants sometimes operate nonstop for years before they are shut down for inspection or repair. The first modern steam turbine was made in 1884. Electric companies now use them to produce electricity. The huge turbines are expensive and have many parts.

Goal

For this activity, you will build a steam generator that will turn a fan-shaped turbine.

Criteria and Constraints

▸▸ Turbine blades should be uniform in size and shape.
▸▸ You must keep a record of turbine speeds and the influence changes in design have on speed.

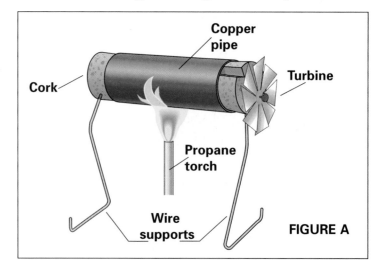

Copper pipe

Turbine

Cork

Propane torch

Wire supports

FIGURE A

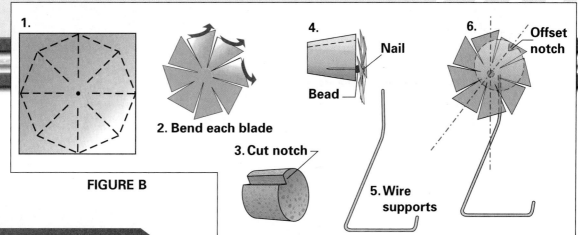

FIGURE B

1.

2. Bend each blade

3. Cut notch

4.
Nail
Bead

5. Wire supports

6.
Offset notch

Design Procedure

❶ Review the drawing in Figure A to get an idea of how the steam turbine will be made.

❷ Use scissors and the template shown in Figure B to cut out an octagon-shaped disk from the beverage can. Punch a hole in the center with a small nail and bend the blades as shown in Figure B. This will be your spinning turbine disk. Color one blade with a permanent marker to help you judge how fast the disk spins.

❸ Use the razor knife to cut a notch in one of the corks as shown in Figure B. The notch will direct steam from the copper tube to the turbine disk and cause the disk to spin.

❹ Slide the turbine disk onto the nail. Then slide on the plastic bead, which will act as a bearing. Push the nail into the notched cork at a spot across from the notch. Place the cork in one end of the copper tube. Place the other cork in the other end of the tube. Your copper tube is now a completed boiler.

❺ Use the heavy wire cutters to make two supports out of clothes hanger wire. See Figure B. Sharpen the ends with a file, and stick one support into the edge of each cork. Your steam turbine is now ready to operate.

❻ Remove a cork and fill the copper tube about two-thirds full of water. Replace the cork.

❼ Devise a method for measuring turbine speed. For example, one person can count rotations while another uses a stopwatch.

❽ Put on the safety glasses and light the propane torch. Carefully and evenly heat the copper tube. Steam will soon come out through the notch you cut in the cork. The force of the steam will spin the turbine.

❾ Measure and record turbine speed.

❿ Try twisting the turbine blades to greater or lesser angles. What effect, if any, does this have on the rotating speed? Record the speeds.

Evaluating the Results

1. Could you tell how fast the disk was rotating? Would you expect it to spin faster when the boiler (copper tube) was two-thirds full of water or only one-third full of water? Why?

2. Would the disk spin faster or slower if the notch you cut in the cork were smaller? Why?

3. What conclusions can you draw about turbine design from this experiment?

SUPERSTARS OF TRANSPORTATION

INVESTIGATION ////

Space travel is probably the most exciting and dangerous form of transportation. Do some research to find out who was the first human to orbit the earth and how long his trip lasted.

Matthias W. Baldwin
(1795-1866)

Trained as a jeweler, Matthias Baldwin built Pennsylvania's first locomotive in 1832. He went on to establish the Baldwin Locomotive Works, which became the largest locomotive manufacturer in the world. He was among the first manufacturers to use standardized parts, and he developed special wheel supports for use on the sharp curves on American rail lines. Baldwin's engines could carry more steam pressure than any others available at the time. Before the company closed in 1955, it had manufactured about one-third of all American locomotives, more than 1,500 of them during Baldwin's lifetime.

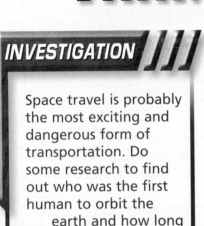

Isambard Kingdom Brunel
(1806-1859)

Isambard Kingdom Brunel was Great Britain's hero of the Industrial Revolution. A designer of bridges and tunnels, he is best known for the construction of several large ocean-going steamships. The *Great Britain,* built in 1845, was the first all-metal, propeller-driven steamship and carried up to 252 passengers on voyages to America and Australia. The *Great Western* made the crossing from Bristol, England, to New York City in fifteen days, a spectacular feat at the time.

Rudolf Diesel
(1858-1913)

For powering boats, trucks, and locomotives, no engine has proved as useful as the diesel engine. It is named for its inventor Rudolf Diesel, who patented the design in Germany in 1892. The gasoline engine had already been invented, but Diesel sought a design that was more efficient for heavy transportation. Early diesel engines used kerosene for fuel and were quite large. The first experimental model, built in 1897, had one cylinder, was about 12 feet tall, and developed 20 horsepower.

Frederick M. Jones
(1893–1961)

Frederick Jones learned mechanics and electronics on the job. He was a garage mechanic in Cincinnati before moving to Minneapolis during the 1920s where he took a job working with movie equipment. There he invented the first process that enabled movie projectors to play back recorded sound. In 1937, he got the idea for a refrigeration unit for trucks that had to carry meat and produce across the country using only ice that melted in the heat. His unit was so successful it was adopted by the U.S. Army during World War II. Jones and a partner then formed the Thermo King Company to manufacture the units. Today, almost all refrigerated trucks and railroad cars use Jones's system. He invented many other devices, including a portable X-ray machine, and earned over 60 patents.

Ivan Getting **(1912–2003)**

Trained as an astrophysicist, Ivan Getting got involved with technology during World War II when he researched the use of radar, ballistic missiles, and microwave tracking systems. Today, he is best known as the architect of the Global Positioning System, the series of satellites orbiting above the earth that is used for mapping and navigation. Early critics thought the system was impractical and too expensive, but it is now considered the most important aid to navigation produced during the 20th century.

Donna Shirley **(b. 1941)**

Donna Shirley's worst subject in school was math, but she worked hard at it anyway, because she wanted to be an engineer. At the age of sixteen, she got her pilot's license, and in 1963 received her first degree in aerospace engineering. In 1966 she began working at NASA's Jet Propulsion Laboratory in California. One of her jobs was to figure out how vehicles could fly safely through the atmosphere of Mars. In the 1980s she worked on space station design and concepts for vehicles used to explore other planets. She is best known as the first woman to head up a project for NASA and for her work on the *Sojourner*, the first rover to explore the surface of Mars.

A transportation system is a way of moving people or products from one place to another. Transportation terminals are locations where people or products enter or leave the system. Airports, bus stops, and docks are terminals.

A transportation system has inputs, processes, outputs, and feedback. For example, inputs to a city bus system include such things as people to operate the buses and fuel to provide energy. Processes include such things as driving the bus and loading passengers. The output is arrival at scheduled stops. Feedback includes comments from satisfied customers.

Transportation systems are interrelated. That means that each one often depends on the others. Buses and cars, for example, take passengers to airports and ship docks. Can you think of other examples? Transportation systems are also embedded within the larger technological, social, and environmental systems that make up our world. This chapter discusses some of those relationships.

 You can see the navigational instruments on the bridge of a large cruise ship.

Objectives

▶▶ **Describe** the different types of land transportation.

▶▶ **Explain** the purpose of transportation subsystems.

▶▶ **Explain** the purpose of a transmission.

Terms to Learn

- **bullet train**
- **driving wheel**
- **four-wheel drive**
- **front-wheel drive**
- **maglev train**
- **mass transportation**
- **rear-wheel drive**
- **tractor-trailer**
- **transmission**

Standards

- Relationships & Connections
- Cultural, Social, Economic & Political Effects
- Use & Maintenance
- Information & Communication Technologies
- Transportation Technologies

mass transportation
transportation that moves many people at one time and is available to the general public

Land Transportation

▶ What are some examples of land transportation vehicles?

When you travel in a car, bus, or train, you are using a land transportation vehicle. Land transportation also includes travel by bicycle, motorcycle, and subway train. Perhaps you can think of other land vehicles.

An automobile is an important part of our land transportation system. However, you need more than just a car to get from place to place. Roads, bridges, and service centers are just a few of the subsystems within a land transportation system that allow you to make good use of your family car.

Government regulations often influence the design and operation of transportation systems. For example, a state government regulates the speed at which you can travel on that state's roadways. This results in another subsystem, the highway patrol. Can you think of others?

Automobiles

▶ Why are automobiles so important in the United States?

The United States is one of the world's largest countries. About half of all the world's automobiles are used in the United States. Why do you think we have so many cars?

Mass transportation moves many people at one time and is available to the general public. However, it is expensive to develop mass transportation systems that can serve a country as large as the U.S. That is one reason why the automobile has become such an important part of our way of life. Automobiles are for personal transportation, not mass transportation. Mass transportation is sometimes slower and less convenient. What other trade-offs can you think of?

Modern automobiles are quite different from early cars. However, your family car has at least two things in common with those early models. Both have transmissions and nearly all have front-mounted engines.

Transmission

▶ **What is the purpose of a transmission?**

A car's **transmission** contains gears and other parts that transfer power from the engine to the axles and wheels. The gears work on the same principles as gears on a bicycle. When you pedal up a hill, you shift into low gear because it takes more effort to pedal up the hill. The rear wheel moves more slowly. On level ground, it takes less effort to pedal and the rear wheel turns more quickly. See Figure 22-1. A car's engine operates best if you use a low gear while climbing a steep hill or starting from a dead stop. You use a car's high gear when driving on a flat highway.

▌**transmission**
vehicle system that contains gears and other parts that transfer power from the engine to the axles and wheels

Figure 22-1 The sprockets on a ten-speed bicycle work just like the gears in a car's transmission. High gear combines the larger crank sprocket with the smallest sprocket at the rear wheel. Low gear combines the smaller crank sprocket with the largest sprocket at the rear wheel.

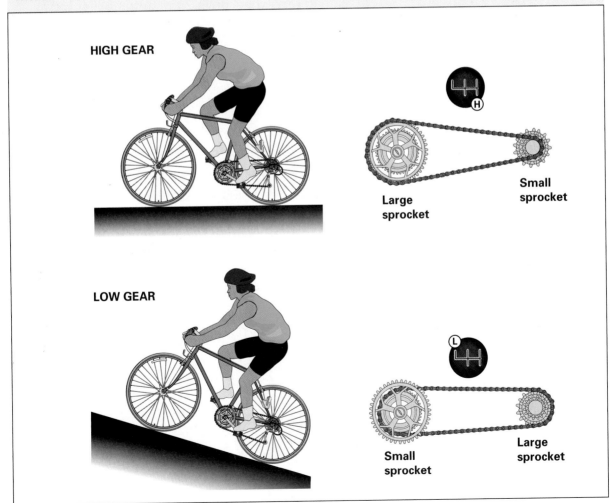

HIGH GEAR

Large sprocket

Small sprocket

LOW GEAR

Small sprocket

Large sprocket

Driving Wheels

▷ **What are driving wheels?**

driving wheel
a wheel that transmits motion from one part of a vehicle to another

rear-wheel drive
transfer of power from the vehicle's engine to the rear wheels

front-wheel drive
transfer of power from the vehicle's engine to the front wheels

four-wheel drive
transfer of power from the vehicle's engine to both front and rear wheels

The power from your legs is transferred to your bicycle's rear wheel with a chain. The rear wheel is a bicycle's **driving wheel**. A car doesn't use a chain. Instead, it transfers power with one or two metal shafts called drive shafts. Some cars transfer power to the rear wheels and are known as **rear-wheel drive** cars. There are also **front-wheel drive** cars. Some send power to all four wheels and are known as **four-wheel drive**, or all-wheel drive, cars. See Figure 22-2.

Four-wheel drive cars can go almost anywhere. Cars that travel on unpaved muddy or snowy roads might need four-wheel drive. Automobile manufacturers make four-wheel drive cars, sport utility vehicles (SUVs), vans, and trucks.

Figure 22-2 Not too many years ago, all cars and trucks had rear-wheel drive. Most automobiles today have front-wheel drive. Some specialty vehicles use four-wheel drive.

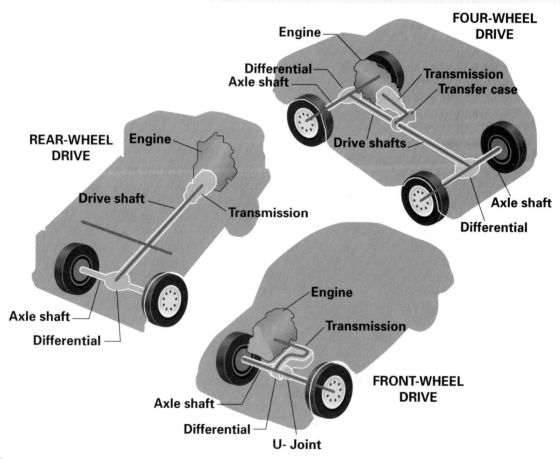

Planning for Change

The world of work is constantly evolving. Planning for change is a good way to stay ahead and be successful.

Customer Service Experts
Growing auto manufacturer needs people at all levels who can help our customers adjust to changes. Applicants must have experience in customer service and must be flexible themselves as job requirements may (change) as the company evolves. Occasional travel is required.

In the past, people often remained at the same job for most of their working lives. This is no longer true. Companies and jobs can change rapidly. People today can expect to make an average of four career moves into four different fields. One of those four fields may not even be in existence today. The chances are also good that they will move to a new neighborhood or city from five to seven times in their lives. You'll need to be prepared and to keep an open mind. Here are some tips for handling change:

✔ Change seldom comes by surprise, so try to be aware. Pay attention to trends and other clues that change may be on its way.
✔ Be positive. Look for the new opportunities change offers.
✔ When change occurs, talk to friends and family. Ask for advice if it's needed.
✔ Be supportive when others must adjust to change. They'll then be more likely to support you when you need it.

Other Subsystems

▷ What other subsystems does a car have?

An automobile has many subsystems. Besides the transmission and driving wheels, some subsystems include those that provide structure and support (frame), propulsion (engine), and guidance (steering wheel). Others provide suspension (springs) and control (brakes). When all these systems work together, the car functions properly.

Many of these subsystems are manufactured by outside suppliers. The anti-lock braking system (ABS), for example, is not usually made by the company that makes the car. The radio, compact-disc player, and windshield may also be manufactured elsewhere. For example, Toyota Motor Manufacturing in Georgetown, Kentucky, uses approximately 350 outside suppliers.

New technology is changing some vehicle subsystems. For example, some cars have guidance systems that can obtain information from the global positioning system (GPS). Maps appear on a screen telling the driver where the car is located.

Buses

▷ How did school buses help provide better schools?

More buses are made in the United States than in any other country. Large and boxlike, buses usually carry thirty or more passengers. They are used for mass transportation between cities. This is called intercity transportation. Buses are also used for transportation within cities and for school transportation. School systems alone use over 400,000 buses.

School buses were being used as early as 1920. They made it possible to gather students together from small rural schools into one large school with improved facilities. About twenty-four million students ride safe and sturdy buses to school each day. See Figure 22-3. These safe, well-designed vehicles are made from strong steel to meet federal manufacturing requirements. Seats and other support systems are specially designed for safety. Many use diesel fuel, which does not burn as easily as gasoline if an unexpected leak should occur.

Intercity buses are also powered by diesel engines. These buses carry up to sixty-four seated passengers and are generally less expensive to ride than trains or airplanes. They also make stops at smaller communities not

Figure 22-3 Almost all school buses are yellow. Do you know why? What do you think is most important to the safety of a school bus?

served by trains and airplanes. They transport nearly 800 million passengers every year in the United States.

Urban or in-city buses are an important part of mass transportation. Urban buses can carry more people than intercity buses because they allow some passengers to stand. Use of urban buses eases traffic congestion and saves fuel. They use about one-third as much fuel per passenger as do automobiles. However, they account for only fifteen percent of all passenger miles traveled in the United States. These are some of the reasons why the federal government provides money to improve a city's bus fleet.

Trucks

▶ **How many different kinds of trucks are there?**

Trucks move much of our country's freight. Cities in the U.S. rely on trucks to supply them with food, fuel, furniture, and other products. They play an important role in transportation because trucks can go directly from the supply location to the customer. Most trains, airplanes, and ships do not share that advantage.

Like other transportation systems, trucks are involved in many processes. They move and deliver cargo. They require places where they can pick up or unload that cargo. People who work in trucking systems must evaluate, market, and manage the services.

We use hundreds of different kinds of trucks. Most are diesel powered, but gasoline engines are also used. Some commercial trucks are as small as pickup trucks, while others are as large as the eighteen-wheel semi-trailer trucks that carry cargo on interstate highways. These large trucks are also called **tractor-trailers**. See Figure 22-4.

There is no such thing as a standard truck, but there are three general types: light duty, medium duty, and heavy duty. Panel and pickup trucks are examples of light-duty trucks. Medium-duty trucks are used locally and include sanitation trucks, soft drink delivery trucks, and fuel trucks. Heavy-duty trucks carry large loads, and an eighteen-wheel tractor-trailer is one type. Only about one in every twenty trucks is an eighteen-wheeler.

TECH CONNECT

LANGUAGE ARTS

The Language Barrier

Large tractor-trailer trucks called eighteen-wheelers in the U.S. are known as articulated lorries in Britain. A dual-carriage way in Britain is a divided highway in the U.S. High-speed, limited-access roads or interstate highways are called motorways in Britain.

ACTIVITY Use a dictionary to look up the words, tram, omnibus, caravan, and the underground. What terms are used in the U.S.?

tractor-trailer
a two-part truck that includes a tractor, or engine and cab, and a trailer that holds the cargo

Figure 22-4 Tractor-trailer trucks can be used for many different needs. This truck is carrying pumpkins to the processing plant.

Figure 22-5 A wind deflector on the top of a truck cab can save fuel costs each year.

The flat front and square shape of many trucks present a large surface that the air can press against. At forty-five miles per hour, one-third of the engine's power is used just to overcome this air resistance. This means that the trucks waste fuel. Manufacturers have tried two ways to help trucks slip through the air more easily. One way is to change the shape of the tractor and the front of the trailer. The other way is to place a wind deflector on the tractor's roof. See Figure 22-5. A properly designed deflector can reduce air resistance by twenty percent. It directs the air around the truck. Next time you ride on the highway, compare the number of trucks you see with deflectors with the number without deflectors.

Locomotives

▷ What kind of engines replaced steam engines in locomotives?

If you lived during the 1800s, the faraway sound of a steam locomotive whistle would have caused your ears to perk up. Trains fired the imagination of everyone, both young and old. By 1960, diesel-electrics had completely replaced steam locomotives on all mainline railroads. There are now approximately 21,000 diesel-electric locomotives in the United States.

Railroads earn most of their money by hauling freight. They deliver bulky items like coal and iron ore. They also carry such things as automobiles and television sets. About 10,000 freight trains roll over the tracks each day. Some are over 200 cars long.

Trains also carry passengers, and the busiest lines are on the east coast. These are operated by AMTRAK, our intercity passenger railroad. Travel between the terminals at Washington, D.C., and Boston, Massachusetts, including cities along the way, is quite popular. The distances are fairly short, and the trains travel directly from downtown to downtown. However, trains carry less than one percent of all U.S. intercity passengers. That is a very low number. In some countries, trains carry up to fifty percent of intercity travelers.

Commuter and subway trains transport workers, tourists, students, shoppers, and others to their daily destinations. They are also part of the mass transportation network. Figure 22-6.

Figure 22-6 The Metro Center subway station is in Washington, D.C.

High-Speed Trains

▷ **How did bullet trains get their nickname?**

The newest all-electric locomotives travel at high speeds and are called **bullet trains**. Their 125-200 miles per hour speed and the pointed shape of the locomotive's nose gave rise to that name. Japan's *Shinkansen* was the first bullet train.

The bullet train in the U.S. is called the *Acela Express*, which was discussed in Chapter 21. It entered service in 2000. Travel time from Boston to Washington, D.C., is about six and one-half hours, which includes stops at intermediate stations.

Maglev Trains

▷ **What drives a maglev?**

One new kind of train doesn't roll on wheels. It doesn't even touch the ground. The train is called a **maglev train**. Maglev stands for "magnetic levitation." The forces of magnetic attraction and repulsion allow the train to float, or levitate, less than one inch above its guideway, or path. The same forces interact to move the trains.

Maglevs are very quiet and produce almost no vibration. See Figure 22-7. However, they are expensive to construct and only a few experimental guideways have been built in the United States. A German company recently won a contract to build a twenty-two-mile long maglev line in China. It will serve Shanghai's international airport. World speed records of over 340 miles per hour have been reached by maglev trains in Japan. Do you think that maglev trains have a place in our transportation system?

> **bullet train**
> high-speed, all-electric locomotive with a bullet-shaped nose

> **maglev train**
> train that is levitated and propelled by the use of electromagnets

TECH CONNECT
SCIENCE

Maglevs on the Move

As you probably know, unlike magnetic poles attract and like poles repel. Special magnets in a maglev train and in its track create magnetism of the same polarity. These like forces repel each other, levitating the train.

ACTIVITY Learn about the principles behind maglev propulsion. Create a poster or display explaining how it works.

Figure 22-7 Japan's high speed maglev train can reach speeds over 340 mph.

Pipelines

▷ **How is cargo transported through pipelines?**

Did you realize that when you turn on a faucet to get a drink of water you are using a transportation system? Some cargo, like water, oil, and natural gas, travels long distances through pipelines. Pipes may be as small as two inches in diameter or as large as fifteen feet in diameter. Most pipelines are buried in the ground. When pipelines are laid out in straight lines, transportation time is reduced. See Figure 22-8.

Pipelines require service facilities such as pumping and control stations. The stations are located along the pipeline and keep the cargo moving. When the cargo is made of particles, such as gravel, it is mixed with liquids to form a slurry. The pumps then force the material through the pipeline. When a pipe is clogged, a tool called a pig is pushed through to clean it.

Figure 22-8 This portion of the Trans-Alaska Pipeline has few bends.

22A SECTION REVIEW

Recall ▸▸

1. What is the purpose of a transmission?

2. How is an intercity bus used? How is this different from the use of an urban bus?

3. Why do some tractor-trailer trucks have air deflectors on top of the tractor?

4. What causes a maglev to "float" above its guideway?

Think ▸▸

5. From a passenger's standpoint, name three advantages to traveling to work by bus.

6. Some people think that high-speed trains instead of planes should be used for distances under two hundred miles, especially in highly populated areas. Why do you think this would be desirable?

Apply ▸▸

7. Systems Make a model or poster that identifies several subsystems or processes in a land transportation system.

8. Research Use the Internet to learn about bullet trains used in Japan, England, France, Germany, and Spain. Report your findings to the class.

Water Transportation

▶ What are the different types of water transportation?

Water has provided transportation routes for centuries. Oceans, rivers, lakes, and other navigable waterways have made natural routes between cities, states, countries, and continents. A **navigable waterway** is a lake or river that is deep and wide enough to allow ships and boats to pass. The five Great Lakes are navigable waterways, as are many rivers, such as the Mississippi and Ohio.

Boats and Ships

▶ What's the difference between a boat and a ship?

A small, open vessel is called a boat. Large, deep-water vessels are called ships, and there are about 24,000 of them in the world.

Water transportation vessels are of three general types. Passenger vessels carry people. Large ocean liners used for vacation cruises are one type of passenger vessel. See Figure 22-9. Cargo ships transport oil, grain, iron ore, automobiles, and many other products. Specialty craft include everything else, such as river barges for transporting coal and other goods, tugboats for pulling large ships into dock, and icebreakers.

Objectives

▶▶ **Discuss** how oceans and inland waterways are used for transportation.

▶▶ **Identify** the different types of ships.

▶▶ **Explain** the concept of intermodal transportation.

▼ Terms to Learn

- containership
- displacement
- intermodal transportation
- navigable waterway
- supertanker

Standards

- Core Concepts of Technology
- Relationships & Connections
- Cultural, Social, Economic & Political Effects
- Transportation Technologies

Used with permission of Royal Caribbean Cruises Ltd.

navigable waterway
a body of water that is deep and wide enough to allow boats and ships to pass

Figure 22-9 Large luxury ships like the *Navigator of the Sea* are often used as vacation cruise liners.

Figure 22-10

Figure 22-10 Modern supertankers transport huge amounts of oil across the seas.

Ships deliver most of the overseas cargo leaving or arriving in the United States. To transport by ship, we need docks with special loading and unloading equipment. We need properly trained people to operate the ship. We also need good communications for weather data and other information. They are all necessary for the entire system to operate efficiently.

For centuries, people used sailing ships to haul cargo and passengers. The ships of 1800 had displacements of about 1,200 tons. **Displacement** is a measure of how much water a ship and its cargo push aside as the ship floats. It's an indication of the ship's size. Today, an average cargo ship that is six hundred feet long might have a displacement of 21,000 tons.

The wind limited how much cargo those older ships could carry. A heavy ship is much more difficult to push than a lighter one and moves more slowly. Today's ships are pushed by powerful engines. Ocean liners and cargo ships use gas turbines. Specialty craft use steam turbine, gas turbine, or diesel engines. It is not unusual for a modern ship to displace 100,000 tons. That means it's about eighty times bigger than a ship of 1800. Very large ships called **supertankers** transport oil across oceans in storage tanks. Their displacements are as high as 500,000 tons. See Figure 22-10. Does this tell you why they are called *super*tankers?

It usually costs less to transport goods by water than by rail, highway, or air. Whenever possible, people try to save as much money as they can by transporting products the cheapest way.

Intermodal Transportation

▷ Can modes of transportation be linked to carry people or products?

Can you think of a time when you used more than one mode (form) of transportation in a single journey? For example, you may have used a bus and then a subway train to reach the airport. Perhaps you then took a plane to a coastal city where you boarded a cruise ship. Different systems were organized to work together to help you. There was a bus stop near the subway station, a subway stop at the airport, and a taxi to take you to the ship. When two or more forms of transportation are used together to move people or cargo more efficiently, it is called **intermodal transportation**.

displacement
the weight of the water that is moved out of the way by a floating ship and its cargo

supertanker
a very large ship that is fitted with tanks for carrying oil across oceans

intermodal transportation
using two or more forms of transportation to move people or cargo more efficiently

Cargo is moved most efficiently when it is packed into large containers. When a product travels overseas, the containers are often loaded on ships called **containerships**. See Figure 22-11. Loading and unloading is easy because the containers are usually the same size and shape.

Containers can then be easily loaded on a train. The products do not have to be unpacked and then re-packed. Some specially designed containers in the form of truck semi-trailers ride on trains to a terminal where they are attached to tractors. They then continue their journey on the highway. Because intermodal transportation is so efficient, it saves both time and money.

containership
large ocean-going vessel designed to carry containers filled with cargo

Figure 22-11 Special cranes are used at this dock to load hundreds of containers onto a containership.

SECTION REVIEW 22B

Recall »

1. Why are the ships that transport oil called *super*tankers?
2. Name the three general types of water transportation vessels.
3. What is the difference between a boat and a ship?
4. Why is intermodal transportation so efficient?

Think »

5. What effects do you think the introduction of engine-powered ships had on the society of the time?

6. What trade-offs do you think take place in developing intermodal transportation systems? Give examples.

Apply »

7. **Research** Look up information and write a short report on how a particular transportation system is a part of the technological, social, and environmental systems that make up our world.
8. **Map Reading** Using a map of the U.S., find out where most navigable waterways are located.

Objectives

▸▸ **Discuss** ways in which air and space transportation are used.

▸▸ **Identify** different types of aircraft and spacecraft.

▸▸ **Explain** how a plane is lifted into the air.

Terms to Learn

- **commercial airplane**
- **helicopter**
- **jumbo jet**
- **lift**
- **lighter-than-air craft**
- **payload**

Standards

- Relationships & Connections
- Cultural, Social, Economic & Political Effects
- Role of Society
- Use & Maintenance

◣ **commercial airplane**
an airplane that carries passengers or freight in order to make money

Air and Space Transportation

▷ What are the main types of air transportation vehicles?

Transportation by air takes place in airplanes, helicopters, and lighter-than-air craft. Hang gliders and sailplanes are primarily used for recreation. Military planes are used for defense. Space vehicles take us beyond the earth. Can you think of other air and space vehicles?

Airplanes are the most important part of our air transportation system. However, many other components are necessary for safe air travel. We need airports, training programs, and radar, for example. Many airplanes are in the air at the same time. This is why air travel is our most complex transportation system. Government regulations specify how air travel is carried out.

Airplanes

▷ What is a jumbo jet?

Many important airplanes were built and flown after the Wright brothers' first flight in 1903. Few were as important as the 1935 DC-3, which had two gasoline engines. The DC-3 was the first **commercial airplane** to make a profit by just carrying passengers. See Figure 22-12. The Boeing

Figure 22-12 The 1935 DC-3 was designed and built in just ten months by the small Douglas Aircraft Company. The plane is affectionately named the Gooney Bird and many are still flying today.

707 was the first American jet, which came out in 1958. It carried 179 passengers, a huge number at the time. In following years, the number of airline passengers increased so much that manufacturers decided to build very large **jumbo jets**. Boeing's 747 was the first; it could carry about 500 passengers.

Because jumbo jets have such powerful engines, they can lift more weight than other planes. This allows them to carry a lot of fuel. As a result, they can stay up in the air for a long time. Jumbo jets fly non-stop from Cincinnati to London and from Detroit to Tokyo. See Figure 22-13.

Smaller jet aircraft, like the two-engine DC-9, are used on shorter flights between smaller cities. The smaller airplanes use less fuel and don't need a long runway to take off. Without such airplanes, smaller U.S. cities might not have good air service.

Some airplanes carry only cargo. However, even the biggest can carry only a fraction of what a ship or train can haul. A Boeing 747 can carry 100 tons of cargo, about the most of any airplane in the world. This is much less than any cargo ship and makes air transportation very expensive. Usually only lightweight items, such as mail and electronics, are shipped by air.

jumbo jet
a very large airplane that carries several hundred passengers at one time

Lift

▷ What lifts a plane into the air?

Did you ever wonder how something as heavy as an airplane could get off the ground? The secret is in the shape of the plane's wing.

As the plane races down the runway, it gathers speed and air rushes over the wing. The shape of the wing causes

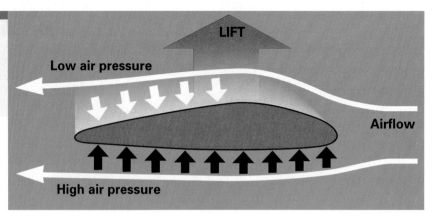

LIFT

Low air pressure

Airflow

High air pressure

lift
the upward movement of an airplane due to reduced pressure above the wing and increased pressure below it

helicopter
an aircraft that is lifted straight up by one or two rotors

the air to travel faster over its upper surface. See Figure 22-14. This reduces air pressure above the wing. It also helps increase the pressure on the wing's lower surface, pushing it upward and creating **lift**. Like magic, the plane rises.

Helicopters

▶ Why do some helicopters have a small tail rotor?

A **helicopter** is an aircraft with one or two rotors that allow it to lift straight up. Helicopters can be as small as a one-person machine with a gasoline engine. They can also be as large as cargo-carrying helicopters that easily lift ten tons. These larger helicopters are powered by one or two gas turbine engines. Some helicopters have a small tail rotor to keep them from spinning out of control. The twirling blades of the rotor are what create lift.

Passenger-carrying helicopters connect downtown New York City with the three major airports in the area. Such services are also available in Chicago, San Francisco, and other large cities. Some helicopters specialize in industrial operations, such as pipeline inspection and delivering parts to construction sites. Others are used to check on automobile traffic and transport people to hospitals. See Figure 22-15.

Figure 22-15 Air ambulance helicopters have saved many lives by delivering medical help quickly to rural areas.

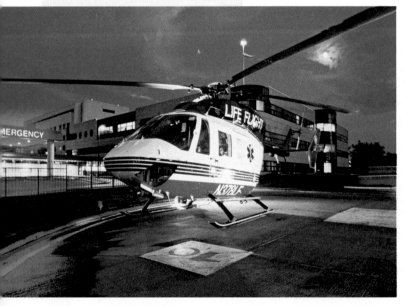

Lighter-than-Air Craft

▷ What causes a hot-air balloon to rise into the air?

Lighter-than-air craft include dirigibles, zeppelins, blimps, and airships. Helium lifts them into the air and gasoline engines turn propellers to move them forward. The engines are located in gondolas, or cars, suspended from the craft. The passenger compartment is located in a separate gondola.

A hot-air balloon is also a lighter-than-air craft. It uses large torches to heat the air inside a huge nylon bag. Hot air weighs less than cooler air, so the balloon rises. Their direction is controlled by ascending or descending into wind currents that push them along. See Figure 22-16.

lighter-than-air craft
an aircraft lifted by being filled with a gas that is lighter than air

Space Vehicles

▷ What does NASA do?

Space transportation is probably the most exciting way to move people and cargo. The National Aeronautics and Space Administration (NASA) is responsible for regulating and directing the entire U.S. space program. This includes the space shuttles and vehicles for exploration.

Figure 22-16 Altitude and air currents are used to navigate in a hot-air balloon.

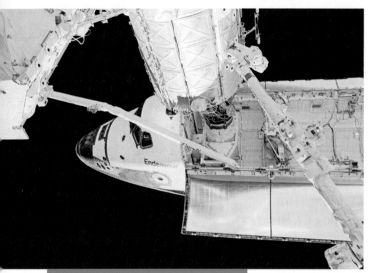

Figure 22-17 Space Shuttle *Endeavor* docks at the Destiny laboratory on the International Space Station.

payload
another term for cargo

Space Shuttles

▷ **What is the purpose of a shuttle?**

John Young and Robert Crippen flew *Columbia* into space in 1981. It was the first space shuttle to go into orbit. Each shuttle flight carries cargo, called a **payload**, in the large cargo bay. The payload can weigh up to 65,000 pounds.

On a typical mission, a space shuttle with four astronauts orbits 115 miles above the earth. Shuttle speed is about 17,000 miles per hour. Once in orbit, an astronaut opens the doors to the huge cargo bay, which usually contains a communications satellite. A fifty-foot mechanical arm, controlled by the astronaut, removes the satellite and releases it a safe distance from the space shuttle. The arm can also grab satellites already in orbit.

The space shuttle can also carry a complete scientific laboratory in the cargo bay. However, many laboratory experiments are now done on the International Space Station. See Figure 22-17.

To return, the astronauts fire small rockets to slow down the space shuttle. It re-enters the earth's atmosphere and glides toward a landing strip.

The destruction of the shuttle *Columbia* in 2003 and the loss of the seven astronauts have made NASA very concerned about safety. New shuttle designs are being considered to replace the aging fleet.

LOOK TO THE FUTURE

Sails in Space

Explorers once crossed the vast oceans in sailing ships. Some future space exploration may also be done with sails—solar sails. Made from thin film into a sheet the size of a football field, a solar sail has a reflective surface, like a mirror. The mirror reflects particles of light (photons) from the sun and this action propels the spacecraft. Top speeds will be about 3,000 miles per *minute*.

Exploratory Vehicles

▷ **What vehicles are used to explore outer space?**

Distances among planets and other points in space are huge. Humans in space must bring along their own life support systems and fuel. However, since Americans first landed on the moon, many exploratory missions have taken place.

Currently, missions are being done without human crews or passengers. The payload—usually cameras and other equipment—sits in the hollow nose of a booster rocket. Once free of Earth's gravity, the nose opens, releasing the payload, which proceeds using its own smaller engines.

In 2001, the *NEAR-Shoemaker* spacecraft traveled a total of 2.3 billion miles to study asteroids that might be dangerous to Earth. In 2004, NASA's planetary probe *Sprint* sent back photographs and other test data from the surface of Mars. In 2005, the *Mars Reconnaissance Orbiter* is scheduled to continue exploration of the red planet. A visit to NASA's Web site will tell you about these and other missions.

TECH CONNECT
SOCIAL STUDIES

The Vastness of Space

In 2003, the European Space Agency (ESA) launched its own mission to explore Mars. China plans to send astronauts to the moon one day.

ACTIVITY Use the Internet to learn about space programs in other countries. Hold a panel discussion on how cultural differences might affect these programs.

SECTION REVIEW 22C

Recall ▶▶

1. Which type of jet carries the most passengers?
2. What jobs do helicopters do?
3. What do the letters NASA stand for?
4. What is the usual type of payload carried by a space shuttle?

Think ▶▶

5. Air travel increased after jets were introduced. What demands, interests, and values in our society do you think contributed to this increase?

6. What value do you think the space program has for the development of technology for use on Earth?

Apply ▶▶

7. **Production** With some of your classmates, produce a videotape documentary showing a typical space shuttle mission.
8. **Research** Do research on new space shuttle designs. Write a paragraph assessing the designs.

Summary Activity

For each numbered blank, pick the answer from the list on the right that makes the most sense in the entire passage. Write your answer on a separate sheet of paper. No answer will be used more than once.

About half of all the world's automobiles are used in the United States. The car's __1__ contains gears for transferring power. The radio in the car is an example of one of the car's __2__. __3__-wheel drive cars can go almost anywhere.

Buses are an important part of __4__ transportation, which is passenger service available to the general public. School buses carry about 24 million students to school each day. Trucks deliver many products. Some larger trucks use __5__ to reduce air resistance.

Most railroads haul __6__. The American bullet train is named the *Acela Express*. A train that floats over a guideway is called a(n) __7__.

There are about 24,000 large deep-water vessels in the world. __8__ is the measure of how much water a vessel moves aside as it floats.

__9__ transportation combines several different kinds. Cargo is often placed in __10__ for efficiency.

The first American-built jet was the Boeing 707 that came out in 1958. Because of increases in passenger air travel, the Boeing 747 became the first of several __11__. Jet airplanes servicing smaller airports don't need a long runway and use less __12__. Helicopters can carry cargo, and some can easily lift ten tons. The force that allows a plane to rise in the air is called __13__.

During a typical mission, a(n) __14__ orbits about 115 miles above the earth. Exploratory vehicles are usually launched using a(n) __15__ rocket.

Answer List

- booster
- containers
- deflectors
- displacement
- four
- freight
- fuel
- intermodal
- jumbo jets
- lift
- maglev
- mass
- space shuttle
- subsystems
- transmission

Comprehension Check

1. Name at least three examples of land transportation.

2. What is a slurry?

3. What is a tractor-trailer?

4. For what types of transportation are oceans and inland waterways used?

5. How is a plane lifted into the air?

Critical Thinking

1. **Infer.** What subsystems do you think are required by locomotives?

2. **Propose.** In addition to exploration, for what purposes do you think space travel should be used?

3. **Judge.** Snowmobiles are a recreational land vehicle.

4. **Compare.** Containerships have made cargo transport more efficient. Do you think the same designs could be adapted for car travel? Why or why not?

5. **Compose.** Write a short paragraph describing your last use of intermodal transportation.

Visual Engineering

Deflecting Head Wind. It is estimated that at least one-third of a conventional eighteen-wheeler's engine power is used to overcome wind resistance. With a wind deflector much of that resistance is reduced. Do a little research to find various designs for deflectors and for the front of the tractor. Make sketches of several of the designs and draw vectors (lines) to show how the wind would be deflected. Do the same for a conventional tractor and trailer with a flat front. Can you see why there is a savings in fuel costs?

TECHNOLOGY CHALLENGE ACTIVITY

Building a Rubber-Band-Powered Vehicle

Equipment and Materials

- one wooden block, 1 inch thick, 2 inches wide, and 6 inches long
- two wooden ready-made wheels, about 2¼ inches in diameter
- two wooden ready-made wheels about 1 inch in diameter
- one dowel rod, ⅜ inch in diameter and about 3 or 4 inches long
- three heavy rubber bands, ¼ inch wide and about 2 inches in diameter
- staples
- about 18 inches of fishing line
- punch
- wire cutters
- hand drill with ⅜-inch drill bit

Background

Modern land transportation vehicles have many basic parts that are the same as those used hundreds of years ago. For example, most have wheels, a body, and a way to provide power. You can learn more about land vehicles by making your own. Modeling, testing, evaluating, and modifying will be used to transform your ideas into practical solutions.

Goal

For this activity, you will make a simple land transportation vehicle powered by a rubber band. You will then experiment with the design to see if you can make the vehicle go faster or farther.

Criteria and Constraints

▶▶ Your first vehicle must follow the basic design.

▶▶ You must keep a record of all speed and distance measurements.

▶▶ You must submit sketches of your design improvements.

FIGURE A

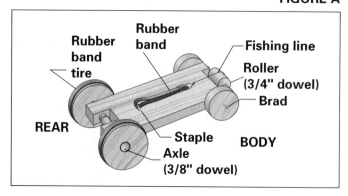

REAR

Rubber band tire
Rubber band
Fishing line
Roller (3/4" dowel)
Brad
Staple
Axle (3/8" dowel)
BODY

⚠ SAFETY

Reminder
In this activity, you will be using tools and materials. Be sure to always follow appropriate safety procedures and rules. Remember, safety is an attitude that you must develop and maintain at all times.

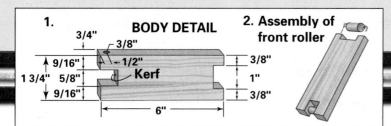

1. BODY DETAIL

3/4"
3/8"
9/16"
1/2"
1 3/4" 5/8" Kerf
9/16"

3/8"
1"
3/8"

6"

2. Assembly of front roller

FIGURE B

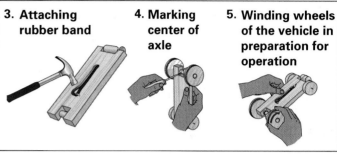

3. Attaching rubber band

4. Marking center of axle

5. Winding wheels of the vehicle in preparation for operation

Design Procedure

❶ Look at the drawing in Figure A to get an idea of what your vehicle will look like.

❷ Drill a ⅜-inch hole at one end of the body. See Figure B. Use the saw to cut notches at the front and rear of the body.

❸ Put the dowel rod through the hole. Cut it to length and attach the 2¼-inch wheels.

❹ Place a rubber band around the outside of each wheel. These will be your car's drive wheels, the wheels that provide the driving power.

❺ Mark the center of the dowel rod with a punch and hammer in a small nail. Leave about ¼ inch sticking out. Later on, you will tie one end of the fishing line to the nail.

❻ Attach the front wheels with small nails. Make sure all four wheels can spin freely.

❼ Make the front roller from a ⅛-inch-long section of the ⅜-inch-diameter dowel rod. Hammer small nails into each end and cut off the heads of the nails. Attach the roller to the body with staples, as shown in Figure B. Don't drive the staples too deep. The roller must turn easily.

❽ Attach a rubber band to the bottom of the body by hammering in a staple. See Figure B. Tie one end of the fishing line to the rubber band. Pass the line over the front roller and under the body of the vehicle. Tie the other end of the fishing line to the small nail on the dowel rod connecting the large wheels.

❾ To provide power for your car, wind the larger wheels backward. See Figure B. Continue until all the fishing line is wound onto the axle and the rubber band is stretched over the front roller.

❿ Place your vehicle on a smooth floor and let go. It will quickly accelerate and then coast to a stop. Use the 25-foot tape measure to measure how far your vehicle travels. Use a stopwatch to measure speed. Repeat a few times to get a good average. Record the results.

⓫ Sketch several improvements for the vehicle and try them out.

Evaluating the Results

1. Did your vehicle travel as fast as you expected it would?

2. How did you change your vehicle to make it go faster or farther?

3. What would happen if you used bigger drive wheels? What would happen if you used smaller drive wheels?

Appendix
Approximate Customary Metric Conversions

	When you know:	You can find:	If you multiply by:
Length	inches	millimeters	25.4
	feet	millimeters	304.8
	yards	meters	0.9
	miles	kilometers	1.6
	millimeters	inches	0.04
	meters	yards	1.1
	kilometers	miles	0.6
Area	square inches	square centimeters (cm2)	6.5
	square feet	square meters	0.09
	square yards	square meters	0.8
	square miles	square kilometers	2.6
	acres	square hectometers (hectares)	0.4
	square centimeters	square inches	0.16
	square meters	square yards	1.2
	square kilometers	square miles	0.4
	hectares (ha)	acres	2.5
Mass	ounces	grams	28.4
	pounds	kilograms	0.45
	tons	metric tons (t)	0.9
	grams	ounces	0.04
	kilograms	pounds	2.2
	metric tons	tons	1.1
Liquid Volume	ounces	milliliters	29.6
	pints	liters	0.47
	quarts	liters	0.95
	gallons	liters	3.8
	milliliters	ounces	0.03
	liters	pints	2.1
	liters	quarts	1.06
	liters	gallons	0.26
Temperature	degrees Fahrenheit	degrees Celsius	0.6 (after subtracting 32)
	degrees Celsius	degrees Fahrenheit	1.8 (then add 32)
Power	horsepower	kilowatts (kW)	0.75
	kilowatts	horsepower	1.34
Pressure	pounds per square inch (psi)	kilopascals (kPa)	6.9
	kPa	psi	0.15
Velocity (Speed)	miles per hour (mph)	kilometers per hour (km/h)	1.6
	km/h	mph	0.6

Glossary

This glossary includes the definitions for all of the "Terms to Learn" that appear in the textbook. The numbers and letters in parentheses indicate the chapter and section where the term is discussed. For example, (16B) means Chapter 16, Section B.

A

abrading Changing the shape of a material by rubbing off small pieces, such as with sandpaper. (16B)

abutment A structure that supports the end of a bridge or dam. (20B)

AC Alternating current; electrical flow that constantly changes direction. (5A)

acid rain A weak sulfuric acid created when sulfur dioxide in the air mixes with rain and oxygen. (4C)

added value The increase in how much a material is worth after it has been processed into a finished product or a part for a finished product. (16B)

advertising Making a public announcement that a product is available for sale. (17C)

agroforestry Turning forests into controlled environments dedicated to the replacement of trees. (15C)

AI Artificial intelligence; computer program that can solve problems and make decisions ordinarily handled by humans. (10B)

alloy A material made by mixing a metal with other metals or materials. (3C)

alternating current See *AC*.

amperage (AM-purr-age) The electrical flow. (5B)

analog signal An electrical signal that changes continuously. (5D)

animation Creating a series of slightly varying drawings or models so that they appear to move and change when the sequence is shown. (13A)

antibiotic In medicine, a substance that kills bacteria. (14C)

aperture (APP-ih-chur) The opening that controls the amount of light that enters a camera. (12A)

aquaculture Growing fish, shellfish, or plants in artificial water ecosystems. (15C)

artificial ecosystem A human-made, controlled environment built to support humans, plants, or animals. (15C)

assembly drawing A drawing that shows how to put parts together to make an item. (8B)

assembly line Series of work stations at which individual steps in the assembly of a product are carried out as the product is moved along. (16A)

audio The recording and reproduction of sound. (13A)

B

balance A state of steadiness or stability; one of the principles of design. (7A)

binary code An electronic code that a computer can understand and which is based on the binary number system. (10A)

biometrics The science of measuring an individual's unique features. (9C)

bionics Creating replacements for human body parts. (14C)

bioremediation Using bacteria and other organisms to clean up contaminated land and water. (15C)

biosynthesis Making chemicals using biological processes. (15C)

brainstorming A group problem-solving technique in which members call out possible solutions. (7B)

browser Software program that provides access to the World Wide Web. (13B)

building code A rule used to control how structures are built. (18B)

building site Location for construction of a building. (19A)

bullet train High-speed, all-electric locomotive with a bullet-shaped nose. (22A)

C

cable-stayed bridge A bridge supported by inclined cables connected to towers. (20B)

CAD Computer-aided drafting or computer-aided design; the use of a computer system in place of mechanical drawing tools to create technical drawings and/or design an object. (8A)

calorie The measure of energy in food. (4A)

CAM Computer-aided manufacturing; a system that uses computers to operate the machinery in a factory. (17B)

cantilever bridge (CAN-tih-lee-vur) A bridge made of beams supported by the ground at only the ends; the beams meet in the middle of the bridge. (20B)

capital Accumulated wealth, which may be money, credit, or property. (2A)

CCD Charge-coupled device; a special microchip inside a digital camera that converts light into an electrical signal. (12A)

central processing unit See *CPU*.

ceramic Material made from nonmetallic minerals that are fused together with heat. (3C)

CIM Computer-integrated manufacturing; the use of one computer system to control the design, manufacturing, and business functions of a company. (17B)

circuit The pathway that electricity flows along; often includes parts such as a wire and a device to which the electricity is being delivered. (5A)

cloning Producing an identical copy of an individual plant or animal. (15B)

closed-loop system A system that has a way of controlling or measuring its product. (2B)

CNC Computerized numerical control; machine tool operation controlled by commands from a computer. (17B)

combining Joining materials together. (3A)

commercial airplane An airplane that carries passengers or freight in order to make money. (22C)

commission A payment made to a salesperson or agent for business they have done. (16C)

communication technology The transfer of messages among people and/or machines through the use of technology. (9A)

composite A material made by combining two or more other materials. (3C and 18B)

computer-aided drafting See *CAD*.

computerized axial tomography See *CT scan*.

computerized numerical control See *CNC*.

computer virus A set of destructive instructions that "infects" a computer system and can cause damage. (10A)

concrete A mixture of cement, sand, stones, and water that hardens into a construction material. (18B)

conditioning Changing the inner structure of a material. (3A)

conductor Material that allows electricity to flow easily through it. (5C)

conservative design Design strategy used to be sure structures can bear more weight than required under normal conditions. (18B)

constraint A restriction on a product. (2C)

containership Large ocean-going vessel designed to carry specially designed containers filled with cargo. (22B)

copy In printing, a graphic message ready for reproduction. (11B)

CPU Central processing unit; the part of a computer that processes information; the "brain" of the computer. (10A)

craft A skilled occupation, usually done with the hands, such as carpentry or sewing. (16A)

crane In construction, a large machine used to lift heavy loads by means of a hook attached to cables. (20A)

crankshaft The part of an engine that changes the reciprocating motion of the pistons to the rotary motion that turns the wheels. (21B)

criteria Standards that a product must meet in order to be accepted. (2C)

CT scan Computerized axial tomography; an X-ray image enhanced by a computer and shown on its screen. (14B)

curtain wall Exterior wall on a skyscraper that does not help support the building. (18B)

D

DC Direct current; the one-directional flow of electrons. (5A)

dehydrate To remove moisture. (15A)

deoxyribonucleic acid See *DNA*.

desktop publishing The use of desktop computers and small printers for publishing. (11B)

developer In photography, a chemical used during processing to reveal the latent image. (12C)

digital compression Reducing the size of a digital file by removing bits of data that can be recreated later. (13A)

digital signal An electrical signal having distinct values. (5D)

dimension A size or location of object parts indicated on a technical drawing. (8A)

direct current See *DC*.

director The person in charge of instructing the performers and guiding the camera work during a video production. (13B)

displacement A measure of the weight of the water that is moved out of the way by a floating ship and its cargo. (22B)

distributed computing Network of computers that researches and analyzes data during computer downtime. (10B)

division of labor System of organizing manufacturing by giving separate tasks to separate workers or groups of workers. (16A)

DNA Deoxyribonucleic acid; the molecules in a gene that carry genetic information. (15B)

drafting The techniques used to make drawings that describe the size, shape, and structure of objects. (8A)

driving wheel A wheel that transmits motion from one part of a vehicle to another. (22A)

drywall In construction, the inside covering of walls and ceilings that is made from plaster and sturdy paper. (19B)

durable good Manufactured product that lasts three or more years. (16A)

dynamic digital printing The use of printing machines that print directly from a computer file instead of a printing plate. (11A)

E

editing Cutting and arranging material in order to decide its final sequence and content in an audio or video production. (13B)

efficiency The ability to achieve a desired result with little effort and waste. (4B)

electricity The movement of electrons from one atom to another. (5A)

electromagnetic carrier wave See *electromagnetic wave*. (5D and 9A)

electromagnetic wave Wave of electromagnetic energy used to carry an electronic signal through the atmosphere; also referred to as an *electromagnetic carrier wave*. (5D and 9A)

electronic device A device that changes one form of energy, such as sound, into an electrical signal that can be transmitted by a sender through a channel to a receiver. (5D)

e-manufacturing Using electronic information in the manufacture of a product. (17B)

embankment The main part of a dam that holds back the water. (20C)

endoscope In medicine, a small, flexible instrument, having a camera attached, that is inserted into a patient through an incision or by some other means. (14B)

energy The capacity or ability to do work. (4A)

energy conservation The management and efficient use of energy sources. (4C)

engineering Designing products or structures so that they are sound and deciding how they should be made and what materials should be used. (1B)

ergonomics The design of equipment and environments to promote human safety, health, and well-being; also called *human factors engineering*. (14A)

excavation In construction, a large hole. (20A)

expert system Form of artificial intelligence in which information from experts is collected and stored in a computer's memory. (10B)

external combustion engine An engine that burns fuel to create energy; its power source is outside the engine. (21A)

F

feedback The part of a closed-loop system that provides control or measurement of the product. (2B)

fertilizer A chemical compound used to restore nutrients to the soil. (15A)

fiber optic cable In communication, thin, flexible, glass strands that transmit light over great distances in order to carry information. (5D)

finishing In manufacturing, the last step in making a product; used to improve appearance. (3A)

flexography A relief printing process that uses a raised, rubber printing plate. (11A)

focus In photography, a term used to describe the sharpness of an image. (12A)

footing The bottom part of a foundation, made of hardened concrete and located under the foundation wall. (19B)

forming In manufacturing, changing the shape of a material. (3A)

fossil fuel Fuel produced from the fossils of long-dead plants and animals. (4A)

foundation The part of a house that rests on the ground and supports the upper structure. (19B)

four-color process printing Combining magenta, cyan, yellow, and black to produce all the other colors used on printed materials. (11B)

four-wheel drive Transfer of power from the vehicle's engine to both front and rear wheels. (22A)

frequency The number of cycles or changes in direction of alternating current; measured in hertz, or cycles per second. (5A)

front-wheel drive Transfer of power from a vehicle's engine to the front wheels. (22A)

fuel cell A device that converts the chemical energy of a fuel, such as hydrogen, into electrical energy. (21C)

G

gable roof A roof with two sloping sides that meet at the ridge and form a triangular shape at either end. (19B)

gene In biotechnology, the factor in cells that carries heredity. (15B)

genetic engineering Altering or combining genetic material in order to treat a disease or modify body characteristics. (14C)

genetic testing Evaluation of a person's genes to discover if the person is at risk for a particular disease. (14B)

geodesic dome In construction, a spherical structure the surface of which is formed by triangles or polygons. (19A)

geothermal energy Heat energy produced under the earth's crust. (4A)

girder bridge A bridge made of girders that rest on the ground on either side of the span. (20B)

graphic communication Sending and receiving messages using visual images and printed words or symbols. (9A)

gravure printing (grah-VYOOR) A printing process in which letters and designs are etched or scratched into a metal plate; ink fills these grooves and is then transferred to paper. (11A)

greenhouse effect Increase in the temperature of the earth's atmosphere caused by a rise in carbon dioxide. (4C)

H

halftone In printing, a photograph reproduced using a series of dots. (11B)

hand tool Tool powered by human muscle. (3B)

helicopter An aircraft that is lifted straight up by one or two rotors. (22C)

horsepower A measurement of power based on lifting 550 pounds one foot in one second. (4B)

HTML Hypertext markup language; the code in which information on the World Wide Web is written. (13B)

human factors engineering The design of equipment and environments to promote human safety, health, and well-being; also called *ergonomics*. (7A)

humanities School subjects having to do with cultural knowledge, such as language and history. (6B)

hybrid *(1)* In biotechnology, an organism bred from two different species, breeds, or varieties. (15A) *(2)* A manufactured product made from a mixture of different elements. (21C)

hydraulic power Fluid power produced by putting a liquid under pressure. (4B)

hydroelectric power Electricity generated by turbines that are propelled by flowing water. (4A)

hydroponics Growing plants in nutrient solutions without soil. (15C)

hypertext markup language See *HTML*.

hypothesis An explanation for something that is used as the basis for further investigation. (6A)

I

ignition system In transportation, an engine system that starts the fuel burning. (21B)

immunization In medicine, a process for making the body resistant to disease, usually by vaccination. (14A)

implant In medicine, an electronic device inserted into the body. (14C)

Industrial Revolution Social and economic changes in Great Britain, Europe, and the United States that began around 1750 and resulted from making products in factories. (16A)

Information Age Period beginning around 1900 in which human activities focused on the creation, processing, and distribution of information. (6B)

ink-jet printing Printing process that uses spray guns to spray ink on the printing surface. (11A)

innovation (in-noh-VAY-shun) The process of modifying an existing product or system to improve it. (7A)

inorganic material Material that comes from minerals that were never alive. (3C)

input Whatever resources that are put into a system. (2B)

insulation Material used to keep heat or cold from entering or leaving a building. (19B)

insulator Material that resists the flow of electricity. (5C)

integrated circuit A tiny chip of semiconducting material, such as silicon, that contains many of the electrical circuits needed to operate an electronic device. (5D and 10A)

intermodal transportation Using two or more forms of transportation to move people or cargo more efficiently. (22B)

internal combustion engine In transportation, an engine in which the fuel is burned inside. (21B)

invention Turning ideas and imagination into new products and systems. (7A)

irradiation In biotechnology, treating products with radiation to destroy pathogens. (14A)

irrigation Bringing a supply of water to crops. (15A)

isometric drawing A pictorial drawing that shows three sides of an object that has been rotated thirty degrees and tilted forward thirty degrees. (8A)

J

jet engine A type of gas turbine engine that pushes a vehicle forward as hot air and exhaust are shot out the back of the engine. (21B)

jumbo jet A very large airplane that carries several hundred passengers at one time. (22C)

just-in-time delivery In manufacturing, a method of scheduling the arrival of materials at the time they are needed so that storage is not necessary. (16C)

L

laser A very powerful, narrow beam of light in which all the light rays have the same wavelength. (5D)

laser surgery In medicine, surgery done with a laser beam instead of a scalpel. (14C)

latent image In photography, the invisible image produced on exposed film. (12B)

layout Arrangement of elements on a page to be printed. (11B)

lens In photography, a piece of glass used to focus and magnify light. (12A)

letterpress printing A printing process that uses raised letters, symbols, or designs that are inked and then pressed against paper. (11A)

lift The upward movement of an airplane resulting from reduced pressure above the wing and increased pressure below it. (22C)

lighter-than-air craft An aircraft lifted by being filled with a gas that is lighter than air. (22C)

line art In printing, drawings and other art elements made of solid lines and shapes. (11B)

lithography A printing process that is based on the principle that oil and water don't mix. (11A)

load The output force of a power system. (4B)

lumber In construction, pieces of wood cut from logs into convenient shapes. (18B)

M

machine A tool with a power system that takes advantage of certain scientific laws that enable the tool to work better. (2A)

machine tool In manufacturing, a machine used for shaping or finishing metals and other materials. (1C)

machine-to-machine communication The transfer of messages from one machine, usually a computer, to another. (9B)

maglev train Train that is levitated and propelled by the use of electromagnets. (22A)

magnetic resonance imaging See *MRI*.

maintenance The process of inspecting and servicing a system on a regular basis to enable it to continue functioning properly, to extend its life, or to upgrade its capability. (21B)

marketing Telling potential customers about products and services in such a way as to make them eager to buy. (16C)

market research The process of getting people's opinions about a product so that a company knows what changes to make or whether to sell the product. (16C)

mass transportation Transportation that moves many people at one time and is available to the general public. (22A)

measuring tool Tool used to identify size, shape, weight, distance, density, or volume. (3B)

mechanical property Way in which a material reacts to a force. (3C)

mode A way of doing something. (9B)

model Replica of a proposed product that looks real but doesn't work. (8C)

monoculture farming In biotechnology, raising only one crop or one species of plant. (15A)

mortar In construction, a mixture similar to concrete used to fasten concrete blocks or bricks together. (19B)

MRI Magnetic resonance imaging; in medicine, the use of a magnetic field to create an image of body structures. (14B)

multiview drawing A technical drawing that shows an object from several different views. (8A)

N

nanotechnology The science of working with the atoms or molecules of materials to develop very small machines. (1C)

National Institute of Occupational Safety and Health See *NIOSH*.

navigable waterway A body of water that is deep and wide enough to allow boats and ships to pass. (22B)

negative In photography, an image produced on exposed film after processing; light areas appear dark and dark areas appear clear. (12C)

NIOSH National Institute of Occupational Safety and Health; the government agency that approves for use protective equipment, such as safety glasses. (17B)

non-durable good A manufactured product that lasts less than three years. (16A)

O

oblique drawing A pictorial drawing that shows one side of the object face-on, without any distortion, and the other sides at an angle. (8A)

Occupational Safety and Health Administration See *OSHA*.

Ohm's law Scientific law expressing the relationship among electrical amperage, voltage, and resistance; it takes one volt to force one ampere of current through a resistance of one ohm. (5B)

open-loop system A system with no way of controlling or measuring its product. (2B)

operating system The program a computer follows that tells it how to operate. (10A)

optimization In manufacturing, creating the most effective and functional product, system, or process. (2C)

organic material Material that comes from something that is or was once alive. (3C)

OSHA Occupational Safety and Health Administration; the government agency that sets safety rules and checks to make sure the rules are being followed. (17B)

outlet works Section of a dam with gates that allow water to flow through. (20C)

output What a system produces. (2B)

P

parallel circuit An electrical circuit having multiple pathways carrying current to individual devices; if one device stops working, the others are not affected. (5C)

pasteurization Heat treatment used to destroy pathogens in food. (14A)

pathogen Organism that causes disease. (14A)

payload Another term for cargo. (22C)

personal privacy The right of individuals to keep certain information away from public view. (9C)

perspective drawing A realistic pictorial drawing in which receding parallel lines come together at a vanishing point. (8A)

pharming In biotechnology, using genetically modified organisms to produce medicines. (15C)

photosite In photography, a tiny light-sensitive cell that converts light into an electrical charge. (12B)

pictorial drawing A drawing that shows a three-dimensional object realistically. (8A)

pier In construction, a concrete column used to add support to a foundation. (20A)

pile In construction, a large shaft driven deep into the soil to support a structure. (20A)

piston A plug that slides inside a cylinder of an engine. (21A)

pixel One of many tiny dots of light used to create a video image; acronym for *picture element*. (13A)

pneumatic power Fluid power produced by putting a gas under pressure. (4B)

portable electric tool Small, portable tool powered by electricity. (3B)

power A measure of work done over a certain period of time when energy is converted from one form to another or transferred from one place to another. (4B)

primary tool A basic tool that is hand held and muscle powered. (2A)

print A printed copy of a graphic message. (11B)

problem statement A statement that clearly defines a problem to be solved. (7B)

process The conversion of a system's input into a useful product, or output. (2B)

producer The person responsible for an entire audio or video production. (13B)

profit Money left over after all the bills for making a product are paid. (16C)

program A set of instructions that a computer follows. (10A)

propellant In transportation, a fuel mixture that causes an explosive thrust. (21B)

proportion The correct relationship between sizes and quantities in a design. (7A)

prototype In manufacturing, a working model of a proposed product. (8C)

Q

quality assurance The process of inspecting products to make sure they meet all standards that have been set. (16C)

R

R&D See *research and development*.

RAM Random access memory; memory in a computer's CPU that stores data temporarily. (10A)

random access memory See *RAM*.

rapid prototyping Using CAD and a special machine to make a three-dimensional model of an object. (8C)

read-only memory See *ROM*.

rear-wheel drive Transfer of power from a vehicle's engine to its rear wheels. (22A)

reciprocating motion (ree-SIP-roh-kay-ting) In an engine, up-and-down or back-and-forth motion that occurs in a straight line. (21B)

recycle To reuse all or part of materials, such as metal, glass, paper, and plastics. (4C)

research and development (R&D) Manufacturing department that searches for and develops new products and processes. (17A)

residential building (rezz ih DEN shul) A building in which people live. (19A)

resin (REZZ-in) A chemical compound used to make plastics. (16B)

resistance Anything that opposes or slows the flow of electrical current. (5B)

resource Something that supplies help or aid to a system; can be a source of information, capital, supply, or support. (2A)

retailer A merchant who buys products from a wholesaler and sells them to consumers. (17C)

robotics The technology involved in building and using industrial robots. (17B)

ROM Read-only memory; permanent memory in a computer's CPU that cannot be deleted or changed. (10A)

rotary motion Circular motion. (21B)

S

sanitation In biotechnology, removal of waste products or contaminants that could cause disease. (14A)

scale drawing Drawing of an object that is not true size but that is in the correct proportions. (8A)

schedule In manufacturing or construction, a plan that includes a list of the work to be done and when it should be finished. (17B)

schematic diagram Drawing that shows the circuits and components of electrical and electronic systems. (8B)

science Knowledge covering general truths or laws that explain how something happens. (1B)

scientific law A theory that has been proven true so often that it is accepted as fact. (6A)

scientific management In manufacturing, the system of developing standard ways of doing particular jobs. (16A)

scientific theory Scientific conclusion carefully developed through experimentation. (6A)

script The written version of an audio or video production that contains a list of characters and their dialogue. (13B)

search engine Software program that helps users find information on the Internet. (13B)

section drawing A drawing that shows the interior of an object. (8A)

semiconductor Material that can be used as either a conductor of electricity or an insulator. (5C)

sensory property Property of a material that we register with our senses, such as taste or texture. (3C)

separating In manufacturing, removing pieces of a material. (3A)

series circuit A single pathway that electrical current flows through to more than one device; if one device along the path stops working, they all stop. (5C)

serigraphy (sih-RIG-rah-fee) A printing process that uses a printing plate made of an open screen of silk, nylon, or metal mesh. (11A)

shadowing program School program in which students spend time in a work environment. (6B)

sheathing In construction, a layer of material between house framing and the outer covering. (19B)

shield In construction, a metal tube that fits inside a tunnel to support the walls. (20B)

shutter A covering that opens to let light into a camera. (12A)

skill The combination of knowledge and practice that enables a person to do something well. (2A)

skyscraper A very tall building. (18A)

solar cell A device that converts sunlight into electrical energy. (4A)

solar heating system A system in which energy from the sun is used to heat a building. (4A)

sound waves Vibrations traveling through air, water, or some other medium that can be perceived by the human ear. (9A)

span (1) Distance between bridge supports. (2) The entire length of a bridge. (20B)

spillway Portion of a dam that allows excess water to spill over the dam. (20C)

standard In manufacturing, a rule or guideline for making a product. (17B)

static electricity A buildup of electrons that are not in motion and are seeking to discharge. (5A)

stick construction Method of building construction in which lightweight pieces of wood or steel are used. (18A)

studs In construction, pieces of lumber, usually 2 x 4s, used for the framework of walls. (19B)

subfloor In building construction, the first layer of flooring, usually made of plywood. (19B)

subgrade In construction, the layer of soil beneath a roadway. (20B)

subsystem A system that is part of another, larger system. (2B)

superconductor Material that has no measurable resistance to electricity. (5C)

supertanker A very large ship that is fitted with tanks for carrying oil across oceans. (22B)

supplier Person or company that provides one or more of the materials or parts for a manufactured product. (16C)

surfacing Applying a material to a roadway that provides a smooth surface. (18A)

suspension bridge A bridge that hangs from large cables across a wide span. (20B)

system A group of parts that work together in an organized way to complete a task. (2B)

T

technical drawing To accurately represent the size, shape, and structure of objects; also called *mechanical drawing*. (8A)

technologically literate Term used to describe someone who is informed about technology and feels comfortable with it. (1A)

technology The practical use of human knowledge to extend human abilities and to satisfy human needs and wants. (1A)

telecommunication Communicating over a distance. (9A)

telemedicine Using communication technology to transmit medical information and advice over a distance. (14C)

thrust The high pressure that pushes a jet engine forward. (21B)

time and motion study In manufacturing, an investigation into the best ways of doing a job; things such as working conditions, wasted time, and unnecessary movement are considered. (16A)

tolerance Respect for others' differences. (9C)

tool An instrument or apparatus that increases a person's ability to do work. (2A)

tractor-trailer A two-part truck that includes a tractor (engine and cab) and a trailer that holds the load. (22A)

trade-off A compromise in which one thing is given up in order to gain something else. (2C)

transformer Device used to change electricity from one voltage to another. (5B)

transgenic organism In biotechnology, an organism into which genes from another organism have been transplanted. (15B)

transistor Small device made of a semiconducting material and used to control electric current. (5D)

transmission Vehicle system that contains gears and other parts that transfer power from the engine to the axles and wheels. (22A)

transmitter The part of a communication system that changes the sender's message into electrical impulses and sends it through the channel to the receiver. (5D)

troubleshooting A problem-solving method used to identify a malfunction in a system. (17B)

truss In construction, a prefabricated triangular framework that supports a roof. (19B)

truss bridge A bridge made from steel beams fastened together in triangular shapes. (20B)

turbine (TUR-bin) A disk or wheel that changes the energy of moving gases or liquids into rotary motion. (21A)

U

ultrasound In medicine, the use of sound waves to create an image of internal body structures on a computer screen. (14B)

uniform resource locator See *URL*.

unity A state in which all parts of a design work together. (7A)

URL Uniform resource locator; an address on the World Wide Web. (13B)

V

vaccine In medicine, a preparation containing dead or weakened pathogens used to stimulate a person's immune system. (14A)

video The recording and reproduction of moving images. (13A)

virtual factory A three-dimensional computer model of a factory. (17A)

visualization software CAD software used to create virtual models. (8C)

voltage The pressure needed to push electricity through a circuit. (5B)

W

wattage A measure of electric power; calculated by multiplying the amperes used times the voltage of the circuit. (5B)

wholesaler A merchant who buys large quantities from a manufacturer and sells smaller quantities to retailers. (17C)

wi-fi A wireless connection to computers on a network and to the Internet. (10B)

wind farm A large collection of windmills located in an area that has fairly constant winds. (4A)

working drawing One of a set of technical drawings needed to make a product or structure. (8B)

X

xerography (zee-RAH-graph-fee) A printing process that transfers negatively charged toner to positively charged paper. (11A)

Credits

AGStockUSA,
   Bill Barksdale, 363, 377
   Russ Munn, 361
American Airlines, 536, 537
Arnold & Brown, 119, 140, 466, 469
Artville, 2, 367
Associated Press, 192, 356, 521
   Ford Motor Company, 382
   Jim Cole, 495
   Suzanne Vlamis, 453
Roger B. Bean, 320, 440
Bentley Motors, 179
Broad Daylight, 247
Bureau of Land Management,
   Edward Bovy, 532
Bureau of Reclamation, 113, 486
Calty Design Research,
   Inc./Toyota, 204
Cincinnati Milacron, 384
Clear Cut Solutions, Inc. &
   Missler Software, 205
Ken Clubb, 76, 102, 103, 170,
   183, 184, 190, 192, 244, 245,
   246, 254, 279, 281, 288, 289,
   329, 355
Colorado Dept. Of
   Transportation,
   Gary Martin, 482
CORBIS 54, 57, 452, 487, 520
CORBIS,
   AFP, 145
   Jean Pierre Amet, 453
   Lester V. Bergman, 331
   James P. Blair, 14, 530
   Regis Bossu, 331
   Matthew B. Brady, 246
   CDC, 173
   Howard Davies, 406
   Randy Duchaine, 9, 283
   ER Productions, 342
   Eyewire, 27
   Firefly Productions, 28
   Owen Franken, 356
   Gehl Company, 57
   Philip Gould, 437
   Charles Gupton, 407

Henry Horsestein, 247
Dave G. Houser, 361, 457
Hulton-Deutsch Collection,
   49, 406, 407, 495, 520
Archivo Iconografico, S.A., 26,
   27
Hanan Isachar, 435
Lester Lefkowitz, 12, 336, 390,
   480, 521
Danny Lehman, 438
Yang Liu, 315
Richard T. Nowitz, 308
Charles O'Rear, 416
Gianni Dagli Orti, 494
Douglas Peebles, 446
Jose Luis Pelaez, Inc., 174, 189
David Pollack, 345
Carl & Ann Purcell, 360, 428
Roger Ressmeyer, 58
Reuters NewMedia Inc., 9, 13,
   240, 260, 368, 435
Charles E. Rotkin, 393, 428
Bob Rowan; Progressive
   Images, 458
SIE Producions, 252
Leif Skoogfors, 12, 415, 425
Lee Snider, 429
Joel Stettenheim, 436
Vince Streano, 534
James A. Sugar, 221
Ramin Talaie, 6, 67
The Cover Story, 521
Peter Turnley, 391
Bill Varie, 394
Doug Wilson, 330
Michael S. Yamashita, 373
Peter Yates, 81
CORBIS/Bettmann, 48, 49, 177,
   193, 220, 221, 246, 247, 353,
   356, 357, 371, 382, 383, 406,
   407, 429, 452, 494, 495, 520,
   521
CORBIS SYGMA,
   Pitchal Frederic, 414
   Hashimoto Noboru, 531
   Eranian Philippe, 411
   Ruet Stephane, 8, 185
   Koren Ziv, 237
David R. Frazier Photolibrary,
   Inc., 13, 442, 461, 462, 464,
   465, 466, 468, 471

Digital Globe, 233
Digital Vision, 7, 60, 110, 116,
   119, 159, 193, 247, 263, 356,
   357, 395, 406, 413, 496, 520
Digital Vision,
   Mark Kemp & Cyberactive,
   497, 517, 523, 543
Dover Elevators, 368
Dreaming Creek Timber Beam
   Homes, 460
Fossil, Inc., 262
Tim Fuller, 383
Ann Garvin, 398, 399, 400
Getty Images, 29, 45, 51, 75, 101,
   105, 127, 133, 150, 157, 169,
   192, 193, 452, 453
Getty Images,
   Theodore Anderson, 419
   Bruce Ayres, 337
   Denis Boissavy, 224, 243
   David Buffington, 431, 449,
   455, 471, 475
   Dennis Cox, 428
   Philip Henry Delamotte, 27
   Bruce Forester, 165
   Robert Frerck, 164
   Garry Gay, 385, 387, 403, 409,
   425
   Alan Hicks, 538
   Hulton/Archive, 26, 49, 173,
   221
   Russell Illiq, 310
   Stephen Johnson, 175, 189,
   195, 217
   Dmitri Kessel, 26
   Kent Knudson/Photolink, 68
   Nick Koudis, 11, 333, 341, 353,
   359, 377
   Christian Laqereek, 522
   Tom Mareschal, 74, 101
   Gjon Mili, 192
   PhotoDisc Collection, 11, 148,
   362
   Andrew Sacks, 408
   Hugh Sitton, 231
   Thinkstock, 223, 243, 249,
   265, 269, 287, 291, 305, 309,
   327, 407
   Time Life Pictures, 172
   V.C.L., 343
   Stan Wayman, 330

Index

References in *italic type* refer to page numbers of activities. References in **boldface type** refer to key terms.

building(s) (*continued*)
dome-shaped, 458
floor plan of, 211 (illus.)
steel-framed, 441, 477, 478
tallest, 428, 434, 476
building codes, 445-446
building site, 457, 460-461
bullet train, 531
business
and ethics, 400
impacts of communication
technology on, 165-166,
240, 284, 420, 423
and Internet, 420, 423
and profit, 401
and scanners for personalized shopping, 399
byte, in binary code, 253

C

cable-stayed bridge, 481
CAD, 204-207, 214
calorie, 108
CAM, 414
camera
on computer, and Internet
chats, 261
digital. *See* digital camera
on endoscope, 343-344, 347
film for, 295-296, 297, 301-
302, *307*
to find victims under wreck-
age, 67
movie, 299
parts of, 292-296
processing images taken
with, 301-303, *307*
cams, 56
canals, 438
cantilever bridge, 481, 482
(illus.)
capital, 58
car(s)
alcohol-fueled, 108
concept, 31 (illus.)
driving wheels in, 526
electric, 513
engines in, 260, 504, 505
and environment, 505, 513,
514
first, 172

front-, rear-, and four-wheel
drive, 526
fuel cell, 514
garbage-fueled, 124
Grease, 38
hybrid, 513-514, *515*
impacts of, 69, 117
manufacture of, 414 (illus.),
416 (illus.)
seatbelts in, *34*
subsystems in, 525-528
transmission in, 525
carbon monoxide, 121
career(s)
in architecture, 468
changing, 527
computer technician, 255
in construction, 445, 468,
483
and creativity, 300
customer service, 527
in design, 33, 178, 283, 468
as drafter, 199
as EMT, 342
in engineering, 143, 178, 506
and ethics, 506
and following directions, 59
in graphic design, 283, 300
in health care, 342, 364
and interests and values, 392
investigating, 166, 364
and listening skills, 230
in management, 112, 392
in medical field, 342, 364
and organizational skills, 167
and personal grooming, 445
as pharmacologist, 364
as photojournalist, 300
and pressure, handling, 143
and responsibility, 112
in sales, 89, 421
as scriptwriter, 319
selecting, 33, 392
and self-discipline, 283
and shadowing program, 166
as sign language interpreter,
230
skills for, 53, 59, 112, 167,
230, 283
as surveyor, 483
as teacher, 167
as Web designer, 33

and willingness to learn, 255
as wind farm manager, 112
Carlson, Chester, 49
Carrier, Willis, 124
carrier wave, 232, 310, 311
Carver, George Washington,
357
casting, 78, *102-103*, 395
CAT scan, 343
caulking, 466
CAVE, 176
CCD, 296, 298
CDs, 254-255, 312-313, 423
cells
body, 340, 341, 367-368
fuel, 514
solar, 111-112
ceramics, 98, 505
change, handling, 527
Channel Tunnel, 483, 484
(illus.)
chemical energy, 106
chemicals
in acid rain, 120-121
biosynthetic, making, 370
in conditioning, 396
to develop photos, 301-302,
306-307
as disinfectants, 335
in manufacturing, 396
as preservatives, 338
chess, AI system for playing,
260
Chrysler, Walter, 36
CIM, 414
circuit, 136
in alarm system, *152-155*
integrated, 149, 221, 250,
251-252
measuring electricity in, 138-
140
overload of, 143, 144 (illus.)
parallel, 143-144
series, 142, *144*
circuit breaker, 143, 144 (illus.)
clamps, 84
cleanliness, 335, 445
climate, structural design for,
444, *447*
cloning, 367-369
closed-loop system, 64-65
CMOS, 296, 298

F

hydroponics, 372-373, *378-381*
hypertext markup language, 322-323, 324
hypothesis, 158

I

ignition system, 501, 506
illness. *See* disease
illustrations, in graphic communication, 280-281
image
    digital, 298-299, 302-303
    and illusion of motion, 299, 311
    latent, on film, 297, 301
    negative, 297
    processing (in photography), 301-303, *306-307*
imaging machines, 341-344
immobots, 261
immunization, 337
impacts
    of airplane, 117
    of antibiotics, 346, 363
    on business, 165-166, 240, 284, 420, 423
    of cars, 69, *117*
    of communication technology, 165, 237-241, 249, 277
    of computer, 249. *See also* Internet
    of consumers. *See* consumers
    cultural, 41, 117, 238-239
    economic, 164, 240, 241, 388, 437, 438
    on education, 163, 165, 238-239
    of energy technology, 117, 120-125
    environmental. *See* environment
    of farming practices, 363, 366-367
    of genetic engineering, 366-367
    of graphic communication, 163, 238-239, 277
    of historical events on technology, 67
    of Industrial Revolution, 41, 164, 361, 387-388

of Information Age, 165
international, 239, 240
of Internet, 420, 423
of microscope, 340-341
of movable type, 27, 163-164
of photography, *295, 296*
of plow, 26, 361
political, 233, 237
of power technologies, 117, 120-125
of printing, 27, 163-164, 238-239
on privacy, 239, 319, 322, 528
social, 41, 117, 163, 164-165, 238-239, 388
of society on technology, *35, 485, 541*
study of, developing, 241
of transportation, 69, 117
weighing, and technological literacy, 69
implants, 348, 349 (illus.)
inclined plane, 56, 57
Industrial Age, 27
Industrial Revolution, 41, 164, 361, 387-388
industrial robots, 91, 383, 416-417, 418
information
    collecting, in problem solving, 185
    as input, 63, 394
    on Internet, finding, 323
    processing of, by computer, 250-253, 254-255
    as resource, 54
Information Age, 41, 164-166, 239, 361
ink-jet printing, 258, 276-277
innovation, 67, 177, 291, 443
inorganic materials, 93
input, 63
    in communication, 225
    in farming, 360
    feedback as, 65
    in manufacturing, 391-394
    in power systems, 117
    in transportation, 523
    types of, 63, 391-394
inspection, and quality assurance, 400, 415

instructions
    for computer, 252, 253-254, 259-260. *See also* software
    destructive (virus), 253-254
    following, 59
    writing, 54, *211, 243*
insulation (house), 466, *469*
insulators (electricity), 141
intaglio printing, 271
integrated circuit, 149, 221, 250, 251-252
intermodal transportation, 534-535
internal combustion engines, 501-511
International Space Station, 438, 488-489
International Standards Organization (ISO), 296
Internet, 322
    advertising on, 420
    browsers for, 323
    cameras for live chats on, 261
    and distributed computing, 262, 263
    and e-manufacturing, 417
    hypertext markup language for, 322, 324
    impacts of, 420, 423
    non-stop, 322
    origin of, 165
    and privacy, 239, 322
    search engines for, 323
    Web pages for, 323-325, *328-329*
    wireless (wi-fi), 262
    and World Wide Web, 322-323
interpersonal skills, 468
Interstate Highway System, 429
invention(s), 177
    building on previous, 31, 40, 146, 149, 177, 220-221, 291
    impacts of, 26, 27, 117, 163
    leading to computer technology, 220-221
    patents on, 38, 177
    by teens, 38
Iron Age, 41
irradiation, 335

processing, in photography, 301-303, *306-307*

producer, role of, 317

product
  advertising, 401, 420, *459*
  custom-made, 423
  designing and developing, 66-68, 398, 410-411
  inspecting, 401, 415, 425
  selling, 401, 420-423
  standards for, 416
  testing, *426-427*

product development, 66-68, 398, 410-411

profit, *160*, 401

program (computer), 252. *See also* software
  artificial intelligence (AI), 259-260
  diagnostic, 260
  expert system, 259-260
  operating system, 252
  viruses in, 253-254
  writing a, for robot, *266-267*

propellant, 508, 510-511

properties, of materials, 93-94

proportion, 180

propulsion, 508-511, 531

protons, 134

prototype, 180, 186, 212, 213

prototyping, rapid 213

pulley, 56, 57

pyramids, Egyptian, 428

Q

quality assurance, 64, 400, *404-405*, 415-416, 425

quality control. *See* quality assurance

R

R&D, 410-411

radiant energy, 106. *See also* solar energy

radio, 310

radioactive waste, 122

rain, acid, 120-121

RAM, 251, 255

rapid prototyping, 213

raw materials, 60

rear-wheel drive, 526

reciprocating motion, 504, 505, *517*

recorders, 312-313

recycling, 96, 123, *241*

relativity, theory of, 160

Renaissance, 41

renderings, 202, 205

renewable energy sources, 107-108, 371

requirements. *See also* criteria
  in construction, 439, 444-447
  in manufacturing, 400, 411, 416
  and product design, 66-67, 411

research and development, product, 410-411

residential buildings, 456. *See also* house

resin, 396

resistance, electrical, 138, 141

resistance of materials, to forces, 94

resources, 52-61
  depletion of, 122-123
  energy, 61, 107-114, 122-123, 371

responsibility, 112

retailer, 422

Revolution, Industrial, 41, 164, 361

Rillieux, Norbert, 192

roads, construction of, 434-436, 479-480, *484*

robot
  ASIMO, 91
  degrees of freedom in, 348
  designing and programming, *266-267*
  flexibility of, 348
  immobile (immobots), 261
  industrial, 91, 161, 383, 416-417, 418
  joints in, 348
  and mathematics, 161
  and medical technology, 347-348, 350
  program for, developing, *266-267*
  safety devices for, 418
  and science, 348
  surgery using, 347-348

robotics, 416-417, 418. *See also* robot

rocket engines, 508-511

Roentgen, Wilhelm, 357

ROM, 251

Roman construction, 27, 434-435, 436

roof construction, 444, 464-465

rotary motion, 504-505, *517*

rubber-band-powered vehicle, *544-545*

S

safety
  and air bags, 34, *35*
  colors for, 90
  in construction, 445, 446-447, 477
  in factories, 418
  with gasoline, 503
  with hand tools, 91
  with hot items, 131

safety restraint system, designing, *354-355*

Salk, Jonas, 356

sanitation, 173, 336

satellites
  energy for, 112
  and GPS, 165, 361
  releasing, into space, 540
  spy, 233
  and telecommunication, 232-233

Savery, Thomas, 172

saws, 86, 87

scale drawings, 198-199, 201

scanner, for high-tech shopping, 399

schedule, 413

schematic diagrams, 209-210

science, 37
  and biotechnology, 333, 359. *See also* agricultural technology; medical technology; medicine
  and effects of lenses, 293
  and electricity, 134-135, 141, 146
  and experimentation, 158-159
  laws of. *See* scientific laws
  and machines, 55-57, 348

Standards for Technological Literacy, 5, 20

static electricity, 133, 135-136

Statue of Liberty, 487

steam engine, 172, 498-499

steam turbines, 499, 500 (illus.), *518-519*

steel
    conditioning, 81
    in construction, 441, 442, 447
    shapes of, 96, 97 (illus.)

stethoscope, 340, *345*

stick construction, 434

storyboard, 318

strength
    of materials, 94
    of structures, testing, *450-451, 492-493*

strokes, in engine cycles, 502-503, 506

Strong, Harriet Russell, 356

studs, 463

subfloor, 463

subgrade, 479

subsystems
    in cars, 525-528
    in communication, 226-227
    defined, 62
    electric circuits as, 136. *See also* circuit
    in houses, 466-467
    in transportation, 524

subway, 530

sun, energy from, 111-112, 444, 467, 489

superconductors, 141

supertankers, 534

supplier, 399, 414, 417, 527

surfacing (road), 435-436, 443, 480

surgery, 346-348

surveyor, 483

suspension bridge, 481, 482 (illus.)

switch, electric, 142

symbols
    in CAD, 204
    recycling, 96 (illus.), 123 (illus.)

on schematics, 210

synthetic materials, 60, 96

system(s), 62-65
    in cars, 525-527
    communication, 225-227, *233*
    computer, 250-258
    constraints on, 67
    construction, 439
    controls for, 68. *See also* quality assurance
    diagramming, 62-64, *233*
    electrical circuits. *See* circuit
    farming as a, 360
    global positioning, 165, 361
    ignition, 506
    interconnected, 65, 139
    maintenance of. *See* maintenance
    malfunctions of. *See* malfunction
    manufacturing, 391-397
    medical technology, 333
    power, 117
    requirements of. *See* criteria; requirements
    solar heating, 111, *128-129*
    sub-. *See* subsystems
    transmission, of car, 525
    transportation, 523

T

Tange, Kenzo, 453

tape recorders, 312

Taylor, Frederick, 388

technical drawing, 197. *See also* drafting

technologically literate, 34, 69

technology, 30, 62-65
    agricultural, 359-381
    as art, 36
    building on prior, 40-42, 53, 54, 146, 149, 172, 177, 220-221, 291
    changes caused by. *See* impacts
    communication. *See* communication technology
    construction. *See* construction

and creativity, 53, 176, 177, 410

and culture, 41, 117, 238-239

and democracy, 42

and economy, 164, 240, 241, 388, 437, 438

effects of. *See* impacts

energy for, 61, 105-120

and ethics, 35. *See also* ethics

farming, 26, 359-381

historical events influencing, 67

history of, 40-42. *See also* evolution

impacts of. *See* impacts

innovations in, 67, 177, 291, 443

international effects of, 239, 240

interrelated, 523, 534-535, 538

inventions. *See* inventions

and language arts, 163-164. *See also* language arts connections

manufacturing, 385-405, 409-424

and mathematics, 160-162. *See also* mathematics

medical, 214, 333-355

nano-, 41-42

and needs and wants, 30, 52, 53, 54, 66

net-zero, 467

political impacts of, 233, 237, 505

power for, 115-119, 497-515

and privacy, 239, 319, 322, 528

resources of, 52-61

and science. *See* science

without scientific knowledge, 37, 158, 362, 373

and skill, 53

and social studies, 164-166. *See also* social studies connections

and society. *See* society

from space exploration, 505, *541*